Lecture Notes in Computer Science   1905
Edited by G. Goos, J. Hartmanis and J. van Leeuwen

W0245942

Springer
Berlin
Heidelberg
New York
Barcelona
Hong Kong
London
Milan
Paris
Singapore
Tokyo

Hans Scholten   Marten J. van Sinderen (Eds.)

Interactive Distributed Multimedia Systems and Telecommunication Services

7th International Workshop, IDMS 2000
Enschede, The Netherlands, October 17-20, 2000
Proceedings

 Springer

Series Editors

Gerhard Goos, Karlsruhe University, Germany
Juris Hartmanis, Cornell University, NY, USA
Jan van Leeuwen, Utrecht University, The Netherlands

Volume Editors

Hans Scholten
Marten J. van Sinderen (CTIT)
Department of Computer Science
University of Twente
P.O. Box 217, 7500 AE Enschede, The Netherlands
E-mail: {scholten,sinderen}@cs.utwente.nl

Cataloging-in-Publication Data applied for

Die Deutsche Bibliothek - CIP-Einheitsaufnahme

Interactive distributed multimedia systems and telecommunication
services : 7th international workshop ; proceedings / IDMS 2000,
Enschede, The Netherlands, October 17 - 20, 2000. Hans Scholten ;
Marten J. van Sinderen (ed.). - Berlin ; Heidelberg ; New York ;
Barcelona ; Hong Kong ; London ; Milan ; Paris ; Singapore ; Tokyo :
Springer, 2000
  (Lecture notes in computer science ; Vol. 1905)
  ISBN 3-540-41130-5

CR Subject Classification (1998): H.5.1, C.2, H.4, H.5, H.3

ISSN 0302-9743
ISBN 3-540-41130-5 Springer-Verlag Berlin Heidelberg New York

Springer-Verlag Berlin Heidelberg New York
a member of BertelsmannSpringer Science+Business Media GmbH
© Springer-Verlag Berlin Heidelberg 2000

Typesetting: Camera-ready by author, data conversion by PTP-Berlin, Stefan Sossna
Printed on acid-free paper   SPIN: 10722662   06/3142   5 4 3 2 1 0

Preface

The first International Workshop on Interactive Distributed Multimedia Systems and Telecommunication Services (IDMS) was organized by Prof. K. Rothermel and Prof. W. Effelsberg, and took place in Stuttgart in 1992. It had the form of a national forum for discussion on multimedia issues related to communications. The succeeding event was "attached" as a workshop to the German Computer Science Conference (GI Jahrestagung) in 1994 in Hamburg, organized by Prof. W. Lamersdorf. The chairs of the third IDMS, E. Moeller and B. Butscher, enhanced the event to become a very successful international meeting in Berlin in March 1996.

This short overview on the first three IDMS events is taken from the preface of the IDMS'97 proceedings (published by Springer as Lecture Notes in Computer Science, Volume 1309), written by Ralf Steinmetz and Lars Wolf. Both, Ralf Steinmetz as general chair and Lars Wolf as program chair of IDMS'97, organized an excellent international IDMS in Darmstadt.

Since 1998, IDMS has moved from Germany to other European cities to emphasize the international character it had gained in the previous years. IDMS'98 was organized in Oslo by Vera Goebel and Thomas Plagemann at UniK – Center for Technology at Kjeller, University of Oslo. Michel Diaz, Phillipe Owezarski, and Patrick Sénac successfully organized the sixth IDMS event, again outside Germany. IDMS'99 took place in Toulouse at ENSICA. IDMS 2000 continued the tradition and was hosted in Enschede, the Netherlands.

The goal of the IDMS series of workshops has been and still is to bring together researchers, developers, and practitioners from academia and industry; and to provide a forum for discussion, presentation, and exploration of technologies and advances in the broad field of interactive distributed multimedia systems and telecommunication services, ranging from basic system technologies such as networking and operating system support to all kinds of teleservices and distributed multimedia applications. To accomplish this goal IDMS remains relatively "small": it has no parallel sessions and a limited number of participants to encourage interaction and discussion.

Although IDMS2000 had tough competition from other conferences and workshops, it received 60 submissions from Europe, Asia, Africa, and North and South America. Every paper was refereed by at least three reviewers. A tedious job, luckily made easier with help from the excellent online conference tool ConfMan, developed in Oslo for the 1998 IDMS workshop. Ultimately, the 26 members of the program committee (PC) and 33 referees selected 24 high quality papers for presentation at the workshop. The main topics of IDMS2000 are: efficient audio/video coding and delivery, multimedia conferencing, synchronization and multicast, communication, control and telephony over IP networks, QoS models and architectures, multimedia applications and user aspects, design and implementation approaches, and mobile multimedia and ubiquitous computing systems. This technical program is complemented with three invited papers: "Energy-efficient hand-held multimedia systems" by Gerard Smit, "Short-range connectivity with Bluetooth" by Jaap Haartsen and "On the failure of middleware to support multimedia applications" by Gordon Blair.

The co-chairs are very grateful for the help they received to make IDMS2000 a successful event. We thank the steering committee for their trust in us to continue the tradition of the IDMS series of workshops, and the PC and referees for the time and effort in selecting the papers for the technical program. We heeded the good advice of the organizers of the previous two workshops Thomas Plagemann, Vera Goebel, Michel Diaz, Philippe Owezarski, and Patric Sénac. Also we would like to acknowledge the cooperation with ACM SIGCOMM and ACM SIGMM, the financial support from KPN Research, Lucent Technologies and Philips, and the financial and organizational support from the Department of Computer Science and CTIT of the University of Twente. Finally, we especially would like to thank the people from the local organization for their help, Marloes Castanada Schlie, Nikolay Diakov, Cléver Guareis de Farias, and Johan Kunnen.

July 2000   Hans Scholten
                                                        Marten van Sinderen

Organization

Program Co-Chairs

Hans Scholten, University of Twente, The Netherlands
Marten van Sinderen, University of Twente, The Netherlands

Program Committee

H. Batteram, Lucent Technologies, The Netherlands
G. v. Bochmann, University of Montreal, Canada
P. Bosch, CWI, The Netherlands
B. Butscher, GMD-FOKUS, Germany
A. Campbell, Colombia University, USA
L. Delgrossi, University of Piacenza, Italy
M. Diaz, LAAS-CNRS, France
H. Eertink, Telematica Instituut, The Netherlands
W. Effelsberg, University of Mannheim, Germany
F. Eliassen, University of Oslo, The Netherlands
L. Ferreira Pires, University of Twente, The Netherlands
V. Goebel, University of Oslo , UniK, Norway
T. Helbig, Philips, Germany
D. Hutchison, Lancaster University, Great Brittain
P. Jansen, University of Twente, The Netherlands
W. Kalfa, TU Chemnitz, Germany
E. Moeller, GMD-FOKUS, Germany
K. Nahrstedt, University of Illinois, USA
P. Owezarski, LAAS-CNRS, France
S. Pink, SICS, Sweden
T. Plagemann, University of Oslo, UniK, Norway
D. Quartel, University of Twente, The Netherlands
P. Sénac, ENSICA, France
G. Smit, University of Twente, The Netherlands
R. Steinmetz, TU Darmstadt / GMD, Germany
L. Wolf, University of Karlsruhe, Germany

Local Organization

Marloes Castaneda Schlie, University of Twente, The Netherlands
Nikolay Diakov, University of Twente, The Netherlands
Cléver R. Guareis de Farias, University of Twente, The Netherlands
Johan Kunnen, University of Twente, The Netherlands

Referees

H. Batteram	C. Kuhmuench
T. Becker	J. Kunnen
P. Berthou	M. Lijding
G. v. Bochmann	K. Lund
P. Bosch	D. Makaroff
B. Butscher	M. Mauve
A. Campbell	E. Meeuwissen
C. Chassot	E. Michiels
R. Denda	E. Moeller
N. Diakov	K. Nahrstedt
M. Diaz	P. Owezarski
J. Dittrich	T. Plagemann
K.-P. Eckert	A. Pras
H. Eertink	D. Quartel
W. Effelsberg	R. Roth
F. Eliassen	P. Sénac
L. Ferreira Pires	G. Smit
C. Fuhrhop	R. Song
N. Georganas	R. Steinmetz
S. Gjessing	D. Thie
V. Goebel	L. Tionardi
C. R. Guareis de Farias	G. Tuquerres
B. Hafskjold	J. Vogel
X. He	C. Wang
T. Helbig	J. Widmer
D. Hutchison	V. S. Wold Eide
P. Jansen	L. Wolf
W. Kalfa	W. Yu
T. Kristensen	

IDMS Steering Committee

Michel Diaz, LAAS-CNRS, France
Vera Goebel, Univ. of Oslo, UniK – Center for Technology at Kjeller, Norway
Eckhard Moeller, GMD-FOKUS, Germany
Thomas Plagemann, Univ. of Oslo, UniK – Center for Technology at Kjeller, Norway
Patrick Sénac, ENSICA, France
Lars Wolf, Universität Karlsruhe (TH), Institute for Telematics, Germany

Supporting and Sponsoring Organizations

ACM SIGCOMM	UT, Computer Science
ACM SIGMM	UT, CTIT
KPN Research	
Lucent Technologies	
Philips	

Table of Contents

Mobile Multimedia and Ubiquitous Computing Systems

Energy-Efficient Hand-Held Multimedia Systems
- Designing the Swiss Army Knife of Computing -

Gerard J.M. Smit

University of Twente, Faculty of Computer Science,
PO Box 217, 7500 AE Enschede, the Netherlands
smit@cs.utwente.nl and http://www.cs.utwente.nl/~smit

Abstract. The trend in wireless terminals has been to shrink a general-purpose desktop PC into a package that can be conveniently carried. Even PDAs have not ventured far from the general-purpose model, neither architectural nor in terms of usage model. Both the notebook and the personal computer generally use the same standard PC operating system such as Windows (CE) or Unix, same applications, use the same communication protocols and use the same hardware architecture. The only difference is that portable computers are smaller, have a battery, a wireless interface, and often use low power components [2].
Even though battery technology is improving continuously and processors and displays are rapidly improving in terms of power consumption; battery life and battery weight are issues that will have a marked influence on how hand-held computers can be used. Energy consumption is becoming the limiting factor in the amount of functionality that can be placed in these devices. More extensive and continuous use of network services will only aggravate this problem since communication consumes relatively much energy.
Another key challenge of mobile computing is that many attributes of the environment vary dynamically. Mobile devices face many different types of variability in their environment [3]. Therefore, they need to be able to operate in environments that can change drastically in short term as well as long term in available resources and available services. Merely algorithmic adaptations are not sufficient, but rather an entirely new set of protocols and/or algorithms may be required. For example, mobile users may encounter a complete different wireless communication infrastructure when walking from their office to the street [4]. A possible solution is to have a mobile device with a reconfigurable architecture so that it can adapt its operation to the current environment and operating condition. Adaptability and programmability should be major requirements in the design of the architecture of a mobile computer.
We are entering an era in which each microchip will have billions of transistors. One way to use this opportunity would be to continue advancing our chip architectures and technologies as just more of the same: building microprocessors that are simply complicated versions of the kind built today. However, simply shrinking the data processing terminal and radio modem, attaching them via a bus, and packaging them together does not alleviate the architectural bottlenecks. The real design challenge is to engineer an integrated mobile system where data processing and communication share equal importance and are designed with each other in mind. Just integrating current PC or PDA architecture with a communication subsystem, is not the solution. One of the main drawbacks of merely packaging the two is that the energy-inefficient

H. Scholten and M. van Sinderen (Eds.): IDMS 2000, LNCS 1905, pp. 1-2, 2000.
© Springer-Verlag Berlin Heidelberg 2000

general-purpose CPU, with its heavyweight operating system and shared bus, becomes not only the center of control, but also the center of data flow in the system and a main cause of energy consumption [1].

Clearly, there is a need to revise the system architecture of a portable computer if we want to have a machine that can be used conveniently in a wireless environment. A system level integration of the mobile's architecture, operating system, and applications is required. The system should provide a solution with a proper energy-efficient balance between flexibility and efficiency through the use of a hybrid mix of general-purpose and the application-specific approaches. The key to energy efficiency in future mobile systems will be designing higher layers of the mobile system, their system architecture, their functionality, their operating system, and indeed the entire network, with energy efficiency in mind. Furthermore, because the applications have direct knowledge of how the user is using the system, this knowledge must be penetrated into the power management of the system.

References

[1]   P.J.M. Havinga, "Mobile Multimedia Systems", Ph.D. thesis University of Twente, February 2000, ISBN 90-365-1406-1, www.cs.utwente.nl/~havinga/thesis.

[2]   Smit G.J.M., Havinga P.J.M., et al.: "An overview of the Moby Dick project", 1st Euro-micro summer school on mobile computing, pp. 159-168, Oulu, August 1998; Moby Dick homepage:

[3]   www.cs.utwente.nl/~havinga/mobydick.html.

[4]   Lettieri P., Srivastava M.B.: "Advances in wireless terminals", IEEE Personal Communications, pp. 6-19, February 1999.

[5]   Weiser M.: "Some computer science issues in ubiquitous computing", Communications of the ACM

Realisation of an Adaptive Audio Tool

Arnaud Meylan and Catherine Boutremans

Institute for Computer Communication and Applications
Swiss Federal Institute of Technology at Lausanne (EPFL)
CH-1015 Lausanne, Switzerland

Abstract. *Real-time audio over the best effort Internet often suffers from packet loss. So far, Forward Error Correction (FEC) seems to be an efficient way to attenuate the impact of loss. Nevertheless to ensure efficiency of FEC, the source rate must be continuously controlled to avoid congestion. In this paper, we describe a realisation of adaptive FEC subdued to a TCP-friendly rate control.*

1 Introduction

The Internet has changed from mainly a file transfer and e-mail tool to a network for multimedia and commercial applications, among others. This change brought up many new technical challenges such as transport of real-time data over non real-time lossy networks, which has been fulfilled by the Real-time Transport Protocol (RTP) [8] and FEC techniques [3].

Unfortunately, FEC is too often used without rate control, what leads to more congestion, loss and then worse audio quality [6]. The purpose of this work is to add adaptive FEC to an existing software: the Robust Audio Tool [7]. FEC will be constrained by a TCP-friendly rate, proposed by Mahdavi and Floyd [16].

In this paper the general problem of optimizing quality at reception is presented, a more specific solution is deduced.

2 State of the Art

2.1 FEC

Definition. Forward Error Correction (FEC) relies on the addition of repair data to a stream, from which the content of lost packets may be recovered at destination, at least in part. Two classes of repair data may be added to a stream [3]: those which are independent of the contents of that stream (e.g. Parity coding, Reed-Solomon codes) and those which use knowledge of the stream to improve the repair process. In the context of real-time audio the most popular scheme, which was standardized by IETF is repetition of audio units on the stream. In the following, we will refer to this scheme as Signal processing based FEC (SFEC).

SFEC. In order to simplify the following discussion, we distinguish a (media) *unit* of data from a *packet*. A unit is an interval of audio data, as stored inter-

H. Scholten and M. van Sinderen (Eds.): IDMS 2000, LNCS 1905, pp. 3–13, 2000.
© Springer-Verlag Berlin Heidelberg 2000

nally in an audio tool. A packet comprises one or more units, encapsulated for transmission over the network.

The principle of FEC used here is to repeat units of audio in multiple packets. If a packet is lost then another packet containing the same unit will be able to cover loss and be played—providing it arrives. This approach has been advocated in [4] and [5] for use on the Mbone, and extensively simulated by Podolsky [6] who calls this framework "Signal-processing based FEC" (SFEC[1]). Redundant audio units are piggy-backed onto a later packet, which is preferable to the transmission of additional packets, as this decrease the amount of packet overhead and routing decisions.

There is a potential problem with apply redundancy in response to loss because when loss occurs the (user) reflex is to add redundancy at the source. This leads to increase the overall rate transmission and congestion, probably producing even worse quality. Effectively, redundancy can be added when loss occurs, but in the same time the source encoding must be changed to use less bandwidth in response to congestion [6]. Our framework is to continuously ensure that the overall rate transmission is smaller or equal than a TCP-friendly rate proposed by Mahdavi and Floyd [16]. It is a combined source rate/redundancy control.

2.2   TCP-Friendly Rate Control

As networked multimedia applications become widespread, it becomes increasingly important to ensure that they can share resources fairly with each other and with current TCP-based applications, the dominant source of Internet traffic. The TCP protocol is designed to reduce its sending rate when congestion is detected. Networked multimedia applications should exhibit similar behavior, if they wish to co-exist with TCP-based applications.

One way to ensure such co-existence is to implement some form of congestion control that adapts the transmission rate in a way that *fairly* shares bandwidth with TCP applications. One definition of *fair* is that of TCP *friendliness* [16] - the non-TCP connection should receive the same share of bandwidth (namely achieve the same throughput) as a TCP connection.

Mahdavi and Floyd [16] have derived an expression relating the average TCP throughput R_{TCP} to the packet loss rate:

$$R_{\text{TCP}} = 1.22 \frac{\text{MTU}}{\text{RTT} \times \sqrt{\pi_1^*}} . \tag{1}$$

where MTU is the packet size being used on the connection; RTT is the round trip time and π_1^* is the loss rate being experienced by the connection.

In our application, we fix the MTU to 576 bytes, the minimum size for TCP. Current values for RTT are obtained using RTCP Sender and Receiver Reports. The packet loss rate π_1^* is computed at the receiver and reported to the sender via the Fraction lost field of the Receiver Reports.

---

[1] "because this approach exploits a signal-processing based model of the audio signal to effectively compute its error-correcting information"

3   The Robust Audio Tool

3.1   Software Presentation

The Robust Audio Tool (RAT) [7] is an open-source audio conferencing and streaming application that allows users to particpate in audio conferences over the Internet.

This software is based on IETF standards, it uses RTP above UDP/IP as its transport protocol, according to the RTP profile for audio and video conference with minimal control [8][9]. RAT features a range of different rate and quality codecs, receiver based loss concealment to mask packet losses, and sender based channel coding.

3.2   Some Useful Concepts about RAT

The channel coder. Let us define the following terms:

A *codec* is defined as an encoder-decoder performing one type of compression/decompression on the audio stream. The encoder part takes a buffer of audio stream in input, performs encoding and outputs a playout buffer of *media units*.

The *channel coder* builds *channel units* which represent the payload of the RTP packets. In RAT the channel coder performs four different kind of channel coding, especially three allowing FEC called *redundancy* (SFEC), *interleaving* and *layered*. For our work, FEC will be added in the form of SFEC, using the redundant channel coder.

The Message Bus. RAT comprises three separate processes: controller, media engine and user interface. Communication between them is provided by the Message bus (Mbus), the sole means to ensure coordination of multimedia conferencing systems.

The Message bus was proposed by Colin Perkins and Jorg Ott [10]. It solves the typical problem of separate tools providing audio video and shared workspace functionality. It maps well on the underlying RTP media streams, which are also transmitted separately

A message contains a header and a body. The former indicates notably the source and destination adresses the latter contains the message having to be delivered to the application. The message is transmitted by UDP in the form of a string. A function maps it into a C function call at destination (unmarshalling).

We considered the following MBus commands:

Command name	Usage
`tool.rat.codec(.)`	Specifies primary codec being used by this source
`audio.channel.coding(.)`	Specifies secondary codec, and its relative offset to the primary
`tool.rat.rate(.)`	Sets the number of audio units (codec frames, typically) placed in each packet when transmitting, assuming a unit is t ms long

These messages provide high level access to the program. The biggest part of implementation was made using these commands.

4 Adaptive FEC Algorithm

4.1 The General Problem

We want to maximize a certain measure of quality at reception. Quality can be estimated according to a non-subjective (loss rate after reconstruction) or subjective measure (perceived quality at destination depending on the quality of each audio unit played). To realize this, the SFEC-channel coder algorithm must choose a range of variables:

- k denotes the number of copies sent for the same audio unit. k is bound to the loss probability of the network and desired apparent loss rate after reconstruction at reception.
- $\mathbf{o} = [o_1, o_2, \ldots, o_k]$ denotes the offset of redundant unit i relative to the first transmitted unit. One always have $o_1 = 0$.

  It has been shown [11] that loss presents some degree of correlation: if packet n is lost, then there is a higher than average probability that packet $n + 1$ will also be lost. This indicates that an advantage in perceived audio quality can be achieved by offsetting the redundancy. The disadvantage of offsetting is the increase of the playout delay. It's a compromise between delay and robustness. Nevertheless in case of half duplex for example, delay is not important and offset can be extensively used.
- The set of codecs available is associated with a finite set of encoding rates r_i: $\mathcal{R} = \{r_i\}_{i=1}^n$. Notice $r_0 = 0$ corresponds to no coding. Codecs are ordered according to their bit rates, namely $i < j \Leftrightarrow r_i \leq r_j$.
- Let $\mathbf{x} = \{x_i\}_{i=0}^{k-1}$ denote the number of codec used to encode the ith redundant unit.
- r is the total payload rate $r = \sum_{i=1}^k r_{x_i}$
- t denotes the frame duration of an audio unit, it is usually 20 ms, 40 ms or 80 ms. It influences end to end delay to increase t and reduces the number of packets sent, so diminishes the header overhead.

Input parameters are:

- The TCP-friendly rate constraint R which is slightly smaller than R_{TCP}. Indeed, R_{TCP} provides an upper bound to the *total* throughput $R_{\text{TCP}} = R + R_{\text{headers}}$ allowed for the application. R only represents the maximum payload (audio) throughput. R_{headers} is the IP/UDP/RTP header throughput: this application sends $(20 + 8 + 12)$ bytes of header each t ms . The payload rate is then

$$R = R_{\text{TCP}} - \frac{40 \cdot 8}{t} \text{ [kb/s]}. \tag{2}$$

- The parameters of the loss process on the network, typically p and q for the Gilbert model.

In this context, it would be useful to maximize a function $f(k, \mathbf{o}, \mathbf{x}, t, R, p, q)$ representing the quality at reception. Our solution will be less general, the (restrictive) hypothesis are presented below.

4.2   Models

Loss Model. The characteristics of the loss process of audio packets are important to determine how to use SFEC. Previous work on the subject [11], [13], propose a two state Markov Chain, the Gilbert model for the loss process (Fig. 1). States 1 and 0 respectively correspond to a packet loss and packet reaching destination. Let q denote the probability of going from state 1 to state 0, and p from state 0 to state 1. The stationary distribution of this chain is $\pi_0^* = \frac{q}{p+q}$ $\pi_1^* = \frac{p}{p+q}$.

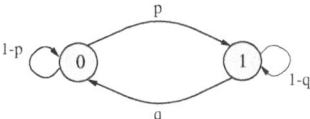

Fig. 1. The Gilbert Model

The actual implementation of the software does not report q. An assumption is then needed. Usually one assumes $p+q = 1$, and the model turns in a Bernoulli of parameter (π_0^*) loss process in which losses are independent. Since this is not realistic [11], we prefer to assume that the average packet loss length is 1.5, therefore $q = 2/3$. This ambitious assumption is based on [14][15], and presents the advantage to reflect the verified correlation of loss process. π_1^* represents the average loss rate, such as reported by RTCP: $\pi_1^* = \frac{p}{p+2/3}$. So p is:

$$p = \frac{2}{3} \cdot \frac{\pi_1^*}{1 - \pi_1^*} \quad . \tag{3}$$

Loss Rate after Reconstruction. We are then interested in loss rates after reconstruction, as a function of k and $\mathbf{o}$. At this stage, a drastic simplification due to actual implementation of RAT appears: k cannot exceed 2, only one level of redundancy can be used. Therefore only o_2 remains to compute.

Let us examine the loss rate after reconstruction for different offsets (we will denote o_2 by n). The probability ρ to lose packet l and $l + n$ is easy to compute using basic probabilities and Markov chains properties

$$
\begin{aligned}
\rho &= P(\text{pack } l \text{ lost} \cap \text{pack } l + n \text{ lost}) \\
&= (\text{pack } l + n \text{ lost} \mid \text{pack } l \text{ lost}) P(\text{pack } l \text{ lost}) \\
&= P(\text{pack } n \text{ lost} \mid \text{pack } 0 \text{ lost}) P(\text{pack } 0 \text{ lost}) \\
&= p_{11}^n \pi_1^* \quad .
\end{aligned}
\tag{4}
$$

Where p_{ii}^n is the probability to return in state i after n steps, i.e. it is the ii^{th} entry of the n step transition matrix P^n.
For this chain one can compute [12] that

$$P^n = \frac{1}{p+q}\begin{bmatrix} q & p \\ q & p \end{bmatrix} + \frac{(1-p-q)^n}{p+q}\begin{bmatrix} p & -p \\ -q & q \end{bmatrix} . \tag{5}$$

hence

$$p_{11}^n = \frac{1}{p+q}(p + q(1-p-q)^n) . \tag{6}$$

Therefore the probability to lose both packets is

$$\rho = \frac{p}{(p+q)^2}(p + q(1-p-q)^n) . \tag{7}$$

The objective is to minimize ρ since it represents the probability to fail to repair the audio stream. The only adjustable parameter is n (positive integer) since p is bound to the network's state, and we assumed a constant value for q, therefore we compute

$$\frac{\partial \rho}{\partial n} = \frac{pq}{(p+q)^2}(1-p-q)^n \ln(1-p-q) . \tag{8}$$

Extrema are obtained for zeros of (8). Three cases are distinguished:

- **$0 < 1 - p - q < 1$** All members of (8) are positive, for $n \to \infty$, $\frac{\partial \rho}{\partial n} \to 0$. $n \to \infty$ leads to a minimum, we do not prove it here. This means that the more you offset for the redundant audio unit the smaller the loss rate after reconstruction. But in the same time playout delay increases since the application must "wait" for the redundant units.
- **$1 - p - q = 0$** All values of n are optimal since ρ is then constant.
- **$-1 < 1 - p - q < 0$** The logarithm function is not defined. We constat on (7) that the smallest value of ρ is obtained for $n = 1$.

Conditions can be simplified in: $p < 1/3$, $p = 1/3$, $p > 1/3$ since $q = 2/3$.

Figure 2 shows plots of ρ as a function of n and p, first one illustrates case $p < 1/3$ second one all three.

On Fig. 2 (a) we can constat that when n increases ρ tends fast to its limit. For $n - 3$, ρ already quite equals the limit. More offsetting probably brings a small reduction of ρ while it increases playout delay. But we do not precisely know which k is optimal since penalty induced by delay on audio quality is not clearly defined. This explains the need for the above mentioned function appraising quality at reception. Without it, a quite arbitrary choice will be to set $n = 3$. Nevertheless it nicely improves the loss rate after reconstruction compared to $n = 1$ and the playout delay introduced seems reasonable[2].

Two last cases are clear. When $p > 1/3$ one clearly chooses $n = 1$ since it guarantees the smallest delay and loss rate after reconstruction. For $p = 1/3$ any

---

[2] extra playout delay due to offset is then 3·frame duration

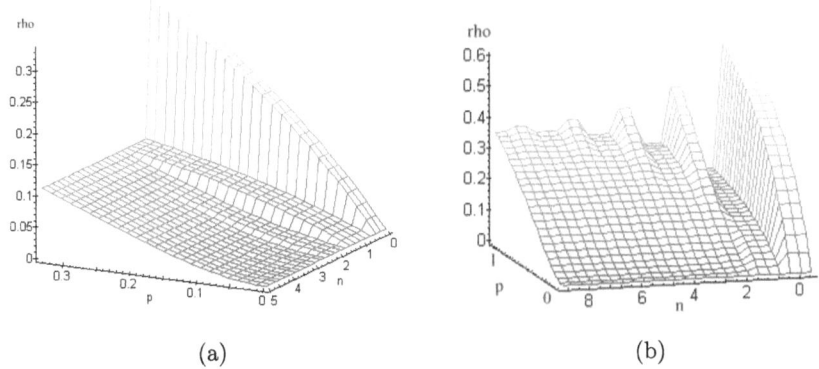

Fig. 2. ρ as a function of n and p. (a) $p \in [0, 0.33]$, (b) $p \in [0, 1]$

value of n is optimal, so we choose $n = 1$ since it minimises delay. We implemented such an algorithm to select offset of redundand unit:

```
p = 2/3 * π₁*/(1 − π₁*); //π₁* denotes loss rate reported by RTCP
if (p < 1/3) n = 3;
else n = 1;
```

These results are specific to this value of q in the general case, the if should be with $p + q$.

Redundancy Level. At this point, we have to decide if redundancy must be used or not, depending on loss rate perceived at reception and the target loss rate at destination τ. In absence of redundancy the loss rate equals π_1^*, if redundancy is used the loss rate after reconstruction is ρ.

It is not very clear how much audio loss can be tolerated at destination, it depends on voice reconstruction techniques, audio unit's length and the subjective tolerance of user. It appears [11] that with packet repetition techniques 5% loss rate after reconstruction can be tolerated. We assume then $\tau = 0.05$ and choice about redundancy becomes:

```
if (π₁* > τ) k = 2;
else k = 1.
```

It is interesting to remark that with this model and one single piece of redundant audio offseted, loss rates on the network up to 22% leads to less than 5% of loss rate after reconstruction. 30% network loss gives 9%. With these theoretical results more than one level of rendundancy seems useless.

Encodings. The last parameters to determine are the codecs used for enco-ding. We use a Free-Phone [13] inspired algorithm which solves the problem assuming a variable number of redundancy levels can be used. Since we have one piece of redundancy at most we derive a simpler algorithm that uses their general result:"The main information[3] should be encoded using the highest qua-lity encoding scheme (among those used to encode the main and the redundant information)".

A set of codecs must be chosen before running the algorithm. We picked one performing encoding in the range of 5.6 kb/s to 64 kb/s with 8 kHz sampling frequency: LPC (5.6 kb/s), GSM (13.2 kb/s), G726-16 (16 kb/s), G726-24 (24 kb/s), G726-32 (32 kb/s), G726-40 (40 kb/s) and A-law (64 kb/s). It repre-sents a quite homogeneous distribution of usable rates, so ensures that available bandwidth will be correctly used.

We ordered the codecs according to their bit rates, this differs slightly from their order of quality. One assumes that choosing a codec with higher bit-rate provides better audio quality.

k denotes the redundancy level ($k \in \{1;2\}$ here) x_i is the codec number for the ith copy, R is the available throughput and r_i is rate used by i^{th} codec assuming N codecs are available. The following algorithm is valid for any $k \geq 1$, it gives the codec number to use for i^{th} redundancy level $i = 1 : k$.

```
x₀ = 1; // ensures minimum encoding
for(i=1:k-1:i++)  xᵢ = 0;  r = rₓ₀;
i = 0;
do{
    if(xᵢ < N)  xᵢ++;
    r=0;
    for(j=0;  j<k;  j++){
        r += rₓⱼ;
    }
    if (r > R){
        if(x₀==1 && x₁==0) break;
        xᵢ--;
        break;
    }
    if(x_{k-1} = N) break; // all codecs have the best quality
    i = ++i % k;
}while(r < R);
```

Frame duration. We chose $t = 20$ ms as frame duration. It enables smaller latency and reduces time to wait to receive offseted packets.

---

[3] the first audio unit sent

5   Evaluation

5.1   Results

We have implemented the above described adaptive channel coding subdued to rate control. Tests were performed between EPFL (Switzerland) and two destinations in Europe: the University College of London and the Faculté Polytechnique de Mons in Belgium.

During one week connections were established with UCL at different times of the day. The loss rates never significantly exceeded 1% so SFEC was not needed and the best possible primary quality was chosen by the algorithm. The absence of loss is certainly explained by the very high speed links available beetween both universities.

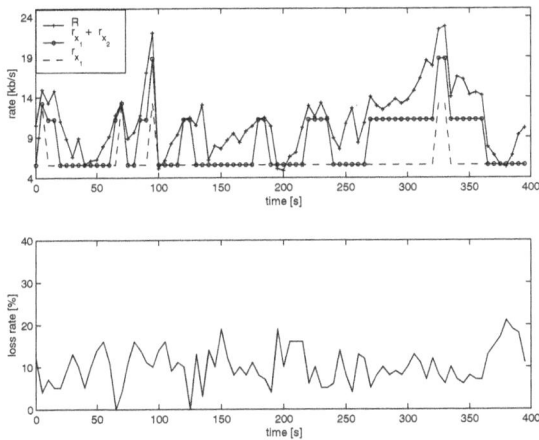

Fig. 3. Source rates and network loss

Connections with the Faculté de Mons were more lossy, the network was in a high degree of congestion since the RTT was about 1.3 sec. and the average loss rate above 10%. Figure 3 shows the TCP-friendly rate constraint R the payload rate $r_{x_1} + r_{x_2}$ and the rate used by the primary encoding r_{x_1} during a 400 sec. connection. This permits to see when redundancy is applied in response to loss on the network which is presented on the bottom.

The rate constraint is respected except when smaller than 5.6 kb/s because our algorithm keeps sending audio at this rate. In this situation the converstion should perhaps be stopped. When redundancy is needed bandwith is shared between primary and secondary encoding. In some cases redundancy should be used but a low rate constraint prevents it.

We could hear a fine quality improvement at destination when using SFEC to attenuate the effect of loss. We do not provide measurements of loss rate after reconstruction here since the efficiency of the method was demonstrated by [13] and [6].

5.2 Future Work

The most important thing to do seems to build a function to estimate quality perceived at reception. Without it, choices must be done statically. With it we could change dynamically offset or frame duration, and probably provide better quality.

Lots of hypothesis where made to build the model, the roughest was certainly to assume $q = 2/3$. It would be useful to have an estimator for this value, computed at destination and reported by RTCP.

In rare cases (loss rates > 22%), one more level of redundancy could be useful. RAT's implementation does not support this at the present time but it seems possible to add this functionality. Notice that then the presented model must be completed to compute the probability to lose all three packets. In our opinion more than three levels of redundancy are not useful.

We stated that a 64 kb/s (A-law) codec provides better quality than a 32 kb/s (G726-32), in our experience the difference is really faint. Instead it could be better to double sampling frequency and keep the G726 codec to get 8 kHz bandpass and 64 kb/s rate. It would have the convicing argument to provide larger bandwidth than Plain Old Telephone System.

6   Conclusion

In this paper we proposed a method to provide adaptive source coding for the Robust Audio Tool. It satisfies fair-rate constraints while improving efficiently quality perceived at reception thanks to adaptation of redundancy level, encoding rates and offset. Nevertheless the proposed algorithms are certainly not optimal because we had to make some suppositions.

Acknowledgements. We would like to thank RAT hackers, Orion Hodson and Colin Perkins for their active support during RAT's code exploration. We are also grateful to Alain Ruelle who allowed us to perform the tests with the Faculté Polytechnique de Mons.

References

1. L. Zhang, S. Deering, D. Estrin, S. Shenker, and D. Zappala. *RSVP:A new resourse ReSerVation Protocol.* IEEE Network Mag.,vol.7, no 5, pp. 8-18, sept 1993.
2. W. Almesberger. *Scalable Resource Reservation for the Internet.* Ph.D thesis number 2051, EPFL, November 1999.
3. C. Perkins O. Hodson V. Hardman. *A Survey of Packet-Loss Recovery Techniques for Streaming Audio.* IEEE Network Magazine, September/October 1998.
4. V. Hardman, M. A. Sasse, M. Handkey, and A. Watson. *Reliable audio for use over the Internet.* In Proceedings of INET'95, June 1995, Honolulu, Hawaii.
5. J.-C. Bolot, and A. Vega-García. *The case for FEC-based eror control for packet audio in the Internet.* to appear in ACM/ Springer Multimedia Systems.

6. M. Podolsky, C. Romer, and S. McCanne. *Simulation of FEC-Based error control for packet audio on the Internet.* In Proceedings of IEEE Infocom'98, April 1998.
7. Robust Audio Tool. *http://www-mice.cs.ucl.ac.uk/multimedia/software/rat/.*
8. H.Schulzrinne. *RTP: A Transport Protocol for Real-Time Applications. RFC 1889.* IETF 1996.
9. H.Schulzrinne. *RTP profile for audio and video conferences with minimal control. RFC 1890.* IETF 1996.
10. C. Perkins, and J. Ott. *A message bus for conferencing systems.* http://www.cs.ucl.ac.uk/staff/c.perkins/confbus/.
11. I. Kouvelas, O. Hodson, V. Hardman, and J. Crowcroft. *Redundancy control in Real-Time Internet Audio Conferencing.* In proceedings of International Workshop on Audio-Visual Services over Packet Networks (AVSPN97), September 1997, Aberdeen, Scotland.
12. P. Thiran. *Processus Stochastiques pour les communications.* Module 6, p. 120 EPFL 2000 http://icawww1.epfl.ch/cours_thi/index.html.
13. J.-C. Bolot, S. Fosse-Parisis, and T. Towsley. *Adaptive FEC-Based Error Control for Internet Telephony.* In Proceedings of IEEE Infocom'99, March 1999, New York.
14. A. Vega-García. *Mecanismes de controle pour la transmission de l'audio sur l'internet.* PhD thesis, Université de Nice-Sophia Antipolis, October 1996.
15. D. Ratton Figueiredo and E. de Souza e Silva. *Efficient Mechanisms for Recovering Voice Packets in the Internet.* In Proceedings of IEEE Globecom'99, December 1999, Rio de Janeiro, Brazil.
16. J. Mahdavi and S. Floyd. *TCP-Friendly Unicast Rate-Based Flow Control.* Draft posted on end2end mailing list, January 1997. http://www.psc.edu/papers/tcp_friendly.html.

A Robust JPEG Coder for a Partially Reliable Transport Service

Johan Garcia and Anna Brunstrom

Karlstad University, SE-651 88 Karlstad, Sweden
{johan.garcia,anna.brunstrom}@kau.se

Abstract. The usage of multimedia applications on the Internet has seen phenomenal growth in recent years. Transport protocols that provide partially reliable service have been suggested as one approach to better handle the requirements of these applications. A partially reliable service provides applications with the possibility of a flexible tradeoff between reliability and delay/throughput. Appropriately designed coders are, however, required to fully utilize a partially reliable service. In this paper we present a JPEG image coder tailored to suit the behavior of a partially reliable byte stream service. With regular JPEG, data loss typically results in severely distorted images. The robust recoder employs three major modifications to standard JPEG in order to adapt to the partially reliable transport: (1) extended resynchronization markers in order to be able to resynchronize effectively, (2) block interleaving in order to spread out the loss of a packet across the image and (3) error concealment in order to minimize the perceived quality loss. The modifications incorporate both new inventions, such as random window interleaving, as well as variations of previously known techniques.

1 Introduction

Internet has proven to be a formidable success, and it has continued to grow with new applications and new communication link technologies never originally conceived. The research community continues to enhance and evolve the Internet and its services. One area of research centers on the concept of partially reliable and partially ordered transport services. The most common Internet transport protocol, TCP [16], provides a fully reliable, ordered transport service. Since the underlying IP layer only provides an unreliable, unordered service the transport layer must in this case provide reliability and reordering mechanisms. The mechanisms used to obtain this incur a user cost in the form of added delay needed for retransmissions and reorder buffering. Even if the application does not need a fully reliable, fully ordered transport service, it experiences the penalty of added delay. Neither TCP nor UDP [15], the other common Internet transport protocol, is suitable for applications that can accept some controlled degree of loss or out-of-order delivery. By using transport protocols such as POC [2] or PECC [8] that provide a transport service with flexible reliability and/or ordering constraints, it is possible to obtain increased application performance by relaxing the reliability and ordering constraints. Increased performance for a video application by relaxed ordering and reliability constraints has been shown in

H. Scholten and M. van Sinderen (Eds.): IDMS 2000, LNCS 1905, pp. 14–25, 2000.
© Springer-Verlag Berlin Heidelberg 2000

[18]. Increased performance can also be obtained for non-streaming applications such as image transfer. One example is the progressive display of GIF images [10]. The JPEG image coding standard [14] is very common on the Internet due to the vast amount of web pages with embedded JPEG images. No reports of the possible benefits of using a partially reliable service for transferring JPEG web images has been found in the literature. This paper describes the design and implementation of a JPEG image coder adapted to work well with a partially reliable transport service. This coder is intended to be used as a component in a test environment where system-wide advantages and disadvantages of using partially reliable transport for web transfers can be evaluated. The robust coder contains both known techniques as well as new ones, such as random window interleaving. The remainder of this document is structured as follows. In Sect. 2 background information and related work is presented. Section 3 gives a brief overview of JPEG, focusing on the details needed to understand the modifications necessary to accommodate for partial reliability. These modifications are described in Sect. 4. Finally, conclusions are given in Sect. 5.

2 Background and Related Work

We have developed PRTP, a partially reliable transport protocol, which allows the required reliability level to be specified by the application. This allows the application to influence the tradeoff between latency and reliability to optimally suit the current operating conditions and user preferences. The PRTP transport protocol allows the reliability to be specified between 0 and 100%. Further details of PRTP and recent performance measurements are presented in [3]. For ease of implementation and simple WWW integration, the first implementation of PRTP is based on TCP. This allows the use of TCPs mechanisms for flow control and retransmissions. It also leads to some disadvantages inherent in TCPs design, such as the inability of performing Application Layer Framing (ALF) [5]. An extensive simulation study [19] shows that PRTP has a more aggressive congestion control behavior than TCP. As discussed in [3], a more aggressive behavior can however be appropriate in certain situations.

When transporting image data, the loss of data may lead to information loss. One characteristic of the human visual system is its ability to extrapolate lost information to a certain degree. In order to ease this extrapolation, the coding of visual data must take the possibility of data loss into account. However, when an excessive amount of data is lost, a severe degradation in image quality will occur. This motivates the use of a partially reliable protocol that puts an upper bound on the amount of data loss possible, and correspondingly ensures a minimum image quality.

Other work examining image transfer over unreliable or partially reliable channels include [4,6,7,11,13,17]. However, none of these proposals present a solution that integrates easily into a web application, codes color images of arbitrary size and is able to handle the packet loss characteristics provided by the PRTP partially reliable transport service.

3   JPEG Basics

JPEG is a relatively complex standard that can be operated in a number of modes. Our coder is primarily aimed at transfer of web-type images, and these are typically coded in the lossy sequential baseline mode and stored in the JFIF [12] interchange format. In order to give a background on how the suggested adaptation differs from the normal approach, a brief, somewhat simplified, introduction to baseline JPEG encoding is provided. The encoding of an image is typically comprised of the following steps:

1. Change the image representation to the YC_bC_r color space and down-sample the chrominance components (C_b, C_r).
2. Split the components into 8x8 sample point blocks and perform a DCT (Discrete Cosine Transform) on each block. The resulting 64 coefficients are divided into one DC coefficient holding the average value of the sample points in the block, and 63 AC coefficients holding the amount of different spatial frequencies.
3. Quantize the coefficients according to the quantization tables. Since higher spatial frequencies are less perceptually important, these are quantized more aggressively.
4. Code the DC coefficient using predictive coding with the DC value of the previous block used as predictor. Run-length encode the AC coefficients in zigzag order.
5. Huffman encode the run-length encoded data and insert JFIF headers.

JFIF places the headers in the beginning of the file together with the tables necessary to perform decoding. In baseline JPEG two quantization tables and four Huffman tables are required. The size of these headers and tables range from 200 to 600 bytes. The header data must be reliably transmitted, which is easily done with PRTP. After the header data follows the Huffman coded data representing the image. The exact data organization is dependent on the down-sampling used. A typical configuration is 2hx2v down-sampling which causes the data to be organized as four Y blocks, then one C_b block and one C_r block. In this case there will be six blocks grouped together, and this is the default configuration in our coder. This group of associated blocks is called a MCU (Minimum Coded Unit) and is used as an atomic unit for resynchronization.

4   JPEG Adaptation

The intended application area is web browsing, and the data loss type considered is packet loss, not bit errors or bit erasures. The coder must be able to handle the loss of one or more packets of variable size, and still resynchronize and conceal the loss as well as possible. The output quality from the coder should degrade gracefully as the loss rate increases. The typical loss rates expected will be lower than 10 %, with the transport service set to guarantee a maximum loss rate of 20%. In order to adapt the JPEG coding for partial reliability a strategy consisting of three steps is employed:

1. Extend the resynchronization capabilities of regular JPEG so that consecutive losses up to several kilobytes can be handled without losing too much data between the end of a loss and the first restart marker.
2. Perform interleaving so that the lost data is not aggregated in one place in the image.
3. Try to conceal the lost data by using redundancies not removed by the source coding.

The steps are further explained below, and their effects are illustrated in Figs. 1-8. The images in the figures were produced by our coder and used an original web image of typical quality as input. The original image, shown in Fig. 1, is a 276 x 185 pixel color image coded at 2.3 bpp. All test images except the original image lose the same amount of data, 1.5 kbyte, at the same positions in the transfered data stream. This corresponds to around 10% packet loss, distributed as three lost packets of 512 bytes each. As can bee seen in Fig. 2, a regular JPEG image becomes considerably distorted by a 10% data loss.

Fig. 1. Original image   **Fig. 2.** Original image after loss

4.1   Decoder Resynchronization

Decoder resynchronization is needed to allow the decoder to come to a known state after a data loss of unknown length. For a web application, the losses are of unknown length since the application have no control over how the transport protocol distributes the data into segments. This also precludes the possibility of using ALF principles which would considerably ease the task of decoder resynchronization. We propose the use of *extended* resynchronization markers due to the specific problems that occur when using a partially reliable protocol that provides a stream abstraction, such as PRTP, in conjunction with JPEG coding. In this case the resynchronization problem can be considered to consist of three subproblems:

- Coder internal state dependence
  Depending on the specific coding technique used, data can be more or less state dependent during decoding. GIF [9] for example is highly state dependent since it builds up a codebook using previous data. One approach to lower the state dependence for GIF images is presented in [1]. The reduction in state dependence has to be paid for with lower compression performance. JPEG has a lower degree of state dependence, in this case originating from the predictive coding used for the DC coefficient.
- Data stream semantics
  A problem occurs if the data stream contains bytes that have different semantic meaning or is comprised of variable length codes such as Huffman codes. When a loss occurs, the decoder must to resynchronize itself with the data stream so that it can apply the correct semantics to the data received after the loss.
- Media positioning
  After a loss of unknown length, the data that comes after the loss must be processed. However, as the amount of lost data is unknown, it is not possible to know how much space in the image the lost data corresponds to. This makes it impossible to map the data received after a loss to a correct position in the image.

The JPEG standard provides optional resynchronization capability. This capability is based on the insertion of restart markers into the data stream. Eight unique restart markers are specified by the JPEG standard. The periodical insertion of restart markers at MCU boundaries solves the coder internal state dependence since the predictive coder used for the DC coefficient is reset at restart markers. The data stream semantics is resolved since restart markers always occur at MCU borders. The media positioning problem is resolved to a degree since the number of lost MCUs can be inferred from the marker number and R, the number of MCUs between resynchronization markers. This is however only true for smaller losses that do not extend over more than eight restart markers. For data losses encompassing more than $8R$ MCUs the decoder cannot unambiguously determine the actual amount of data that has been lost, and hence cannot position the incoming data correctly in the image. The value of R can be increased, hence allowing for larger data losses without loss of decoder synchronization. Increasing R however also increase M, the mean amount of correct MCUs received after a data loss that cannot be decoded due to lack of decoder synchronization ($M = R/2$). This creates the optimization problem of choosing a large enough R to ensure resynchronization while minimizing M. Rather than performing this optimization, we use extended restart markers that instead of eight unique markers provide 190 markers. This allows R to be set to 1 and still guarantees unambiguous media positioning for all practical image sizes. Since the extension of the restart markers uses unused JPEG marker space, the extended restart markers provide enhanced resynchronization without having to use longer markers. The effect of using extended resynchronization markers instead of regular JPEG markers is illustrated in Figs. 3 and 4, where Fig. 3 shows regular JPEG resynchronization markers inserted at each MCU row and Fig. 4 shows the effect of using extended restart markers inserted at each MCU.

Fig. 3. JPEG regular restart   **Fig. 4.** Extended restart

When a data stream is adapted for resynchronization it becomes larger than the original. This data expansion occurs since resynchronization markers are inserted and the compression efficiency is lowered by the periodical resets of the predictive coding used for the DC coefficients. The data expansion can be expressed as an additional increase α of the resynchronized data over the original. In [11], the value of α is said to be approximately $\alpha = k\frac{\rho}{R}$, where ρ is the compression ratio and k is a constant that has been experimentally determined to 0.04 for black and white pictures. The value of k for color images is dependent on the amount of down-sampling of the chrominance components. Preliminary measurements indicate a data expansion lower than 10% for color images.

4.2   Interleaving

Interleaving is performed with the objective of distributing data losses as evenly as possible in the image in order to facilitate error concealment by some kind of interpolation. In order to achieve this distribution two problems must be addressed:

1. How to ensure that a data loss of a typical size is as evenly distributed as possible over the image.
2. How to ensure that the interleaved blocks lost in a second data loss is distributed as far away from the previously lost blocks as possible.

In addition to distributing the blocks, the coefficients can also be distributed [4]. The use of coefficient distribution is decoupled from the interleaving method used for block interleaving. Three block interleaving strategies will be presented, and these can all be used either with or without coefficient interleaving.

In the following the length of a typical data loss is assumed to be L bytes. Such a loss should be as evenly distributed as possible over an image. To do this, the size of the image has to be known to ensure that the blocks will not be placed above each other by the interleaver. The pixel width and height of the image are denoted by W and H. The compression ratio expressed as ρ (bits/pixel) is also interesting since it determines the average number of lost blocks that maps to a data loss of size L.

Fig. 5. Max. spreading interleaved   **Fig. 6.** Max. spreading concealed

Maximum spreading interleaving. A simple interleaver would just use the values discussed above to space the interleaved blocks D blocks apart according to a formula such as

$$D = \left\lceil \frac{8F_x\rho \left\lceil \frac{W}{S_x^W 8} \right\rceil \left\lceil \frac{H}{S_x^H 8} \right\rceil}{L} \right\rceil \qquad (1)$$

The value F_x is used to specify the relative amount of sample points present in component x, and $S_x^{W,H}$ is the amount of down-sampling used. For the typical 2hx2v down-sampling the values for the Y component are $F_Y = \frac{4}{6}$ and $S_Y^{W,H} = 1$. For the color components the values are $F_{Cb,Cr} = \frac{1}{6}$ and $S_{Cb,Cr}^{W,H} = 2$. If D is selected so that $W \bmod D \neq 0$ and $D \bmod W \neq 0$, a satisfactory solution is achieved. This approach spreads a data loss evenly over the whole image, and for losses of length L a maximum spreading is achieved[1]. However, if a second data loss occurs so that the deinterleaved blocks neighbor the previously lost blocks, then many lost blocks will have lost a neighboring block as illustrated in Fig. 5. This occurs since both losses are interleaved with a fixed interblock distance D. The repeated absence of a neighboring block severely hampers the performance of error concealment algorithms. This is the main disadvantage of this method.

Random block interleaving. In order to minimize the probability of multiple data losses incurring a stride of lost neighboring blocks it is possible to use random block interleaving. This interleaving uses a seeded random number generator to obtain the interleaved position of each block. This method requires that a common random number generator is implemented in the adapted JPEG coder and decoder, or that a generator is available from the system. The reported experiments used the random number generator present in Linux. The algorithm for random block interleaving is as follows:

1. Place the sequence number of each image block in an array A. The array will have B elements, where B is the number of blocks in the component under interleaving.
2. Feed the initial seed to the random number generator $G(\ )$ and set the counters N and M to 0. The function $G(y)$ returns a value between 0 and $y - 1$.
3. Assign the array index I as: $I = N + G(B - N)$
4. Switch the values at $A[I]$ and $A[N]$. Increment N by one.
5. Update the output array O as follows: $O[M] = A[I]$. Increment M by one.
6. Goto step 3 until M equals B.

The above algorithm will produce an interleaving that is (pseudo-)randomized and hence will not produce strides of lost blocks which also have lost neighbors as is shown in Fig. 7. However, for the case with only one data loss, this algorithm may produce a few lost blocks which also has a lost neighbor. This is a disadvantage compared to the maximum spreading method which will not interleave so that neighbor blocks are lost for one data loss.

---

[1] If an image contains areas which differ greatly in the amount of encoded data produced, this will cause imbalance in the spreading.

Fig. 7. Random block interleaved   **Fig. 8.** Random block concealed

Windowed random interleaving. In order to improve the random block interleaving we have devised a window-based random interleaving which is capable of alleviating the neighbor block loss problem for single data losses. The algorithm described above is modified as follows. First a window-size P is calculated. The value P signifies the number of MCUs that map to a loss of size L. If a given block after interleaving has any of its neighboring blocks within P MCUs, then a data loss could cause a neighbor block loss. In order to avoid this, a test is performed in step 4 to see if the block to be interleaved has its above or left neighbor located within P MCUs from the MCU to which $O[M]$ belongs. This could cause a loss of size L to include a neighboring block, which is undesirable. If the block has a neighbor within the window, a new counter is initialized $N' - N + 1$ and steps 3 and 4 above are repeated with N' and a new test is performed. This is repeated until a sequence number producing a block without any neighbor within the P window is obtained or N' equals B, in which case a placement of the block leading to a neighbor within the window is inevitable, and N is used. The value P is dependent on the down-sampling used and can be calculated as

$$P = \frac{L}{max\left\{S_Y^W, S_{Cb}^W, S_{Cr}^W\right\} max\left\{S_Y^H, S_{Cb}^H, S_{Cr}^H\right\} 8\rho} \qquad (2)$$

4.3   Error Concealment

When a data loss has occurred there will be a number of blocks with missing information. If coefficient interleaving has been used, then there will be many blocks which have one or a few coefficients missing. If coefficient interleaving was not used there will be a few blocks where all coefficient values are lost. The error concealment scheme employ the same technique regardless of whether coefficient interleaving was used or not. Figures 5 to 8 were produced with coefficient interleaving, and Figs. 6 and 8 show the effect of the described error concealment.

DC coefficient. The DC coefficient contains the average color of the block. If this value has been lost it is important to try to interpolate it in a way that produces the least visible difference. If the DC coefficient differs too much from the correct value, the edges around the block may become noticeable. These edges are especially perceptively important since they are straight line artifacts, which are not well masked by the visual system.

We base our reconstruction on a method mentioned in [17], which we refer to as the least delta method. This method compares the difference between the two vertical neighbor blocks with the difference from the two horizontal neighbor blocks. The two blocks which have the lowest difference are averaged to obtain the interpolated value.

AC coefficients. The AC coefficients contain information on the presence of different spatial frequency contents in the block. This fact is used when performing the interpolation as described in [4]. The coefficients can be divided in three categories: (1) containing mainly horizontal frequency information, (2) containing mainly vertical frequency information or (3) containing both. Accordingly, coefficients containing information about vertical patterns are interpolated from the blocks directly above and below. Correspondingly, coefficients with horizontal patterns are interpolated from blocks to the left an right of the missing block. This can be further enhanced by checking that the maximum difference between the two blocks used for interpolation is not too large, in which case the interpolation instead use four or eight neighboring blocks. For coefficients that contain both horizontal and vertical frequencies simple four-way interpolation is used.

4.4   Measurements

Numerical results that quantify the quality difference visible in Figs. 2-8 are given in Table 1. All values are relative to the original web-quality image shown in Fig. 1 and were computed for grayscale versions of the images. The peak signal-to-noise ratio (PSNR) is a mathematically derived value often used for image comparisons. In addition to the PSNR a perceptual metric based on just-noticeable-difference (JND) [20] is also presented. For PSNR high values are desirable whereas for JND low values are better. The values are image-dependent and other images will produce different results. In order to fully evaluate the performance of the proposed methods more comprehensive tests must be performed.

Table 1. Error concealment performance

Image	PSNR	JND
Fig. 2: 10% loss	12.437	243.487
Fig. 3: 10% loss, JPEG regular restart	19.066	159.023
Fig. 4: 10% loss, extended restart	21.802	134.521
Fig. 5: 10% loss, maximum spreading interleaved	23.022	118.748
Fig. 6: 10% loss, maximum spreading interleaved concealed	28.888	32.254
Fig. 7: 10% loss, random interleaved	22.138	128.095
Fig. 8: 10% loss, random interleaved concealed	29.733	52.807

5 Conclusions

By coupling a partially reliable transport protocol to suitable image coding improved image transfer performance can be achieved in lossy networks. The partially reliable transport protocol guarantees that at least a specified fraction of the data is delivered to the application, and hence a lowest acceptable image quality will be enforced. We have designed and implemented a modified JPEG coder suitable for such applications. Some of the required modifications, such as random window interleaving are new, while others are adoptions of known techniques. Our recoding results show that JPEG can be adapted to handle packet losses in a graceful way. For future work we intend to integrate a compressing text coder capable of handling a partially reliable service with the JPEG coder and a proxy system.

References

[1] P. Amer, S. Iren, G. Sezen, P. Conrad, M. Taube, and A. Caro. Network-conscious GIF image transmission over Internet. *Proc 4th Int'l Workshop on High Performance Protocol Architectures (HIPPARCH)*, June 1998.

[2] P. D. Amer, C.Chassot, T. Connolly, M. Diaz, and P. Conrad. Partial order transport service for multimedia and other applications. *IEEE/ACM Transactions on Networking*, 2(5), October 1994.

[3] A. Brunstrom, K. Asplund, and J. Garcia. Enhancing TCP performance by allowing controlled loss. *Proceedings SSGRR 2000 Computer & eBusiness Conference*, L'Aquila, Italy, August 2000.

[4] E. Y. Chang. An image coding and reconstruction scheme for mobile computing. In T. Plagemann and V. Goebel, editors, *LNCS 1483: Proceedings of IDMS*, Oslo, Norway, Sept. 1998. Springer.

[5] D. D. Clark and D. L. Tennehouse. Architectural considerations for a new generation of protocols. *Proceedings of ACM SIGCOMM*, 1990.

[6] P. Cosman, J. Rogers, P. G. Sherwood, and K. Zeger. Image transmission over channels with bit errors and packet erasures. *Proceedings of the 32nd Asilomar Conference on Signals, Systems, and Computers*, Monterey, November 1998.

[7] J. M. Danskin, G. M. Davis, and X. Song. Fast lossy internet image transmission. In *Proceedings of ACM Multimedia*, November 1995.

[8] B. J. Dempsey. *Retransmission-based error control for continuous media.* PhD thesis, University of Virginia, 1994.

[9] Graphics Interchange Format, version 89a. Technical Report, Compuserve Incorporated, Columbus, Ohio, July 1989.

[10] S. Iren, P. D. Amer, and P. T. Conrad. Network-conscious compressed images over wireless networks. In T. Plagemann and V. Goebel, editors, *LNCS 1483: Proceedings of IDMS*, Oslo, Norway, Sept. 1998. Springer.

[11] L. Jianhua, M. L. Liou, K. B. Letaief, and J. C.-I. Chuang. Mobile image transmission using combined source and channel coding with low complexity concealment. *Signal Processing: Image Communication*, 12(2):87–104, April 1998.

[12] The JPEG file interchange format. *Maintained by C-Cube Microsystems Inc., ftp://ftp.uu.net/graphics/jpeg/jfif.ps.gz.*, 1998.

[13] A. E. Mohr, E. A. Riskin, and R. E. Ladner. Graceful degradation over packet erasure channels through forward error correction. *Proceedings of the 1999 Data Compression Conference (DCC)*, March 1999.

[14] W. B. Pennebaker and J. L. Mitchell. *JPEG Still Image Data Compression Standard.* Van Nostrand Reinhold, New York, NY, USA, 1993.

[15] J. Postel. RFC 768: User Datagram Protocol, Aug. 1980.

[16] J. Postel. RFC 793: Transmission Control Protocol, Sept. 1981.

[17] J. Ridge, F. W. Ware, and J. D. Gibson. Image refinement for lossy channels with relaxed latency constraints. *IEEE Wireless Communications and Networks Conference 1999 (WCNC'99)*, 2:993–997, September 1999.

[18] L. Rojas-Cardenas, E. Chaput, L. Dairaine, P. Sénac, and M. Diaz. Transport of video over partial order connections. *Computer Networks*, 31(7):709–725, April 1999.

[19] S. K. Schneyer. The effects of PRTP on the congestion control mechanisms of TCP. *Master's Thesis 1999:05, Karlstad University*, June 1999.

[20] A. B. Watson. et al. DCTune 2.0. *http://vision.arc.nasa.gov/dctune/* , 2000.

A Dynamic RAM Cache for High Quality Distributed Video

Nicholas J. P. Race, Daniel G. Waddington, and Doug Shepherd

Distributed Multimedia Research Group,
Computing Department, Lancaster University,
Lancaster, UK
{race, dan, doug}@comp.lancs.ac.uk

Abstract. As technological advances continue to be made, the demand for more efficient distributed multimedia systems is also affirmed. Current support for end-to-end QoS is still limited; consequently mechanisms are required to provide flexibility in resource loading. One such mechanism, caching, may be introduced both in the end-system and network to facilitate intelligent load balancing and resource management. This paper introduces new work at Lancaster University investigating the use of transparent network caches for MPEG-2. A novel architecture is proposed, based on router-based caching and the employment of large scale dynamic RAM as the sole caching medium. Finally, the architecture also proposes the use of the ISO/IEC standardised DSM-CC protocol as a basic control infrastructure and the caching of pre-built transport packets (UDP/IP) in the data plane. The work presented in this paper is in its infancy and consequently focuses upon the design and implementation of the caching architecture rather than an investigation into performance gains, which we intend to report in future publications.

1 Introduction

The delivery of real time continuous media such as digital audio and video is becoming increasingly important in today's computing environment. However, the high data rates and strict delivery constraints that continuous media imposes, have proven to be difficult to meet in high demand situations. A wealth of research has been carried out over the past ten years to solve these problems, combining developments in efficient filing systems, highly optimised scheduling policies, admission control and resource management [1]. Whilst research has led to high performance servers, there are still complex issues surrounding the end-to-end delivery of audio and video across the Internet. The large data size not only places a substantial load on the network, but also represents a high cost for video distribution, particularly if expensive backbone links are involved in the delivery process. One approach to reducing this cost is the use of caching. By inserting cache nodes in the local network, popular videos or clips can be serviced from local caches, resulting in a reduced network load over the greater distance. Caching also brings additional benefits, in that videos streamed from the cache decrease the load on the server,

H. Scholten and M. van Sinderen (Eds.): IDMS 2000, LNCS 1905, pp. 26-39, 2000.
© Springer-Verlag Berlin Heidelberg 2000

allowing it to service other requests. As caches are located close to clients, there is also reduced latency in establishing connection set-ups.

Advances in hardware have brought about increased dynamic RAM capacities at a reduced cost; a pattern that is expected to continue for years to come. Coupled with the move to 64-bit architectures, there is now more potential than ever to use RAM as a medium for video storage. RAM brings many advantages to the real-time delivery of continuous media; not least the elimination of disk I/O bottlenecks and increased overall bandwidth, which offer the potential to support a very large number of concurrent accesses. When serving a large number of streams, it is suggested that RAM provides a more economic solution than disks due to its higher bandwidth capabilities [2].

In this paper we present a cache node using main memory as the primary caching medium. In order to maximise throughput performance between RAM and the network interfaces, the node uses pre-built IP, which stores video data in main memory in a pre-packetised network/transport format, to be inserted into the IP stack with minimal processing overhead. Much of the implementation focuses upon the use of MPEG-2 as a streaming media type. This is the generally accepted format for broadcast quality digital video and is generic in its deployment capabilities. The paper is divided into the following sections. Section 2 gives an overview of the caching system from an architectural perspective, describing envisaged deployment scenarios and integration into the control environment. Continuing, section 3 discusses some of the implementation aspects and the realisation of the caching node in Windows 2000. Finally, sections 4 and 5 overview some related work in the area and outline some directions of interest for future work.

2 Caching Architecture

This section provides an overview of the caching architecture, describing envisaged deployment scenarios and integration into the control and data environment.

2.1 Transparent Caching

Traditional web caching is achieved through the use of a proxy server, which is physically located between a client web browser and a server. The proxy intercepts all packets, and examines each one in order to determine whether it can service the request itself, or whether an additional request has to be made to the server. Proxies generally need to be explicitly configured within the web browser for each client, which presents a large cost and an unscalable solution for service providers.

A fundamental objective of the caching node is that it should be completely transparent to both video client and server so that no additional cost of ownership is incurred, hence making the node suitable for wide-scale deployment. A number of approaches to transparent caching were considered, the most common of which use L4 switches or policy-based routers. In this case, user requests for web pages are diverted by a router/switch to a local cache, and all other network traffic forwarded as normal. If the requested data is not available in the local cache, a separate TCP connection is established to the web server in order to retrieve (and store) it. The data is then returned to the client, with any subsequent requests for the same data serviced from the local storage of the cache.

Our proposed approach includes IP routing functionality within the cache node itself (bearing in mind the cache node is generally positioned at the edge of the network and not in the core). This ensures that all requests pass through the node, enabling it to filter out requests and cache incoming data. Both the server and client are unaware of the presence of the cache, and there are no configuration changes required within the router in order to forward requests to the cache. Cache nodes are deployed within the local area network close to clients, and do not perform the intensive routing operations associated with core routers.

2.2 Topology

The basic networking architecture is shown in Fig. 1. A continuous media server that can support a number of concurrent stream accesses is connected via a high-speed interconnect to a backbone router. The cache node is installed in each LAN, acting as an IP router and is directly attached to the backbone router connecting the LAN to the WAN. The proposed architecture uses UDP/IP over an ATM-based IntServ/DiffServ infrastructure, and assumes negligible packet loss on such connections. We believe that UDP is the preferable choice over TCP since it does not provide error correction and control, and is connection-less and therefore ideally suited to interception and masquerading.

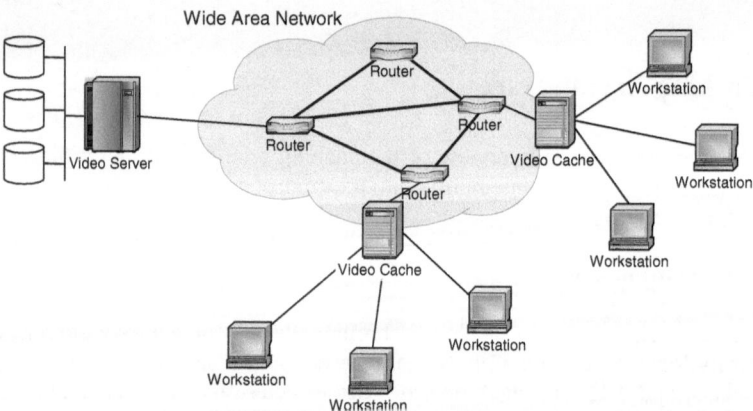

Fig. 1. Network Topology

The caches are able to use pre-built transport units, in this case UDP packets, as a unit of caching. This avoids both the need for user level processing and the need for re-assembly of data leaving the node. Because of the simplistic and connection-less nature of UDP, packets can be easily sent from the cache node in a form of masquerading whereby the client is unable to determine the originality of the data and thus the caching appears to be transparent.

2.3 Integration with Multimedia Data Architecture (MPEG-2)

The MPEG-2 (Moving Pictures Experts Group) ISO/IEC 13818 standard [3] is designed to provide high quality video by exploiting spatial and temporal redundancies in the input source in order to achieve compression. MPEG-2 has been widely adopted as the standard for digital video, and is used by DVB (Digital Video Broadcast), DAVIC and DVD.

The MPEG-2 systems layer (ISO/IEC 13818-1) specifies two mechanisms for the multiplexing and synchronisation of elementary video and audio streams to form a single data stream that is suitable for storage or transmission. Each mechanism is tailored for a different operating environment. The first scheme, known as a *Program Streaming,* uses large packets of variable sizes and is designed for use in a largely error-free environment. The scheme is similar to the MPEG-1 multiplexing standard and can support only one program (a number of elementary video/audio streams with a common time base). The second scheme, known as a *Transport Streaming*, can combine multiple programs with independent time bases into a single stream. Transport Streams use fixed length 188-byte packets, with additional error protection and incorporate timestamps within the packets to ensure correct synchronisation. They are intended for use in error-prone environments. However, because of their additional complexity, Transport Streams are more difficult to create and de-multiplex than Program streams, and therefore the initial caching architecture focuses upon the handling of Program Streams only.

Program Streams are constructed from one or more Packetised Elementary Stream (PES) packets. A PES packet consists of a header and a payload. The header contains important timing information in the form of a *Presentation Time Stamp* (PTS) and a *Decoding Time Stamp* (DTS) which is used to ensure correct synchronisation at the decoder. The header also contains a *stream_id* field in order to distinguish one elementary stream from another within the same program. The payload consists of the video and audio data bytes that have been encoded from the original source stream. In a Program Stream, PES packets are arranged into logical groups known as 'packs'. A pack consists of a pack header, an optional system header and any number of PES-packets. The pack header also contains important timing information in the form of a *System Clock Reference* (SCR).

2.4 Integration with Multimedia Control Architecture (DSM-CC)

In designing a caching architecture, integration into the media control architecture is essential. In this work we have adopted the ISO standardised Digital Storage Media – Command and Control (DSM-CC) protocol [4]. This protocol is a specific application protocol, intended to provide the basic control functions and operations to manage digital storage bit streams akin to MPEG-2. It is designed for the command and control of retrieval/storage applications such as video-on-demand, interactive video services and electronic publishing. In the proposed caching architecture, DSM-CC is used in stand-alone mode (i.e. it is not embedded within the data stream). It is encapsulated in PES packets and transmitted over UDP/IP. To initiate simple playback a client issues a bit stream select command to a given DSM server. This select command carries a bit stream identifier corresponding to an ISO/IEC 13818 stream. Providing the server holds the appropriate stream, a select acknowledgement is returned (to the sender's UDP port). The client can now control the stream (play, stop, pause, resume and jump) through subsequent retrieve commands.

Within the caching topology, the cache server intercepts DSM-CC requests and if it is able to provide the desired bit stream from cache, then masquerades as the server itself (this is possible because of the simplicity of the control architecture), otherwise the DSM command is forwarded to the appropriate server. In some scenarios the cache may be able to provide a portion of the bit stream (termed a partial hit), in which case it is necessary for the cache to send a request to the server itself and then 'hand over' at the appropriate time. Providing that this request is made at the appropriate time, the retrieved stream can be optionally cached and then forwarded.

To determine which UDP packets make up a given bit stream, the caching node must assume that all PES packets originating from the DSM server, to a given port and address, constitute the same bit stream until an MPEG-2 end code is received. It is therefore necessary for the cache to 'snoop' the UDP payload during the caching process. Where IPv6 is deployable, the cache is able to use the flow label instead, in which case the need for snooping is avoided.

2.5 Caching Behaviour

Because of the large size of MPEG-2 video objects (typically in the order of a number of megabits per second, per film), it is intended that the node cache complete video clips and popular portions of large video objects. The behaviour of the cache is controlled by a set of policies that dictate what to do in the event of a cache miss, a cache hit or a lack of available cache memory. A range of cache replacement policies are under consideration for use within the caching architecture. These include strategies from traditional caching research such as LRU (Least Recently Used), LFU (Least Frequently Used) and LRU-k [5]. However, it is clear from an analysis of logged accesses to video data [6] that the initial portion of a video is often used to determine a users interest. The importance of a replacement policy in maintaining the initial portion of a video within the cache is also affirmed in [7], who propose a prefix caching technique that stores the initial frames of popular clips. It is suggested that this hides any latency, throughput and loss effects between the server and the proxy, and if combined with work ahead smoothing can reduce the variability of network resources between the proxy and client.

3 Design and Implementation

This section presents an overview of the engineering and implementation of the caching architecture. A significant contribution of this work is in the realisation of the system within a prototype environment. Although the prototype is still in its infancy, already the design and implementation have provided a number of additional insights into the problems faced in network caching strategies.

The prototype cache node is based on extensions to the Microsoft Windows 2000 Advanced Server operating system and the hardware consists of an Intel L440GX+ motherboard, which is specifically designed for enterprise servers. The existing prototype supports 2Gb of main memory (74bit registered ECC DIMMS). In terms of network interfaces, the L440GX+ motherboard incorporates an Intel 82559 Fast Ethernet (100Mbps) adapter. This is coupled with an additional Intel PRO/100B PCI Ethernet adapter providing a secondary routing interface. Finally, the cache node's processing power includes dual Intel Pentium III 450MHz processors.

In terms of client/server networking infrastructure, the prototype cache node is connected as a gateway router between two private IP sub-networks. Consequently, communications between given client/server nodes are forced to pass through the cache. The video distribution application used within the experimental environment is a proprietary application providing simple MPEG-2 streaming over UDP/IP. The application streams according to in-line transmission rates (pack header *program_mux_rate*), assembling PES packets into transport payloads. The server application is not designed for serving a large client base and does not employ techniques such as striping, commonly found in commercial video distribution applications.

At the client side, the payload from the UDP packets is passed from the socket layer into a Creative Labs Dxr2 MPEG-2 decoder card, supported by a Linux 2.2 open source driver. This client is able to decode and render program streams to both overlay and external analogue output.

3.1   Kernel Mode Processing

The Windows 2000 operating system is a micro-kernel architecture, based on ideas originally founded within the Mach operating system from Carnegie Mellon University [8]. This design means that privileged instructions, which are potentially dangerous to system stability, are restricted to a very small subset of functions that reside within the kernel. Furthermore, access to these is strictly controlled by the privilege level of given process or application. In the Intel Processor Architecture [9] this protection is managed by one of four privileges levels (also known as ring 0 – ring 3). These determine which code and data segments are accessible by the calling program, and thus what instruction sets may be accessed. Conventionally, ring 0 represents the most privileged level, reserved for kernel routines for memory control, error handling, task switching etc. Rings 1-3 are reserved for drivers, operating system extensions and applications respectively. However, the Windows 2000 architecture only uses ring 0 for the operating system and ring 3 for applications; rings 1 and 2 are not used.

In the Intel implementation of Windows 2000, system calls to kernel services (also known as system services) are dispatched through software interrupts, invoked by the *int2E* instruction. This causes a system trap which allows the executing thread to transition into kernel mode and execute the respective trap handler. When this interrupt occurs, the processor first examines the appropriate descriptor (interrupt gate) in the Interrupt Descriptor Table (IDT), and then pushes onto the stack the flags register, current code segment and current instruction pointer. Providing that the protection constraints of the caller and the gate are agreeable, the appropriate handler is called. The handler (in this case *_KiSystemService*) verifies and copies the user mode stack, and then uses an argument in the register EAX to index into the system service dispatch table and execute the respective service. On completion of the services an IRET instruction is issued which pops the instruction pointer, code segment and flags, and then resumes execution in the caller code.

This approach to software protection is effective, but does nevertheless incur some overhead (in the order of thousands of clock cycles) since each time a system call is made a context switch must occur. Furthermore, Windows 2000 does not use hardware multi-tasking (such as found on the Intel processors) and thus context switches are executed by software alone and are consequently processor intensive. To avoid this overhead, all processing by the cache node is executed within the kernel. No interactions with user mode code are made, and thus context switching is avoided. To achieve this engineering goal, the caching mechanisms are implemented as a kernel module (device driver) which interacts directly with the communications protocol stack.

3.2 IP Interception and Injection

The Windows 2000 networking support, from the lowest level, consists of a series of chained kernel device drivers for physical/network/transport layer processing, followed by a number of user level Winsock2 layers. The Winsock2 layers consist of the API itself (provided as ws2_32.dll) in conjunction with one or more network Service Provider (SP) layers. These interact with the kernel transport drivers, via the operating system's I/O manager, to provide a proprietary user level transport API to the Winsock layer. Generally these layers do not provide any of the core implementation of any individual protocol. They do however provide functionality that enables the services offered by the kernel drivers to be 'projected' into the user space, including support for asynchronous and shared I/O.

As previously discussed in Section 0, the cache node is designed to cache pre-built IP packets. Because the core IP protocol processing is provided by the kernel, we are able to execute the necessary caching interactions without entering the upper layers. To intercept packets, a hook routine is attached to the TCP/IP driver (tcpip.sys). Once an IP packet has been assembled from NDIS packets, it is passed to the hook routine which then determines, in accordance to caching policies, whether or not to cache the packet. This process is carried out on all packets passing through the node. Packets which are being forwarded by the node, which in fact make up the majority, are passed to the outgoing interface in the form of the NDIS buffers used to assemble the original packet. This optimisation eliminates the need to re-fragment into network transport units. In theory it would be possible to re-write the network adapter so that DMA is used to place data directly into the caching area (this would also require an adapter capable of 64bit addressing) and hence avoid the need to carry out any data copying (zero-copy). However, the prototype implementation does not support this and executes exactly one copy on all cached packets, which is made during the interception, rather than injection, process.

In the reverse direction, packets which have been previously cached must be passed back out onto the network as required. This process of 'injecting' packets into the network is also done solely in the kernel. Before cached packets can be sent back out into the network a process of adaptation or 'moulding' must take place. The primary purpose of this is re-addressing and the adjustment of any checksum fields. New address fields for the IP packets are determined from the intercepted DSM-CC packets. The initial prototype implementation uses IPv4, whilst adoption of IPv6 is underway. To support IPv4, checksums for both the IP header and the UDP header must be re-calculated from the adjusted field, whilst IPv6 only requires re-calculation of the UDP checksum (since IPv6 headers do not have a checksum).

In the case of IPv6, it is also necessary to adjust the packet's flow label. This is a 28-bit value made up of a traffic class field (4 bits) and a flow identifier (24 bits). In the initial prototype implementation the flow identifier is randomly generated by the cache node and the traffic class fixed at zero. Nevertheless, better use of the traffic class and flow identifier are envisaged in future work. Possibilities include selecting network QoS according to the requirements of the cached video and using the flow identifier to differentiate across multiple streams being received on the same UDP port.

3.3   Cache Entries/Indexing

This section discusses the general internal organisation of the cache, and the adopted strategy for the storage and retrieval of cache blocks. An entry in the cache, known as a cache block, is made up of an informational header followed by a number of pre-built IP packets, made up of an IPv4 or IPv6 header and a series of MPEG PES packets within a UDP payload. Each cache block corresponds to one second of presentation time for a given program stream. This interval represents the minimal required granularity of media access (seeking to a finer granularity serves no useful purpose in Video-on-Demand scenarios). By caching data in blocks of one second severe fragmentation is avoided and the scope required by the indexing scheme is reduced.

Cache blocks are uniquely identified for storage and retrieval through a combination of *bitstream_ID* and *presentation time* (each 32 bits wide). The *bitstream_ID* uniquely identifies a given media stream and is assigned by the DSM-CC server which maps them to more meaningful names. The assumption is made that within the context of a distributed server architecture, identifiers are co-ordinated accordingly and are globally unique to a given stream content. The presentation time, in units of a second, is derived from the Presentation Time Stamp (PTS) of the first PES packet in the first IP packet of the cache block. In some streams PES packet PTSs may not be sequential. This is because different elementary streams, within a program stream, may have different decoding latencies, and therefore the presentation-decode time mappings are different. To avoid this problem, cache blocks are stored in relation to the PTS fields of a specific elementary stream which acts as a reference stream for time stamp information. Finally, within the MPEG-2 Systems specification, the PTS field is defined as optional, and therefore in its absence (across all elementary streams in the program), time stamps are generated by the cache node using either the *program_mux_rate* from the packet header or the local time at which the block is cached.

The proposed caching architecture uses a two level indexing scheme to hold the mappings between the bitstream_ID/presentation time and the physical address of the cache block. The scheme is strikingly similar to that found in hardware virtual memory support such as found in the Intel Architectures. The first index, termed the cache index, provides a mapping between the stream identifier and the address (virtual) of the cache table. This index has n entries, where n is the total number of unique streams distributed by the server(s). Each cache table, per entry in the cache index, consists of an array of physical addresses corresponding to each cache block stored in the cache area for the given stream. An overview of the indexing scheme is shown in Fig. 2. The overhead of the indexing scheme is given as:

$$total\_size\_in\_bytes = (8 * n) + \Sigma(8 * t_n)$$

; where n is number of unique streams and t_n is total presentation time for stream n.

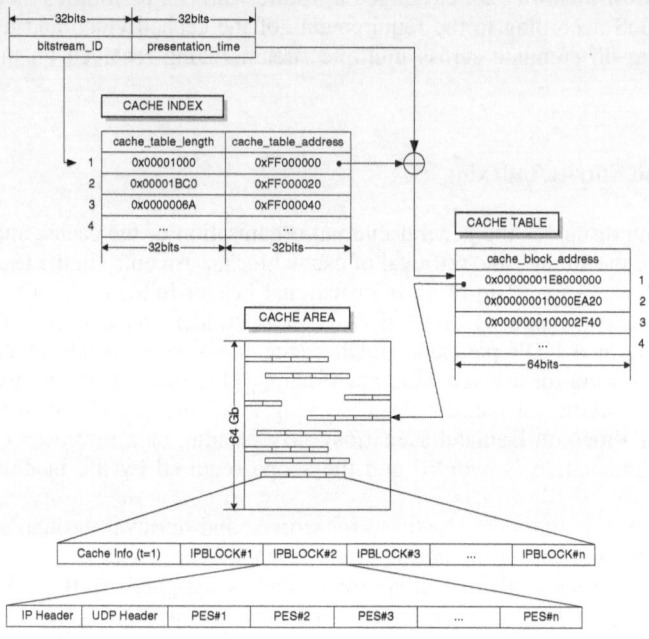

Fig. 2. Cache Indexing Scheme

Because the scheme provides direct mappings, the resources required by the indices may be potentially large. Nevertheless, we are making the assumption that the reducing factor is the total number of unique streams concurrently handled by the system (in a more dynamic media environment, where content is frequently changed and therefore a large number of different streams may exist, a process of re-mapping is required). Consequently, the cache node need only maintain a cache index capable of addressing the total number of unique streams that may exist in the cache at any given time. Furthermore, hardware and cost limitations means that this is constricted by the 'capacity' of the node. Our initial prototype provides ~2Gb of caching memory, capable of storing approximately 68 minutes (4096 seconds) of media. Thus the maximum total overhead of the cache tables, assuming an average stream rate of 4Mbps, is only 32Kb.

As mentioned, the proposed architecture is similar to virtual memory mapping techniques, offering a minimal direct mapping approach. In further development, we envisage the replacement of the soft indexing scheme with 64 bit hardware virtual memory management. An alternative to the chosen approach could be to use hash tables. However, because the required indexing scope (number of unique streams and total length of cached segments) varies significantly, the use of hash tables would result in substantial redundancy.

3.4 Caching Memory Sub-system

The nature of the caching architecture and its demands of memory management are significantly different than the requirements found in general purpose operating systems. Allocations, corresponding to individual cache blocks are relatively large (ranging from 256K – 1024K). Blocks may exist outside the range of the virtual address space, at least within 32-bit architectures, which only provide 2^{32} virtual addresses (4Gb). Furthermore, this problem is exacerbated by the operating system's use of reserved areas, which within the Windows 2000 operating system, only leaves somewhere in the region of 512Mb of virtual addresses available for kernel modules. In order to fulfil these requirements, the cache node uses a memory sub-system to manage the caching area. The prototype memory sub-system includes the following features:

- *Large Scale Addressing* - support for addressing beyond 4Gb of physical memory. The prototype is based on the 32-bit Intel Pentium III architecture which includes support for the Physical Address Extension (PAE). PAE enables an extension of physical address from 32 bits to 36 bits, thus allowing up to 64Gb of main memory. It is supported in the linear address translation scheme through the incorporation of an additional indexing table, the page-directory-pointer table. This table provides support for up to 4 page directories.

- *Dynamic Address Mapping* - as a side effect of large scale addressing there are generally less virtual addresses available than accessible physical addresses. Consequently, the proposed sub-system maps virtual address 'on demand'. This means that when the caching engine wishes to access the caching area, to store or locate a block, the sub-system must temporarily map a virtual address to the cache block for the duration of the required access. Once the engine has finished with a block the virtual address is released. The mapping process can also be optimised so that virtual address are only released when there are no more available, although it is more efficient to release addresses as soon as possible.

- *Modular Sub-System* - The memory sub-system is abstracted as a kernel module and offers a cleanly defined API. Future developments of the cache node are expected to support Intel's 64-bit (IA64) architecture. The initial sub-system implementation uses Intel's 36 bit addressing, with the intention of its seamless replacement with 64 bit support.

- *Lazy Defragmentation* - To avoid fragmentation and thus redundancy occurring, the sub-system employs a low-priority worker thread to defragment the memory area when blocks are not being accessed. The cache blocks are defragmented according to a least recently used metric.

The memory sub-system uses a simple address ordered free list and a first fit policy [10]. Although a single free list is not as efficient as a segregated free list, it does provide ample performance for the prototype implementation. We also chose not to use a buddy memory management system due to its complexity in coalescing and its apparently poor fragmentation.

3.5 Caching Engine

The cache engine is activated through the receipt of data on a network port, i.e. via a hardware interrupt, which instantiates a kernel thread. In turn this up-calls into the IP hook routine which then snoops the payload and caches accordingly. Traffic which does not contain PES packets and is not part of stream being cached, is forwarded out. Traffic which does contain PES packets, either DSM-CC or data, is passed into the caching engine. Here, DSM-CC requests are interpreted and the necessary masquerading/proxy actions performed. Alternatively, data PES packets are written into the cache according to the system's caching policies. In the current implementation a policy is defined by an individual hook routine. We envisage future implementations supporting more dynamic policy specification and loading mechanisms. Other than the kernel threads generated by network interrupts; the current prototype implementation uses low priority worker threads to execute the de-fragmentation. When no fragmentation exists these of course sleep. Currently, the caching engine is executed solely in the kernel. Nevertheless, we do envisage the use of user-mode processes for general policy management and policy specification.

4 Related Work

Research into multimedia caching in the context of buffering and caching within multimedia servers exploits the fact that multimedia objects are typically accessed in a sequential manner. Blocks retrieved for one client can therefore be reused for subsequent requests within a short time interval. One technique to exploit this is *interval caching*, whereby intervals between successive streams (viewing the same data) are cached, in order that the subsequent stream avoids disk access [11, 12, 13]. Similarly *distance caching* [14] replaces blocks of data based on the distance between successive clients.

Research into caching multimedia streams within the network stems from the work in managing a distributed hierarchical video-on-demand system. The Berkeley VOD System [15] is designed to provide transparent access to large amounts of video material. Continuous media objects are stored on tertiary storage devices, and only copied to a file server when required. The MiddleMan architecture [6] is a collection of cooperative proxy servers that, as an aggregate, cache video files within a local area network. The design incorporates a coordinator that is responsible for managing the video files stored at each proxy, controlling the storage and replacement of files and redirecting requests accordingly.

In addition to minimising start up latency, caching multimedia streams can also smooth the playback of variable bit rate (VBR) video streams. Video streams exhibit burstiness due to the encoding scheme and variations within and between frames, which can be a problem in terms of buffer management and network utilisation (ensuring that a high level of utilisation is achieved). Proxy prefix caching [7] overcomes this by caching the initial frames of popular audio/video clips and performing work ahead smoothing of variable bit-rate streams in order to reduce the resource requirements from the proxy to the client. [16] also proposes a prefix caching scheme, but extends this with a selective caching mechanism that caches intermediate frames based on the encoding properties of the video stream and the users buffer size. It attempts to store frames within the cache that are more critical to maintaining the robustness of the stream. [17] describes a technique called video staging that pre-fetches and stores selected portions of VBR streams in a proxy. The aim is to reduce backbone bandwidth requirements by storing the bursty portions of a VBR stream within the proxy and combining them with a constant bit rate (CBR) video stream from the server for playout. [18] considers an end-to-end architecture for the delivery of layered-encoded streams in the Internet using proxy caches to smooth out variations in quality by pre-fetching segments that are missing from the cache.

Much of the work on proxy cache replacement strategies is tailored to HTML documents and images, and does not consider the impact of multimedia streams [19]. However, [20] provides an investigation into multimedia streaming and cache replacement policies, introducing a caching algorithm based upon the resource requirements of an object. Finally, [21] provides one of the few investigations into the use of main memory for caching, using trace-driven simulations to evaluate the performance benefits of main memory caching for web documents.

5   Further Work

Work to date is focused on the use of large-scale RAM as a caching medium. It is proposed that this be extended to incorporate fast access disk as an additional level to a more hierarchical caching approach. However, the cache node's disk storage will only be used as an intermediary before cache content is totally dropped. It will not be used as a source medium for streaming.

Future work will also examine the implications of other media types, in particularly ISO/IEC MPEG-4 [22]. This is a multimedia format that is object-oriented and lends itself to partial caching. We envisage using media 'objects' as a unit of caching, and the distributed gather of such objects to form a scene. The adoption of such media types also opens up other areas of research interest. One such area is co-operative caching, where by caching nodes within the network use some proprietary inter-nodal protocol to handoff cache requests to other nodes in the event of a local cache miss.

6 Concluding Remarks

This paper presents ongoing work examining the initial design and implementation of a network caching architecture for high quality video, and more specifically ISO's MPEG-2. The discussed architecture is based on the notion of a transparent caching node, situated at the edge of the network, which is able to masquerade as a remote video server to its clients. This transparency lends itself to simple management and straightforward integration into the majority of existing network topologies.

The cache node, based on the Windows 2000 operating system, exploits high-speed main memory to cache pre-built UDP/IP packets containing MPEG-2 Program Stream PES packets. This technique of using transport/network level data units avoids the transport layer re-fragmenting the video data, thus increasing performance. In order to deploy such a cache, it is essential that the cache is able to integrate and cooperate with the video control architecture. In our prototype implementation, we have chosen to support DSM-CC as the basic control protocol. The cache behaves as a DSM proxy to the unaware client, intercepting control requests and forwarding/handling this as necessary.

The work presented is in its early stages and concrete evidence of its potential success is yet to be gained. Nevertheless, since its commencement, the work has highlighted some important issues in the design and implementation of such a scheme and further outlined the advantages which may be gained. In our next phase of work we hope to carry out more extensive testing of the prototype and give further indications of its effectiveness. We also wish to address some unresolved issues such as what are the constraints on caching policies and how partial hits and other scenarios should be handled.

Acknowledgements

We acknowledge the kind support of the Microsoft Research Labs, Cambridge, UK in funding this research under the LandMARC Research Project.

References

1. D. Gemmell, H. Vin, D. Kandlur, P. Rangan and L. Rowe, "Multimedia Storage Servers: Tutorial", IEEE Computer, Vol. 28, No. 5, May 1995.
2. M. Kumar, "Video-server designs for supporting very large numbers of concurrent users", IBM Journal of Research and Development, 1998, Vol. 42, No. 2, pp. 219-232.
3. ISO/IEC 13818-1, "Generic Coding of Moving Pictures and Associated Audio Information: Part 1 - Systems", Information Technology Specification, International Standard, 1996.
4. ISO/IEC 13818-6, "Generic Coding of Moving Pictures and Associated Audio Information: Part 6 - Extension for DSM-CC", Information Technology Specification, International Standard, 1996.
5. E. O'Neil, P. O'Neil and G. Weikum, "The LRU-k page replacement algorithm for database disk buffering", in Proceedings of the International Conference on Management of Data, May 1993.

6. S. Acharya, "Techniques for improving multimedia communication over wide area networks", Ph.D Thesis, Department of Electrical Engineering, Cornell University, January 1999.
7. S. Sen, J. Rexford and D. Towsley, "Proxy prefix caching for multimedia streams", Proceedings of the IEEE Infocom, 1999.
8. R. Rashid, D. Julin, D. Orr, R. Sanzi, R. Baron, A Forin, D. Golub, M. Jones. "Mach: A System Software kernel", in Proceedings of the 34th Computer Society International Conference COMPCON 89, February 1989.
9. Intel Architecture Software Developer's Manual, Volume 3: System Programming Guide (Order Number 243192), http://developer.intel.com/, May 1999.
10. P.R. Wilson, M.S. Johnstone, M. Neely and D. Boles, "Dynamic Storage Allocation: A Survey and Critical Review", in Lecture Notes in Computer Science, Vol. 986, Ch. 18, pp.1-116, 1995.
11. A. Dan, D. Dias, R. Mukherjee, D. Sitaram and R. Tewari, "Buffering and caching in large scale video servers", in Proceedings of the IEEE COMPCON, pages 217-224, 1995.
12. A. Dan, D. Sitaram, "A generalized interval caching policy for mixed interactive and long video environments", in Proceedings of the IS&T SPIE Multimedia Computing and Networking Conference, San Jose, CA, January 1996.
13. A. Dan, D. Sitaram, "Multimedia caching strategies for heterogeneous application and server environments", Multimedia Tools and Applications, pp.279-312, 1997.
14. B. Ozden, R. Rastogi, A. Silberschatz, "Buffer replacement algorithms for multimedia storage systems", in Proceedings of the International Conference on Multimedia Computing and Systems, pp. 172-180, June 1996.
15. D. W. Brubeck and L. A. Rowe, "Hierarchical Storage Management in a Distributed VOD System", IEEE Multimedia, Fall 1996, Vol. 3, No. 3.
16. Z. Miao and A. Ortega, "Proxy caching for efficient video servers over the Internet", in Proceedings of the 9th International Packet Video Workshop (PVW '99), New York, April 1999.
17. Y. Wang, Z.-L. Zhang, D. Du and D. Su, "A network-conscious approach to end-to-end video delivery over wide area networks using proxy users", in Proceedings of the IEEE Infocom, April 1998.
18. R. Rejaie, M. Handley, H. Yu and D. Estrin, "Proxy caching mechanism for multimedia playback streams in the Internet", in Proceedings of the 4th International Web Caching Workshop, San Diego, California, March 31-April 2, 1999.
19. M. Abrams, C. R. Standridge, G. Abdulla, S. Williams, E. A. Fox, "Caching Proxies: Limitations and Potentials", in Proceedings of the 4th International World-Wide Web Conference, pp. 119-133, December 1995.
20. R. Tewari, H. Vin, A. Dan and D. Sitaram, "Resource based caching for web servers", in Proceedings of SPIE/ACM Conference on Multimedia Computing and Networking (MMCN), San Jose, 1998.
21. E.P. Markatos, "Main Memory Caching of Web Documents", Computer Networks and ISDN Systems, 1996, Vol. 28, No. 7-11, pp. 893-905.
22. ISO/IEC 14496-1, "Coding of Audio-visual Objects – Part 1: Systems", Information Technology Specification, International Standard, 1999.

Fast and Optimal Multicast-Server Selection Based on Receivers' Preference

Akihito Hiromori[1], Hirozumi Yamaguchi[1], Keiichi Yasumoto[2],
Teruo Higashino[1], and Kenichi Taniguchi[1]

[1] Graduate School of Engineering Science, Osaka University
1-3 Machikaneyamacho, Toyonaka, Osaka 560-8531, JAPAN
{hiromori,h-yamagu,higashino,taniguchi}@ics.es.osaka-u.ac.jp
[2] Faculty of Economics, Shiga University
1-1-1 Bamba, Hikone, Shiga 522-8522, JAPAN
yasumoto@biwako.shiga-u.ac.jp

Abstract. In this paper, we propose static and dynamic server selection techniques for multicast receivers who receive multiple streams from replicated servers. In the proposed static server selection technique, if (a) the location of servers and receivers and shortest paths between them on a network and (b) each receiver's preference value for each content are given, the optimal server for each content that each receiver receives is decided so that the total sum of the preference values of the receivers is maximized. We use the integer linear programming (ILP) technique to make a decision. When we apply the static server selection technique for each new join/leave request to a multicast group issued by a receiver, it may cause server switchings at existing receivers and may take much time. In such a case, it is desirable to reduce both the number of server switchings and calculation time. Therefore, in the proposed dynamic server selection technique, the optimal server for each content that each receiver receives is also decided so that the total sum of the preference values is maximized, reducing the number of server switchings, by limiting both the number of receivers who may switch servers and the number of their alternative servers. Such restrictions also contribute fast calculation in ILP problems. Through simulations, we have confirmed that our dynamic server selection technique achieves less than 10 % in calculation time, more than 90 % in the total sum of preference values, and less than 5 % in the number of switchings on large-scale hierarchical networks (100 nodes), compared with the static server selection.

1 Introduction

Multicast is a useful way for saving bandwidth consumption by simultaneous transmission of a data stream such as WWW pushing of contents and live video streaming to multiple receivers [1,2]. However, due to the limited bandwidth that can be used for multicast traffics, when multiple streams of live video are transfered, we need efficient bandwidth control of network resources used by each stream. For this purpose, we have proposed bandwidth control techniques to

H. Scholten and M. van Sinderen (Eds.): IDMS 2000, LNCS 1905, pp. 40–52, 2000.
© Springer-Verlag Berlin Heidelberg 2000

maximize the quality requirements of receivers for unicast and multicast streams [3,4].

Regardless of unicast or multicast communication, bandwidth shortage caused by multiple streams is due mainly to path length between servers and receivers, since competition occurs among different streams at some common bottleneck links. To overcome this problem, it may be useful to place some replicated servers at remote nodes [5] where video sources are transmitted to the replicated servers through high-speed links (backbone), and each receiver selects one of these servers depending on network traffics, server loads and so on. This technique improves network utilization without changing underlying routing protocols.

In recent years, such multi-server techniques have been researched [5,6,7]. [6, 7] have proposed unicast server selection techniques based on metric information such as packet delay, hop count and server load. [5] has proposed a multicast server selection technique where the optimal server assignment for each receiver to minimize the total link cost is formulated as a mathematical problem on a graph. [5] also gives a heuristic for the dynamic server selection problem where the server switching cost caused by join/leave requests to multicast groups is considered. However, in multi-media applications using multicast communication such as video-conferences at multiple locations, each receiver requires to receive more than one stream and his/her preference for each stream may differ from others. In general, when a receiver receives a stream which other receivers would not like to receive and their path from the server is quite long, the bandwidth used by the stream may reduce the benefit of the whole receivers. Therefore, on networks where available bandwidth is limited, it is desirable to consider each user's preference for each stream and to maximize the benefit of the whole receivers.

Such optimization can statically be calculated if the set of receivers and their preferences to all streams are known in advance. However, in general, join/leave requests are repeatedly issued by receivers. In such a case, re-optimization should be done dynamically. If we do such optimization for every request, the following problems arise.

1. Calculation time: the problem to select servers is a combinatorial optimization problem. Therefore, large amount of calculation time may be required in large-scale networks (we show calculation time against the number of nodes in Section 5).
2. Switching frequency: optimization may force existing receivers to switch the current servers of its receiving streams to others even if they do not want overhead caused by multicast join/leave requests (join/leave latency and so on).

Therefore, it is desirable to apply the optimization technique to a part of network where such dynamic changes happen, reducing the number of server switchings at receivers as well as keeping the sum of the preference values higher than a reasonable threshold.

In this paper, we propose static and dynamic server selection techniques for multicast streams transfered from multiple replicated servers. In the proposed static server selection technique, if several replicated servers and each receiver's preference value for each content are given, the optimal server for each content that each receiver receives is decided so that the total sum of the satisfied preference values is maximized. In the proposed dynamic server selection technique, when a new join/leave request to a multicast group is issued by a receiver, the number of server switchings at existing receivers should be reduced, and the total sum of the satisfied preference values should be increased. Therefore, in our dynamic server selection technique, we limit the number of receivers who may switch servers and their alternative servers so that the number of server switchings is drastically reduced and the sum of preference values is increased.

In our static optimization technique, in networks with certain link capacities, from given (1) the location of servers and receivers, (2) shortest paths between them and (3) each receiver's preference value to each stream, we construct a logical conjunction of linear inequalities which represent the bandwidth constraint on each link used by streams. The objective function is set to maximize the total sum of preference values of all receivers to streams. Thus, we use the integer linear programming (ILP) technique to make an optimal server selection. Here, we assume that streams with lower preference values may not be received in case of bandwidth shortage.

In our dynamic optimization technique, for each join/leave request issued by a receiver, we also construct linear inequalities, in order to obtain a solution where the sum of the preference values is maximized, reducing the total number of server switchings. Here, for fast calculation in the target ILP problem, we add some constraints to restrict receivers to ones who may suffer server switchings and also restrict their alternative servers to the two servers whose multicast trees are the closest of all the servers from the requested receiver.

We have simulated our static and dynamic server selection techniques and measured (a) calculation time, (b) the sum of preference values, and (c) the number of server switchings in a random topology and a hierarchical Internet topology called Tiers topology [8] consisting of LAN, MAN and WAN, where SPF (Shortest Path First) routing protocol is supposed. As a result, we have confirmed that our dynamic server selection technique achieves less than 10% in calculation time, more than 90 % in the sum of preference values, and less than 5 % in the number of server switchings, compared with the static server selection.

2   Preliminaries

A network is modeled as an undirected graph with capacity $CAP(e)$ for each link e. Replicated multicast servers (or just servers hereafter) $S = \{s_1, ..., s_m\}$ and receivers $R = \{r_1, ..., r_n\}$ exist on the network. Each server forwards contents $C = \{c_1, ..., c_p\}$ sent from source nodes to the receivers. Therefore, from the receivers, each server can be regarded as a multicast server which has these contents. Each receiver can receive each content from one of these servers, and specifies a value called a *preference value* to each content. A preference value

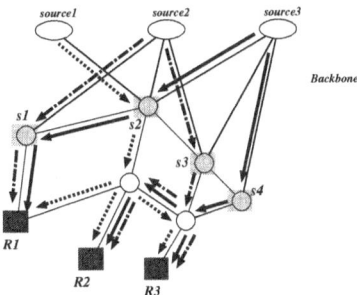

Fig. 1. Network with Replicated Multicast Servers

means how eagerly the receiver wants to receive the content. An example of the network model is shown in Fig. 1.

In Fig. 1, *source*1, *source*2 and *source*3 are the source nodes of contents c_1, c_2 and c_3, respectively. These contents are delivered to some of the replicated servers s_1, s_2, s_3 and s_4 through the connections with large capacities. s_1, s_2, s_3 and s_4 are the candidates of multicast servers, and each receiver selects one of them to receive each content. In the figure, R_1 receives c_2 from s_1, and both of c_1 and c_3 from s_2.

In this paper, for given (1) location of servers and receivers, (2) the shortest path between each pair of server s_i and receiver r_j and (3) the preference value of each receiver to each content, we formulate a problem to decide a server for each pair of a receiver and a content so that the total sum of satisfied preference values is maximized. We call this problem *static server selection problem*. Note that each receiver may not be able to receive all the contents that she/he required, due to bandwidth constraints.

Then we consider the case that receivers dynamically start or stop receiving contents, that is, receivers join/leave multicast groups. For such a case, we can optimize the total sum of preference values by solving the static server selection problem when every join/leave request is issued. However, we cannot avoid suffering (a) the exponential growth of computation time and (b) the overhead of server switching at almost all receivers. Regarding (a), we have experienced simulation on networks with 100 nodes, 50 receivers, 10 servers and 5 contents, and it took 200 seconds in average and more than 400 seconds in the worst case to get an optimal solution on an average machine (Pentium III, 500MHz). Such durations are allowed if, for example, we design the total layout of multicast trees before a new continuous service is started, however, not feasible for each small change of receiver status. Regarding (b), if we consider that the overhead increases proportional to the number of server switchings, it is much better to reduce it. For these reasons, in this paper, we propose another optimization technique called *dynamic server selection* for each join or leave request of a receiver. In Sections 3 and 4, we describe the static and dynamic server selection problems, respectively.

Let us define the following terms.

- $path(s_i, r_j)$: the set of links on the shortest path from s_i to r_j
- $uplink(e_l, s_i, r_j)$: the up-link (next to e_l) on $path(s_i, r_j)$
- $endlink(s_i, r_j)$: the bottom-link (attached to r_j) on $path(s_i, r_j)$
- $pref(r_j, c_k)$: the preference value of r_j given to c_k
- $bw(c_k)$: the transmission rate of c_k
- (s_i, c_k): the multicast group where c_k is delivered from s_i

3   Static Server Selection Problem

In order to formulate the static server selection problem, we define the following two types of boolean variables. Each variable in one type represents the fact that a receiver can receive a content from a server. Each variable in another type represents the fact that the multicast tree of a content from a server uses a link.

- $rcv[s_i, r_j, c_k]$: its value is one only if receiver r_j receives content c_k from s_i, otherwise zero.
- $deliver[e_l, s_i, c_k]$: its value is one only if content c_k from server s_i is delivered through link e_l, otherwise zero.

Using these variables, the static server selection problem can be formulated as the following integer linear programming (ILP) problem.

$$\max \sum_i \sum_j \sum_k pref(r_j, c_k) \cdot rcv[s_i, r_j, c_k] \tag{1}$$

subject to:

$$\sum_i rcv[s_i, r_j, c_k] \leq 1, \quad \forall j, \ k \tag{2}$$

$$deliver[e_l, s_i, c_k] \leq deliver[uplink(e_l, s_i, r_j), s_i, c_k], \quad \forall i, \ j, \ k, \ l \tag{3}$$

$$rcv[s_i, r_j, c_k] \leq deliver[endlink(s_i, r_j), s_i, c_k], \quad \forall i, \ j, \ k \tag{4}$$

$$\sum_i \sum_k bw(c_k) \cdot deliver[e_l, s_i, c_k] \leq CAP(e_l), \quad \forall l \tag{5}$$

Objective function (1) represents the total sum of all receivers' preference values. Constraint (2) states that one receiver selects at most one server for each content. Constraints (3) and (4) concern the form of multicast trees and indicate that if r_j receives c_k from s_i, the multicast tree of c_k from s_i must contain the shortest path from s_i to r_j. Constraint (5) is a bandwidth constraint on each link.

4   Dynamic Server Selection Problem

Due to the dynamic behavior of receivers in multicast communication, members in a group are not unique throughout a session. Therefore, fast re-optimization for each join/leave request of a receiver is desirable.

Let $rcv'[s_i, r_j, c_k]$ denote the fact that receiver r_j is currently joining the group (s_i, c_k) (its value is one if r_j is joining the group, zero otherwise). The join/leave behavior of a receiver is described as follows.

- **join:** a receiver r_q where $\sum_i rcv'[s_i, r_q, c_k] = 0$ wants to join one of the groups $(s_1, c_k), (s_2, c_k), \ldots$ and (s_m, c_k).
- **leave:** a receiver r_q where $rcv'[s_i, r_q, c_k] = 1$ wants to leave the group (s_i, c_k).

We limit the number of receivers who are forced to switch their servers to others, in order to prevent a receiver's join/leave behavior from affecting all the receivers spread in wide-area networks. We also limit the number of possible alternative servers (servers to be switched) when a receiver switches its server of a content so that the receiver does not select servers far from him/her. On assuming such restrictions, we formulate the dynamic server selection problem.

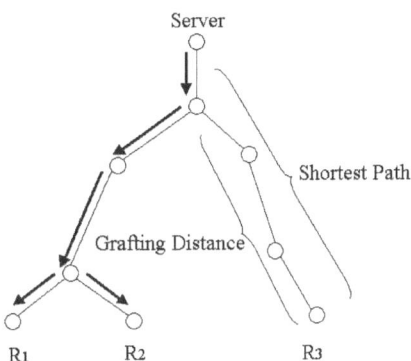

Fig. 2. Grafting Distance

[join] We define a *grafting distance* as the number of links on the shortest path from a server to a receiver through which a content c_k from the server has not been delivered yet (*i.e.*, the number of links where the content would be started to deliver when the receiver joins the group). An example is shown in Fig. 2. We adopt the following policy.

- receiver r_q selects one of the two servers (say s_{i_1} and s_{i_2}) with the shortest two grafting distances.
- For each $path(s_{i'}, r_{j'})$ which shares some of links with $path(s_{i_1}, r_q)$ or $path(s_{i_2}, r_q)$, if receiver $r_{j'}$ is receiving content c_k from server $s_{i'}$, $r_{j'}$ is one of the receivers who may switch servers. As an alternative server to receive c_k, $r_{j'}$ may select one of the two servers with the shortest two grafting distances.

Intuitively, each receiver whose receiving streams may compete with r_q's new stream may have to switch servers. Furthermore, we limit the number of their possible alternative servers to only two for each pair of a receiver and a content.

The dynamic server selection problem by each receiver's join request is an ILP problem with the same objective function (1), the same constraints (2)–(5) as in Section 3 and the following constraint to fix the status of receivers who should not switch servers:

$$rcv[s_i, r_j, c_k] = 1, \quad \forall (i, j, k) \neq (i', j', k') \tag{6}$$

where (i', j', k') is a tuple satisfying the following constraint. Note that all the paths are given and $rcv'[s_i, r_j, c_k]$ in the constraint have already been decided. Therefore such tuples (i', j', k') are uniquely determined.

$$(path(s_{i_1}, r_q) \cap path(s_{i'}, r_{j'}) \neq \emptyset \vee path(s_{i_2}, r_q) \cap path(s_{i'}, r_{j'}) \neq \emptyset)$$
$$\wedge rcv'[s_{i'}, r_{j'}, c_{k'}] = 1$$

[leave] We adopt the following switching policy when r_q leaves group (s_i, c_k).

– For each $path(s_{i'}, r_{j'})$ which shares some of links with $path(s_i, r_q)$, $r_{j'}$ can select $s_{i'}$ as the server to receive content c_k which $r_{j'}$ has not received.

The dynamic server selection problem by each receiver's leave request is an ILP problem with the same objective function (1), the same constraints (2)–(5) and the following constraint to fix the status of receivers who should continue to receive streams.

$$rcv[s_i, r_j, c_k] = 1, \quad \text{if } rcv'[s_i, r_j, c_k] = 1 \tag{7}$$

In Section 5, we have measured the performance of dynamic server selection for a receiver's join request compared with the static server selection, in terms of the computation time, the total sum of satisfied preference values and the total number of server switchings.

5   Simulation

We have used two types of networks based on (a) Tiers Model [8] (Fig. 3) and (b) Random model (Fig. 4). Tiers is a hierarchical model organized by three domains, LAN, MAN, and WAN. For Tiers model, we randomly decided the number of nodes contained in LAN, MAN and WAN. We also decided the link capacities of LAN, MAN and WAN so that they are in the ratio of $1 : 10 : 100$. Then we simulated 5 times on the networks varying the number of nodes $|N|$. For Random model, we randomly decided each link capacity based on Gaussian distribution, and simulated 20 times on the networks varying $|N|$. Also, we had $|S| = 0.1|N|$ servers, $|R| = 0.5|N|$ receivers and $|C| = 5$ contents in the networks and selected receivers' preference values from 25, 16, 9, 4 and 1, randomly. We simulated the dynamic server selection for a receiver $r_{|R|}$ who tried to join a group in the situation that the server selection had been already optimized for

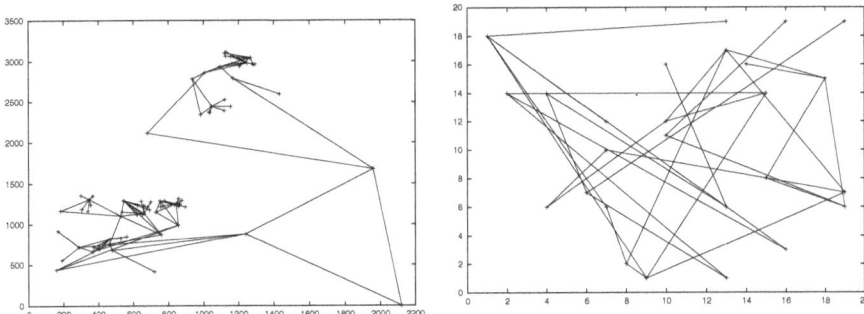

Fig. 3. Network Topology (a) Tiers Model

Fig. 4. Network Topology (b) Random Model

the receivers $r_1, ..., r_{|R|-1}$ by the static server selection (this simulation is denoted by (d)), and the static server selection for the receivers $r_1, ..., r_{|R|}$ (denoted by (s)). Then we have measured the calculation time, the total sums of satisfied preference values and the numbers of server switchings of (d) and (s). The results are shown in Section 5.1. Also in order to examine the validity that the number of alternative servers is limited to 2, we have also measured these values in the dynamic server selection with the different number of alternative servers $|AS|$, (d-1) $|AS| = 0.25|S|$, (d-2) $|AS| = 0.5|S|$ and (d-3) $|AS| = 0.75|S|$. The results are shown in Section 5.2.

5.1   Comparison of Static and Dynamic Server Selection

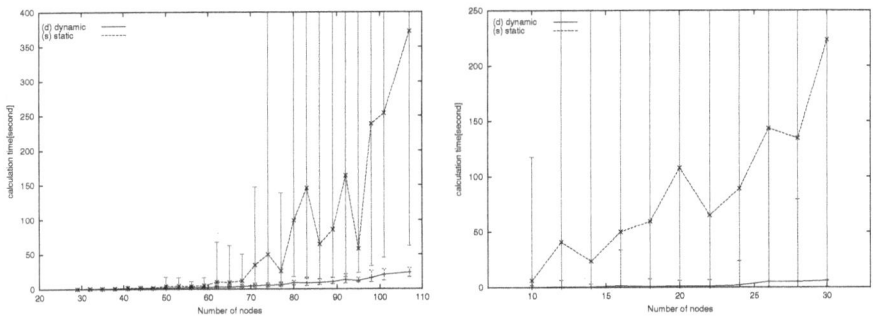

Fig. 5. Number of Nodes vs. Calculation Time: (a) Tiers Model

Fig. 6. Number of Nodes vs. Calculation Time: (b) Random Model

Calculation Time We show the calculation time of the static server selection and the dynamic server selection. We varied $|N|$ (the number of nodes) from 28 to 107 by every 3 nodes on Tiers model (Fig. 5) and from 10 to 30 by every 2 nodes on Random model (Fig. 6). In these graphs, the plots of the average time of (s) the static and (d) dynamic server selections are connected by dashed and solid lines, respectively. For each number of nodes, the range between the worst time and the best time is also shown by a vertical line.

We find the exponential increase of the calculation time in the static server selection around 90 nodes or higher on Tiers model and around 28 nodes or higher on Random model. On the other hand, we find the linear increase of the calculation time in the dynamic server selection. From these results, we can say that the proposed dynamic server selection can solve the problem within a reasonable time. Especially on Tiers model, the calculation time is just 10 seconds on 70 nodes in the worst case. On Random model, it took more calculation time in the static server selection. This is due to the distribution of distances between servers and receivers. On Tiers model, since the distances to servers largely differ from each other, the number of the candidate servers may be reduced in the calculation process. On the other hand, the distances are very close to each other on Random model.

The Total Sum of Preference Values. We have measured the total sum of preference values of the dynamic and static server selections. The ratios of the former to the latter on Tiers model and Random model are shown in Fig. 7 and Fig. 8, respectively. On Tiers model, even in the worst case, the ratio is 89% and the average ratio is 95%. They are good enough to consider the tradeoff between the calculation time and optimality of satisfied preference. On Random model, there are a few worst cases that the ratios are in the range of 50%∼60%. This is because many receivers' alternative servers are converged to a few servers. However, the average ratio is kept more than 95%, therefore our dynamic server selection can keep high optimality compared to the static server selection on both models.

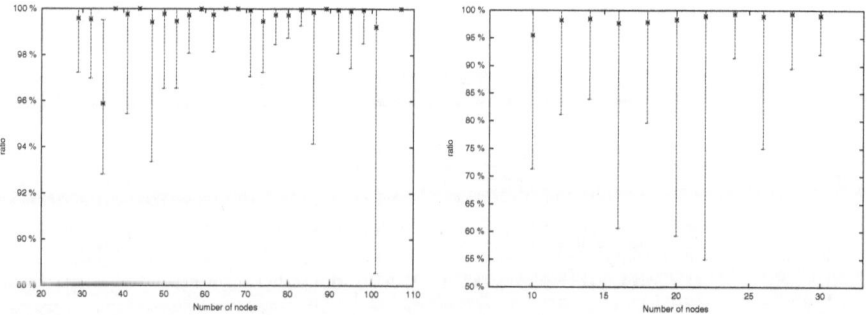

Fig. 7. Number of Nodes vs. Total Sum of Preference Values: (a) Tiers Model

Fig. 8. Number of Nodes vs. Total Sum of Preference Values: (b) Random Model

The Number of Server Switchings We define a new variable $switch[s_i, r_j, c_k]$ for each set of variables s_i, r_j and c_k where $rcv'[s_i, r_j, c_k] = 1$ as follows.

$$switch[s_i, r_j, c_k] = 1 - rcv[s_i, r_j, c_k] \qquad (8)$$

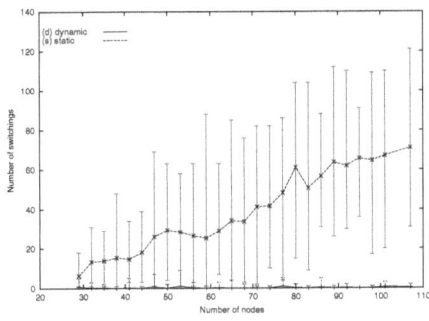 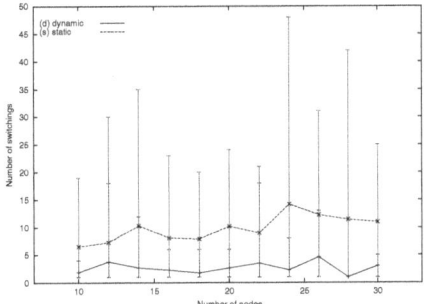

Fig. 9. Number of Nodes vs. Number of Server Switchings: (a) Tiers Model

Fig. 10. Number of Nodes vs. Number of Server Switchings: (b) Random Model

$switch[s_i, r_j, c_k]$ represents the fact that r_j stops receiving c_k from s_i. Thus we can represent the number of server switchings as follows.

$$\sum_i \sum_j \sum_k switch[s_i, r_j, c_k] \qquad (9)$$

We have measured the number of server switchings. We show the results on Tiers model (Fig. 9) and on Random model (Fig. 10).

In the static server selection, the server switchings occurred 60 times on average on Tiers model with 100 nodes. Since we decided $|R| = 0.5|N|$, each receiver has at least one server switching in estimation. It is too much overhead in consideration of multicast join/leave latencies. On the other hand, the maximum number of server switchings is largely reduced in the dynamic server selection.

5.2 Effect of Number of Alternative Servers

In order to examine the effect of the number of alternative (selectable) servers to the calculation time, the total sum of satisfied preference values and the number of server switchings, we have measured these values on Tiers model in the dynamic server selection with the different number of alternative servers $|AS|$, (d-1) $|AS| = 0.25|S|$, (d-2) $|AS| = 0.5|S|$ and (d-3) $|AS| = 0.75|S|$.

Calculation Time We show the average of the calculation time in Fig. 11. Compared with (d) where $|AS| = 2$, we find the feature of divergence in (d-3), not so much as (s). Therefore we can say that our policy to limit the number of alternative server is adequate enough.

The Total Sum of Preference Values We have measured the average and worst of the total sum of preference values and shown the ratios of (d), (d-1), (d-2) and (d-3) to (s) in Fig. 12 (average case) and Fig. 13 (worst case), respectively. The ratios of (d-1), (d-2) and (d-3) are greater than (d) and are kept more than 95% in the worst case. However, in the average case, (d) achieved almost the same values as (d-1), (d-2) and (d-3).

50   A. Hiromori et al.

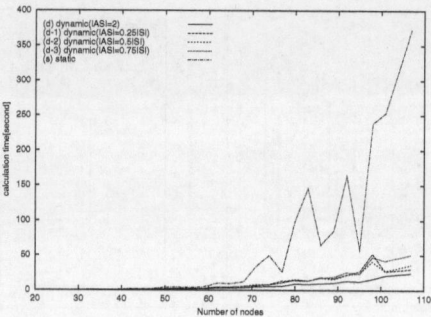

Fig. 11. Number of Nodes vs. Calculation Time on Tiers Model

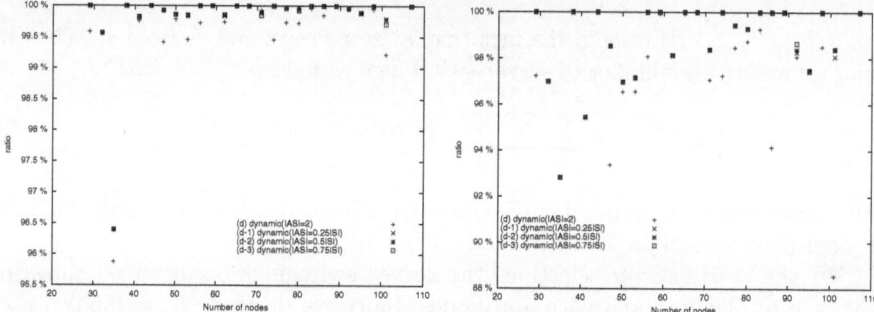

Fig. 12. Number of Nodes vs. Total Sum of Preference Values: Average Case

Fig. 13. Number of Nodes vs. Total Sum of Preference Values: Worst Case

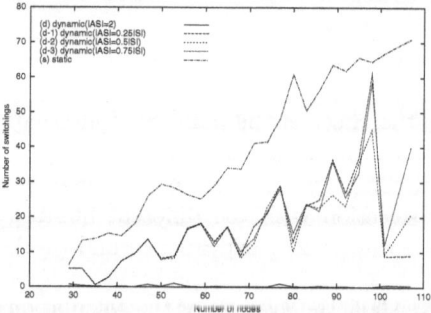

Fig. 14. Number of Nodes vs. Number of Server Switchings on Tiers Model

The Number of Server Switchings We show the number of server switchings in Fig. 14. The behavior of (d-1), (d-2) and (d-3) is similar to (s), while it is kept low in (d). From the results above, we can say that out policy to limit the number of alternative servers is adequate enough.

6 Conclusion

In this paper, we have proposed static and dynamic replicated server selection techniques for multiple multicast streams. In the proposed static server selection technique, if the location of servers and receivers and the shortest path between each pair of a server and a receiver on a network and each receiver's preference value for each content are given, the optimal combinations of the servers and the contents for each receiver are decided so that the total sum of the preference values of the receivers is maximized. We use the integer linear programming (ILP) technique to make a decision. Furthermore, in our dynamic server selection technique, the combinations of the servers and contents for each receiver are decided so that the number of server switchings is reduced and the total sum of the preference values is kept high, by restricting receivers who may suffer server switching and also alternative servers to be switched. Such restrictions also contribute fast calculation in ILP problems. Through simulations, we have confirmed that our dynamic server selection technique achieves less than 10 % in calculation time, more than 90 % in the sum of preference values and less than 5 % in the number of switchings, compared with the static server selection.

As our future work, we plan to design and implement an architecture to let receivers select optimal servers in existence of replicated multicast servers, based on the proposed method. We consider that our technique can be incorporated into application-layer anycast[9]. Application-layer anycast is an implementation of anycast at an application level, and is organized from ADN (Anycast Domain Name). An ADN server provides location service and replies to client's request with the list of servers that can provide the requested service. The ADN server can select those servers based on certain metrics. Therefore if the ADN server knows certain network information needed for our server selection technique, calculating the optimal allocation of servers will be possible on the ADN server. However, we have to consider the following two problems. The first one is that ADN servers reply to only the receivers who requested services. Therefore, we need an additional mechanism to let the other receivers switch their servers. The second one is how to collect the information. We are now investigating an efficient way to realize these requirements.

Moreover, in order to show the feasibility of the receiver/server limitation policy adopted in our dynamic server selection, we will try to analyze the upper bound of the solution (that is, the optimality of the solution of dynamic server selection) compared with the static server selection on Tiers model.

References

1. C. Diot, W. Dabbous and J. Crowcroft, "Multipoint Communication: A Survey of Protocols, Functions, and Mechanisms," *IEEE Journal on Selected Areas in Communications*, Vol. 15, No. 3, pp. 277-290, 1997.
2. X. Li, M.H. Ammar and S. Paul, "Video Multicast over the Internet," *IEEE Network Magazine*, Vol. 13, No. 2, pp46-60, 1999.

3. H. Sakate, H. Yamaguchi, K. Yasumoto, T. Higashino and K. Taniguchi, "Resource Management for Quality of Service Guarantees in Multi-party Multimedia Application," *Proc. of 1998 Int. Conf. on Network Protocols (ICNP'98)*, pp. 189–196, 1998.
4. H. Yamaguchi, K. Yasumoto, T. Higashino and K. Taniguchi, "Receiver-Cooperative Bandwidth Management for Layered Multicast," *Proc. of 1999 Int. Conf. on Network Protocols (ICNP'99)*, pp. 43–50, 1999.
5. Z. Fei, M.H. Ammar and E.W. Zegura, "Optimal Allocation of Clients to Replicated Multicast Servers," *Proc. of 1999 Int. Conf. on Network Protocols (ICNP'99)*, pp. 69–76, 1999.
6. R.L. Carter, M.E. Crovella, "Server Selection Using Dynamic Path Characterization in Wide-area Networks," *Proc. of INFOCOM'98*, 1998.
7. Z. Fei, S. Bhattacharjee, E.W. Zegura and M.H. Ammar, "A Novel Server Selection Technique for Improving the Response Time of a Replicated Service," *Proc. of INFOCOM'98*, 1998.
8. K.L. Calvert, M.B. Doar and E.W. Zegura, "Modeling Internet Topology," *IEEE Communications Magazine*, Vol. 35, No. 6, pp. 160–163, 1997.
9. S. Bhattacharjee, M.H. Ammar, E.W. Zegura, V. Shah and Z. Fei, "Application-Layer Anycasting," *Proc. of INFOCOM'97*, 1997.
10. G. Riley, M. Ammar and L. Clay, "Receiver-Based Multicast Scoping: A New Cost-Conscious Join/Leave Paradigm," *Proc. of 1998 Int. Conf. on Network Protocols (ICNP'98)*, pp. 254–261, 1998.
11. E. Amir, S. McCanne and R. Katz, "Receiver-driven Bandwidth Adaptation for Light-weight Session," *ACM Multimedia*, 1997.

Mcast: A Multicast Multimedia Communication Software Development Platform

Chung-Ming Huang[1] and Hsu-Yang Kung[2]

[1] Dept. of Computer Science and Information Engineering, National Cheng Kung University, Tainan, Taiwan, R.O.C.
huangcm@locust.iie.ncku.edu.tw
[2] Dept. of Management Information Systems, National Pingtung University of Science and Technology, Pingtung, Taiwan, R.O.C.
kung@mail.npust.edu.tw

Abstract. The Internet explosion is driving the need for distributed multimedia presentations (DMPs), which provide multiple users with QoS-controlled multimedia services such as media distribution and virtual classroom. The provision of DMPs is usually based on the multicast communication. However, the jitter phenomenon over the best-effort Internet always disturbs the orchestration of multimedia presentations. Furthermore, the characteristics of multiple media streams combining the multicast delivery complicate the network and system deployment. In this paper, we describe the major considerations and techniques of designing multiple-stream multimedia presentations in a multicast communication environment. Based on the proposed control schemes, we develop a communication engine named Mcast. Mcast (i) provides the flexible authoring tool to allow users to author a multiple-stream multimedia presentation in a multicast environment and (ii) achieves smooth multimedia presentations with the well-designed temporal control mechanism.

1 Introduction

With the advances of computer and communication technologies, distributed multimedia presentations (DMPs), e.g., video distribution and distance learning, becomes more and more popular. DMPs can be characterized by the integrated multicast communication and presentation of multiple continuous and static media streams [7]. Multicast communication simultaneously transmits media to multiple recipients, each of whom has the same multicast address [3]. A continuous medium, such as video or audio, is a time-dependent medium that possesses temporal relations between media units; a static medium, such as text or still image, is a time-independent medium that has no temporal relation between media units [1]. Temporal relations of a multimedia presentation can be pre-defined and be scheduled in a pre-orchestrated multimedia presentation. The

H. Scholten and M. van Sinderen (Eds.): IDMS 2000, LNCS 1905, pp. 53–64, 2000.

goal of a DMP is to present all composed media streams according to the temporal presentation schedule.

Clients of a multicast group, which scatters on separated LANs, also possess diverse capabilities of processing multiple media streams and have distinct available bandwidth in the paths leading to them. Multiple streams mean that media streams are retrieved from media bases and are transmitted via different communication channels with diverse QoS (Quality-of-Service) requirements [11]. For transmitting multiple streams, continuous media are usually based on UDP due to the tolerance of media loss and the real-time requirement; static media are usually based on TCP or RMTP (reliable multicast transport protocol) due to the reliability requirement [8]. Heterogeneous communication protocols induce serious temporal anomalies between media streams. That is, different network devices and links, which have diverse communication capabilities and network traffic situations, always disturb the pre-defined temporal relations and severely degrade the QoS satisfaction degrees of the clients in a heterogeneous multicast environment [6]. In order to compensate for temporal anomalies, multimedia synchronization is the essential requirement to achieve smooth coordination and cooperation among various media.

Multimedia synchronization is used to coordinate, schedule, and present media units in the distributed environment [5]. Two types of temporal synchronization are the intra-medium synchronization and inter-media synchronization [1]. Intra-medium synchronization ensures intra-medium temporal relation of a medium stream and compensates for jitter, which is the asynchronous anomaly between consecutive media units of a medium stream [9]. Inter-media synchronization ensures inter-media temporal relations among related media streams and compensates for skew, which is the time difference between related media streams [9]. With temporal synchronization mechanisms, DMPs achieves smooth presentations.

A well-designed DMP essentially consists of two components that are the authoring system and the temporal control system [2]. The authoring system is the generating mechanism of behavior specifications. Behavior specifications represent (i) the media attributes of related streams, which includes involved media streams, temporal and spatial attributes of each medium stream, and (ii) the temporal relationship between related media streams. The authoring system allows a user to specify behavior specifications of the corresponding multimedia presentation. The temporal control system is the synchronization and presentation mechanisms that achieve temporal relations specified in the corresponding behavior specifications. In order to accurately specify the behavior specifications of multiple streams and to perceptively achieve multimedia synchronization/presentation, a communication engine named Mcast is proposed and developed in this paper.

Mcast provides an authoring tool to specify the media behavior specifications and the required multicast environment. The temporal control system of Mcast is composed of (i) the synchronization mechanism to achieve multiple-stream multimedia synchronization based on the proposed synchronization schemes, and

(ii) the presentation mechanism to achieve a perception presentation based on the proposed presentation schemes. System developers can use Mcast to develop DMPs over a multicast communication environment.

The rest of the paper is organized as follows. Section 2 describes the network and system architecture of Mcast. Section 3 describes the proposed control schemes for resolving multiple-stream multimedia synchronization/presentation. Section 4 evaluates the performance of Mcast. Finally, Section 5 concludes this paper.

2   Network and System Architecture of Mcast

The proposed multicast multimedia network, which is called Multicast Multimedia Communication Network (M^3CN), is a two level hierarchical architecture that spans a distributed environment. The M^3CN consists of a WAN and a lot of LANs that are attached with the WAN. Each LAN is composed of a local Multicast MultiMedia Server (M^3 server) and clients. An M^3 server transmits media units to hosts via LAN or (and) WAN. Clients of a presentation group, which scatter over different LANs, present the same multimedia resource simultaneously.

In M^3CN, the concept of a "virtual server" is adopted. A virtual server receives media units from the "physical server", which owns the presentation resource, and re-transmits them to end clients. The virtual server is a local server of a LAN and compensates for WAN anomalies by means of pre-depositing some media units and suitable synchronization schemes. The concept of virtual servers can simplify the overhead of synchronization control in clients because WAN anomalies are compensated and media streams are synchronized at virtual servers. Clients become simpler and low-end, e.g. a Set-Top-Box, a diskless networking PC, or a networking TV.

Figure 1 depicts the network and system architecture of Mcast. The underlined network communication is based on MBone [4], which provides multicast transmission across Internet. The authoring tool is the authoring system. The temporal control mechanism is composed of the presentation information file, the code generator, and the multicast presentation system. An author uses the authoring tool to specify temporal and spatial attributes of media and to author his multimedia presentation. Temporal attributes can denote thirteen temporal relationships [10]. In order to represent the temporal relationships, we propose a temporal definition to specify temporal attributes of a multimedia presentation in Section 4. After specifying media attributes, the authoring tool generates the presentation information file accordingly. The presentation information file contains a set of data structures that record the spatial and temporal attributes of the corresponding multimedia presentation. Based on the presentation information file, the code generator generates the corresponding C codes to compose the multicast presentation system.

The multicast presentation system is composed of the physical server system (PSS), the virtual server system (VSS) and the client system (CS). Main func-

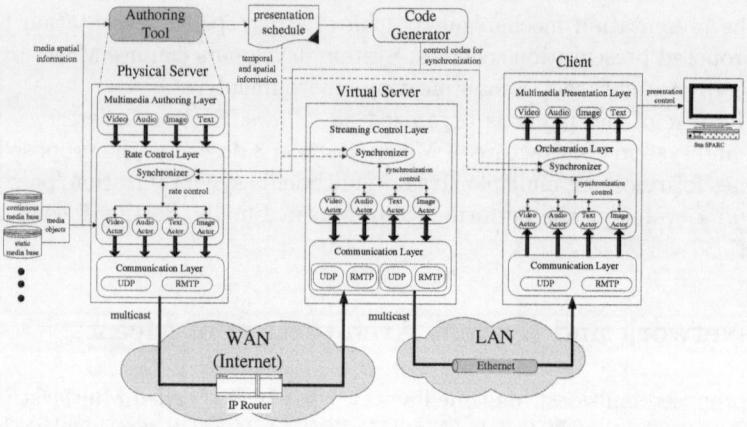

Fig. 1. The abstract network and system architecture of Mcast.

tions of the PSS are (i) to store media resources that are requested by the virtual servers, (ii) to specify the presentation schedule that contains temporal and spatial relations of the related media streams, and (iii) to multicast requested media to the virtual servers. The PSS is composed of three system layers. The multimedia authoring layer provides the authoring tool to allow users to author the multiple-stream multimedia presentation and to generate the corresponding presentation schedule. A system manager can specify the communication configuration that contains the multicast group address, and communication ports. The rate control layer is responsible for retrieving media from the media bases and transmitting these media. The rate control layer is composed of two kinds of components, which are Actors and Synchronizer components. An Actor controls a medium stream. An Actor retrieves media units and multicasts these media units to the corresponding Actors of the virtual servers. The Synchronizer coordinates the rate control among Actors. The communication layer achieves (i) UDP multicast, which provides more efficient transmission and is used to transmit continuous media, and (ii) RMTP (Reliable Multicast Transport Protocol) multicast, which provides reliable multicasting and is used to transmit static media and the presentation schedule.

The VSS is a proxy. After receiving media units, the VSS stores them in the media buffer temporarily. According to the presentation schedule, the VSS multicasts media units to the designated group members using the proposed synchronization control, which compensate for WAN network anomalies. The VSS is composed of two system layers. The streaming control layer is responsible for multiple-stream synchronization and streaming transmission. The streaming transmission achieves continuous and steady multicast transmission. An Actor receives and transmits the media units of a medium stream and controls the medium flow. That is, an Actor controls intra-medium synchronization. The Synchronizer achieves the inter-media synchronization to compensate for skew anomalies and to coordinate Actors' behaviors. The communication layer transmits the continuous and static media via the UDP and RMTP connections, respectively.

The main function of the CS is to achieve a smooth multimedia presentation using proposed synchronization and presentation controls. The CS is composed of three system layers. The presentation layer is responsible for presenting a multiple-stream multimedia presentation. The orchestration layer compensates for jitter and skew anomalies. The Actor achieves the intra-medium synchronization and the presentation control. The Synchronizer achieves the inter-media synchronization among Actors. The communication layer is responsible for receiving media from the corresponding connection channels.

In brief, the steps of preparing a multicast multiple-stream multimedia presentation using Mcast are as follows. (i) A user constructs the desired presentation schedule using the authoring tool. (ii) The code generator generates C codes for part of synchronization/presentation control according to the presentation schedule. (iii) The system manager compiles the generated C codes with the kernel of Mcast to generate the complete executable C codes for the desired multimedia presentation. (iv) The complete executable C codes contain the physical server codes, the virtual server codes and the client codes.

3   Resolutions of Multiple-Stream Multimedia Synchronization and Presentation

A multiple-stream multimedia presentation always contains several kinds of media, such as video, audio, text, and image. Each medium stream owns its presentation schedule and may have related temporal relations with other media streams. In order to have a consistent presentation, clients of a multicast group have to present media streams according to the presentation schedules as much as possible. However, the diversity and heterogeneity of multicast environments inevitably disturb the temporal relations of media streams, i.e., disturb the presentation schedules of media streams. Therefore, clients synchronize media streams with suitable synchronization schemes. The issues of achieving multiple-stream multimedia synchronization at the presentation layer are as follows. (i) Designers have to specify the related synchronization points between/among multiple streams. According to these specific synchronization points, accurate presentation schedules that define temporal relations of multiple streams are derived. (ii) Based on the presentation schedules, designers adopt suitable and practical synchronization/presentation schemes to achieve a perception DMP as much as possible.

In this Section, we specify the temporal distinction of a multiple-stream multimedia presentation in order to induce adequate synchronization points. Then, we describe the proposed synchronization and presentation schemes, which are suitable for diverse media streams.

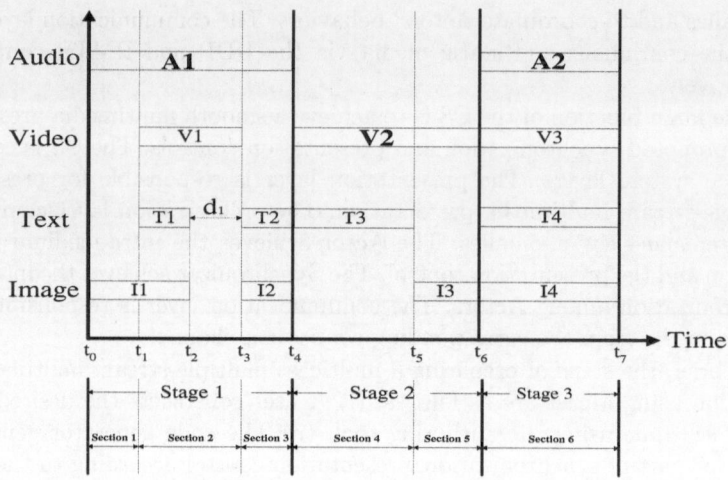

Fig. 2. The presentation time bar chart of an illustrative multimedia presentation.

3.1   Types of Synchronization Points

Figure 2 depicts a presentation example with a time-bar chart. In the presentation example, there are four kinds of media streams, including video, audio, text, and image. Each medium stream is presented or is idle during some time periods and may be related with other streams. In order to identify the temporal relations among multiple media streams, we propose three kinds of temporal synchronization points, which are the stages, sections, and segments.

1. Stage. A presentation stage is a semantic cut of a multimedia presentation. For example, let the multimedia presentation be CNN news broadcast about the chess race between world chess champion Gary Kasparov and supercomputer Deep Blue. Figure 2 depicts the presentation as follows. (1) The news reporter reports the news about a chess race between Gary Kasparov and Deep Blue. The news reporter's audio, Gary Kasparov's video, and the related news texts and images are presented. (2) Gary Kasparov thinks and moves a piece. Then, the video of chess explanation, the texts, and the images about the introduction of Gary Kasparov are presented. (3) An agent moves the piece according to Deep Blue's determination. The background music and some auxiliary texts and images are always presented. Thus, the presentation depicted in Figure 2 is divided into three stages. At the commencement of each stage, inter-media synchronization among related media streams is required to achieve a consistent presentation.
2. Section. A presentation section represents that some media objects have temporal relations. Steinmetz specified the temporal relations with thirteen different relations, which are the equal, start, before, meet, during, overlap, finish relations, and their reversed relations [10]. Based on these possible

temporal relations, a presentation stage can be specified into several presentation sections. One medium object's presentation in a section depends on another medium object's presentation status and a cut point between two presentation sections is a synchronization point. For example, the text of news and the video of Gary Kasparov appear when a specific audio is presented. As depicting in Figure 2, the presentation of the audio object A1 starts the presentations of V1 and T1 at synchronization point t_1, i.e., at the end (commencement) of section 1 (section 2). The object $V1$ finishes and the objects $T2$ and $I2$ start at synchronization point t_3, i.e., at the end (commencement) of section 2(section 3). At the commencement of each section, an inter-media synchronization control is required to achieve a consistent presentation.

3. Segment. In a presentation section, it is possible that a medium stream has no medium object presented during some time periods. A presentation period is denoted as an active segment; an idle period is denoted as an idle segment. In Figure 2, the text medium has two segments in section 2. The first segment displays the text object $T1$ and the idle segment lasts d_1 time units. With the help of presentation segments, the presentation section can be resolved.

Based on the concept of stages, sections, and segments, we (i) clearly specify the temporal relations among multiple streams and (ii) develop the authoring tool and the temporal control system of Mcast.

3.2 Synchronization Control Scheme

Based on the concept of different synchronization point types, we propose the stage-master-based synchronization scheme to solve the multiple-stream multimedia synchronization problem. The stage-master-based synchronization is a refinement of the master-based scheme, which is adopted by Yang et al. [12]. Yang demonstrates that the audio stream is always the master stream since humans are more sensitive to variations in audio. However, in some presentation examples, the audio stream is not always available during the whole presentation, e.g., a piece of silent news. During this silent period, the inter-media synchronization control can not be achieved due to the absent master stream.

In order to solve the problem of the absent master stream, each presentation stage is associated with a master stream to coordinate the presentation and the master stream is changeable between stages. Based on the stage-master-based synchronization scheme, the master stream dominates the commencement and finish of media presentations within the presentation stage. (1) If a slave stream finishes its presentation earlier than the master stream at a synchronization point, the slave stream has to block or extend its presentation until the master stream finishes its presentation. (2) When the master stream finishes its presentation at a synchronization point, the late slave streams have to discard media units to keep pace with the master stream. (3) The master stream is changeable from one stage to the other stage.

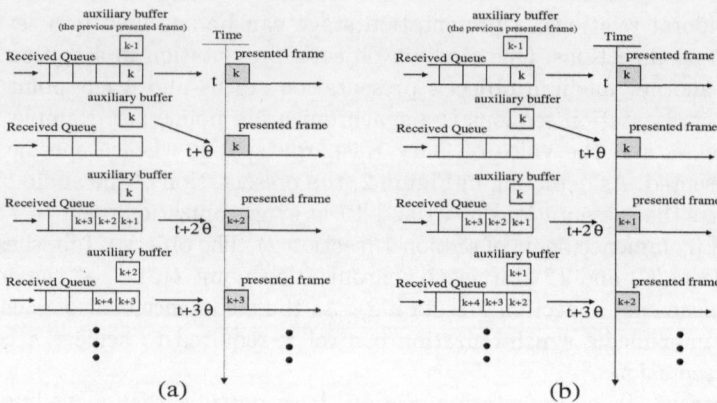

Fig. 3. An example of (a) the time-oriented presentation control, and (b) the content-oriented presentation control for a video stream.

In Mcast, the criterion of selecting a master medium in a stage is based on the human sensitivity about the media. The general principle is as follows. (1) If audio exists in a stage, an audio stream is the master; (2) if audio is absent in a stage, a video stream becomes the master; (3) if there is no continuous medium stream, one of the static media streams is selected to be the master. In Figure 2, the audio stream A1, the video stream V2, and the audio stream A2 are the masters of the stages 1, 2, and 3, respectively.

3.3   Presentation Control Scheme

In order to compensate for jitter anomalies, one can adopt the blocking scheme for the audio medium and the non-blocking scheme for the video medium to achieve intra-medium synchronization. Figure 3(a) depicts an illustrated example for the video stream. When medium unit k was presented at time t, the next medium unit $k+1$ should be presented at time $t+\theta$, where θ is the presentation duration of a medium unit. Unfortunately, medium unit $k+1$ does not arrive on time. Hence, medium unit k is re-presented at time $t+\theta$ according to the non-blocking scheme. During the time of re-presenting medium unit k, medium units $k+1$ and $k+2$ arrive before time $t+2\theta$. At time $t+2\theta$, the "expected" medium unit is presented. Should the "expected" medium unit be unit $k+1$ or unit $k+2$? In order to solve the above problem, two presentation schemes that are considered: (i) the time-oriented and (ii) the content-oriented schemes.

If the main concern is (i) to satisfy time-related temporal relations and (ii) to keep the actual presentation time length equal to the nominal presentation time length as much as possible, the time-oriented presentation scheme can be adopted. The "expected" medium unit should be the one that is closest to the nominal one. In Figure 3-(a), medium unit $k+2$ is presented at time $t+2\theta$ and medium unit $k+1$ is discarded. The mathematic formula for obtaining the expected medium unit is as follows. At time $t + i\theta$, we assume that (1) the last presented medium unit is medium unit $k + j$, and (2) the received queue contains media units $k + r_1$, $k + r_2$, ..., and $k + r_n$. Let the expected medium unit be $k + m$ at time $t + i\theta$. Then, m is the maximum number of the subset of $(j, r_1, r_2, ..., r_n)$, in which all of the elements in the subset are less than or equal to i, i.e., $m = MAXIMUM\{x \in A \mid x \leq i\}, A = \{j, r_1, r_2, ..., r_n\}$. The time-oriented scheme is suitable for the continuous media. The drawback of the time-oriented scheme is that there may be some flickers at the synchronization point when several delayed media units are discarded at the same time.

If the main concern is to keep the completeness of a medium presentation as much as possible, the content-oriented presentation scheme can be adopted. Each medium unit is presented as much as possible. In Figure 3-(b), medium unit $k+1$ is presented at time $t+2\theta$ and medium unit $k+2$ is presented at time $t+3\theta$. The mathematics formula for obtaining the expected medium unit is as follows. At time $t + i\theta$, we assume that (1) the last presented medium unit is medium unit $k+j$, and (2) the received queue contains media units $k+r_1$, $k+r_2$, ..., and $k + r_n$. Let the expected medium unit be $k + m$ at time $t + i\theta$. Then, m is the minimum number of the subset of $(j, r_1, r_2, ..., r_n)$, in which all of the elements in the subset are less than or equal to i, i.e., $m = MINIMUM\{x \in A \mid x \leq i\}, A = \{j, r_1, r_2, ..., r_n\}$. The content-oriented scheme is suitable for the static media and the content-critical continuous media. The drawbacks of the content-oriented scheme are twofold. (i) The total presentation time may become longer than the nominal presentation time, and (ii) more inter-media asynchronous anomalies may exist until an inter-media synchronization is achieved.

4   Performance Evaluation

The presentation quality of a DMP can be evaluated by some essential QoS parameters, which are the presentation jitter and skew. In order to reveal Mcast's efficiency, we compare the performance evaluation of Mcast with that of the traditional client/server architecture based on the essential QoS parameters. Both of Mcast and the client/server architecture are implemented according to our proposed synchronization and presentation schemes. The number of total presented frames is four thousands and the whole presentation contains two stages. For the reason of comparison, the pre-deposited audio/video frames at the virtual server are fixed to 20. The pre-deposited frames of a Mcast's client are 5; the pre-deposited frames of a client in the traditional client/server architecture are 5 or 25. The evaluation results that are depicted from Figure 4 to Figure 6 are arranged as follows. Case (a) and case (b) are the evaluation results based on the traditional client/server architecture. The pre-deposited frames in cases (a) and (b) are 5 and 25, respectively. Case (c) are the evaluation results of Mcast and the pre-deposited frames are 5. We note that the pre-deposited frames in case (b) equal the summation of the pre-deposited frames of a virtual server and a client in case (c).

62   C.-M. Huang and H.-Y. Kung

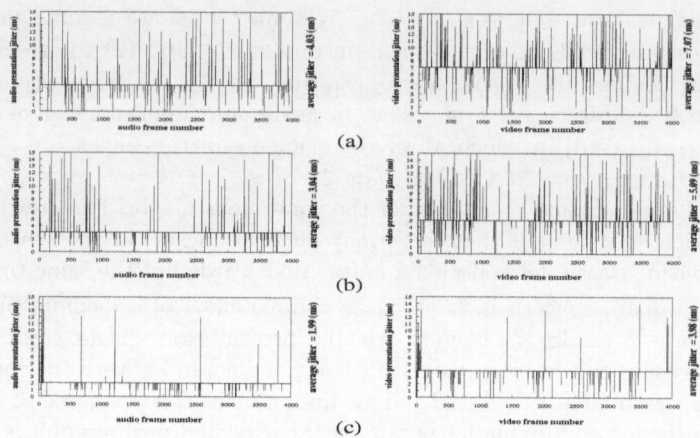

(a)

(b)

(c)

Fig. 4. Presentation jitter of the audio and video streams based on the content-oriented scheme. Cases (a) and (b) are based on the traditional client/sever architecture and case (c) is based on Mcast.

Figure 4 depicts the presentation jitter of the audio and video streams at the client site based on the content-oriented scheme. Having a longer pre-deposited length, the curve and the presentation jitter of case (b) are more stable and less than those of case (a). Although the pre-deposited frames in case (b) equal the total pre-deposited frames of case (c), the curve and the presentation jitter of case (b) are more oscillatory and larger than those of case (c). It is because the presentation overhead of clients in case (b) includes not only the media transmission overhead but also includes the synchronization overhead, which compensates for the WAN jitter. On the other hand, clients of case (c) presents synchronized media frames, which are synchronized by the virtual server. Since the volume of the video stream is much larger than that of the audio stream, the presentation jitter value of the video stream is larger than that of the audio stream.

Figure 5 shows the presentation jitter based on the time-oriented scheme. We observe that the average presentation jitter based on the time-oriented scheme is greater than the average presentation jitter based on the content-oriented scheme. It is because more media frames are skipped based on the time-oriented scheme in order to catch up the nominal presentation.

Figure 6 depicts the presentation skew of the audio and video streams at the client site based on the content-oriented and the time-oriented schemes. The curve and skew value of case (c) are more stable and less than those of cases (a) and (b). We note that the average presentation skew based on the time-oriented scheme is less than that based on the content-oriented scheme. It is because, based on the time-oriented scheme, each medium stream is presented according to the nominal presentation schedule as much as possible. Thus, the skew between media streams can be reduced.

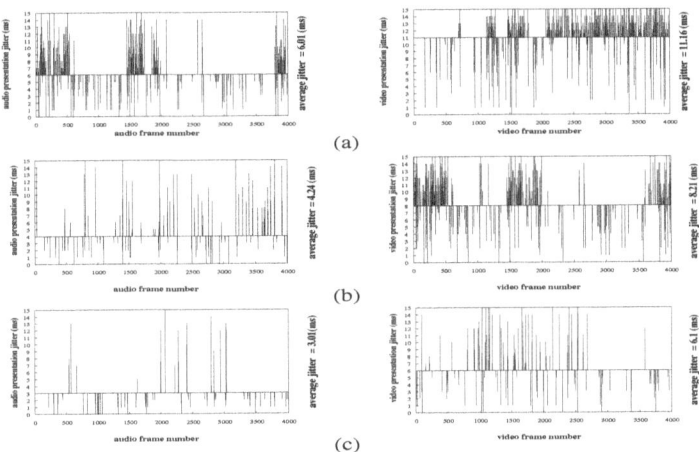

Fig. 5. Presentation jitter of the audio and video streams based on the time-oriented scheme. Cases (a) and (b) are based on the traditional client/sever architecture and case (c) is based on Mcast.

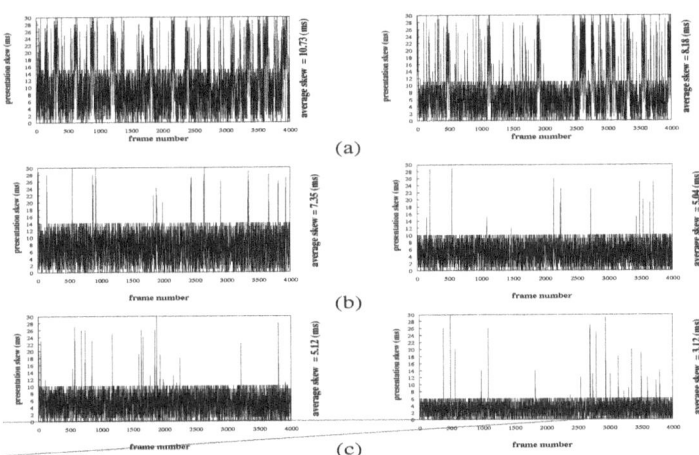

Fig. 6. Presentation skew between the audio and video streams. Cases (a) and (b) are based on the traditional client/sever architecture and case (c) is based on Mcast. The left evaluation results of cases (a), (b), and (c) are based on the content-oriented scheme; the right evaluation results of cases (a), (b), and (c) are based on the time-oriented scheme.

5   Conclusion

This paper describes the major considerations, resolutions, and techniques that are involved in designing and developing a multicast multiple-stream multimedia presentation system. Based on the design and development considerations and the proposed synchronization/presentation schemes, the Mcast communication engine has been implemented. Mcast provides (1) the authoring tool to specify presentation's appearance, and (2) the temporal control mechanism to achieve smooth presentations. Evaluation results reveal that the Mcast's performance is acceptable and Mcast is more efficient than the traditional client/server architecture.

Acknowledgment. The research is supported by the National Science Council of the Republic of China under the grant NSC 88-2219-E-006-012.

References

1. Blakowski, G., Steinmetz, R.: A Media Synchronization Survey: Reference Model, Specification, and Case Studies. IEEE Journal on Selected Areas in Communications **14** (1) (1996) 5–35
2. Coulson, G.: A Configurable Multimedia Middleware Platform. IEEE MultiMedia Magazine **6** (1) (1999) 62–76
3. Diot, C., Dabbous, W., Crowcroft, J.: Multicast Communication: Survey of Protocols, Functions, and Mechanism. IEEE Journal on Selected Areas in Communications **15** (3) (1997) 277–290
4. Eeriksson, S.: MBone: The Multicast Backbone, Communication of the ACM **37** (8) (1994) 54–60
5. Huang, C.M., Lo, C.M.: An EFSM-based Multimedia Synchronization Model and The Authoring System. IEEE Journal on Selected Areas in Communications **14** (1) (1996) 138–152
6. Metz, C.: IP QoS: Traveling in First Class on The internet. IEEE Internet Computing **3** (2) (1999) 84–88
7. Pasquale, J.C., Polyzos, G.C., Xylomenos, G.: The Multimedia Multicasting Problem. ACM/Springer-Verlag Multimedia Systems Journal **6** (1) (1998) 43–59
8. Paul, S., Sabnani, K.K., Lin, J.C., Bhattacharyya, S.: Reliable Multicast Transport Protocol (RMTP). IEEE Journal on Selected Areas in Communications **15** (3) (1997) 407–421
9. Steinmetz, R.: Human Perception of Jitter and Media Synchronization. IEEE Journal on Selected Areas in Communications **14** (1) (1996) 61–72
10. Steinmetz, R.: Synchronization Properties in Multimedia Systems. IEEE Journal on Selected Areas in Communications **8** (3) (1990) 401–412
11. Xiao, X., Ni, L.M.: Internet QoS: A Big Picture. IEEE Network **13** (2) (1999) 8–18
12. Yang, C.C., Huang, J.H.: A Multimedia Synchronization Model and Its Implementation in Transport Protocols. IEEE Journal on Selected Areas in Communications **14** (1) (1996) 212–225

Enforcing Multipoint Multimedia Synchronisation in Videoconferencing Applications

Philippe Owezarski

LAAS-CNRS
7, Avenue du Colonel Roche
F-31077 Toulouse cedex 4

Abstract. This paper deals with the problem of multimedia synchronisation in fully meshed multipoint videoconferencing applications. This work relies on previous work dealing with multimedia synchronisation in point to point videoconferencing: Up to now, the two main approaches are based on (1) timeline or on (2) temporal intervals composition. But, the first approach is not suited for multimedia data, and the second not suited for multipoint synchronisation. This paper then proposes a hybrid synchronisation solution taking advantage of the two preceding approaches. A multipoint videoconferencing tool based on this hybrid solution, called *Confort*, has been designed, developed, implemented and evaluated. This paper then deals with implementation details, as well as evaluation measurements.

1. Introduction

Videoconferencing is nowadays a mandatory tool for many real time collaborative working environments. A lot of videoconferencing tools already exist and among them it is possible to find industrial and commercial products as Netmeeting from Microsoft, or academic tools as VIC from LBL or Rendez-vous (the IVS successor) from INRIA. The market of videoconferencing tools is increasing rapidly, pushed by strong needs in concurrent engineering (in industrial domains as aeronautic or automotive manufacturing, electronic design, etc.). But the first projects dealing with concurrent engineering evaluated videoconferencing tools in actual experiments and showed that no collaborative working environment, and in particular, no videoconferencing application fulfil collaborative working requirements.

In particular, the main problem with videoconferencing tools is their lack in term of QoS (Quality of service) guarantees. Among all the QoS parameters the most meaningful seem to be the audio quality that can currently range in a single work session from correct to incomprehensible (while it has to be constant), the end to end delay that has to be as short as possible to increase the interactivity level between users, and all temporal synchronisation constraints. In fact, users are really disturbed when the temporal features of the audio and video streams are not enforced, as voice can be hard or impossible to understand, and the lack of lips synchronisation makes the audio / video correlation disappear. Besides, multimedia synchronisation is currently the key point to address and the most difficult to solve for designing multimedia systems [3].

H. Scholten and M. van Sinderen (Eds.): IDMS 2000, LNCS 1905, pp. 65-76, 2000.
© Springer-Verlag Berlin Heidelberg 2000

In related work, multimedia synchronisation is most of the time not addressed, and when it is, the selected solution is based on timestamps. This is the case for all videoconferencing applications using RTP [12] as VIC, VAT or Rendez-Vous, even if it is now well known that the semantics of timestamps does not match the requirements of multimedia synchronisation. Based on this observation, some work lead to another synchronisation approach based on temporal intervals composition. This approach has been selected and improved at LAAS, and lead to the design of a point to point videoconferencing tool: PNSVS (for Petri Net Synchronised Videoconferencing System). So, this paper deals with the extension of PNSVS toward a multipoint videoconferencing application enforcing multipoint multimedia synchronisation. This new videoconferencing tool is called "*Confort*".

The remainder of this paper is as follows: first the point related to multimedia synchronisation is described (section 2). In particular, this section focuses more specifically on the problem of multiple users synchronisation. Then section 3 describes the principle of the two main synchronisation approaches that were spoken about just before. Given the analysis of these two approaches, and based on the experience gained from the PNSVS design and development, a hybrid synchronisation is presented (section 4). Using this hybrid approach, section 5 describes the implementation principles of *Confort*, and section 6 gives some experimental results and measurements. Finally section 7 concludes this paper.

2. Synchronisation Problematics

2.1. Multimedia Synchronisation

The point of multimedia synchronisation is closely related to the characteristics of the media to synchronise. In fact, in a videoconferencing tool, media (audio and video) are dynamic with strong temporal requirements. The problematics of multimedia synchronisation consists then in ensuring both the intra and inter streams temporal constraints. Ensuring the intra-stream temporal constraints consists in controlling the jitters on each audio or video object in order to keep it under acceptable maximum values. It means that the temporal profile of media objects has to be respected on the receiving site. For instance, a 25 images/s video stream has to be played back with a rate of 25 images/s. The only tolerance is called jitter.

Ensuring the inter-streams synchronisation consists in controlling the drifts between the audio and video streams (due to the cumulative effects of jitters) within a maximum value (sounds to pictures in lip synchronisation for example where lips movement has to be related to the voice stream).

2.2. Multipoint Synchronisation

Multipoint synchronisation can be seen as multiple users synchronisation. This synchronisation task aims to avoid semantics incoherence in the dialogue between all users. In fact, delays can change significantly in distributed systems, and it is mandatory to avoid that one user receives the response to a question before the question itself. In collaborative systems, this kind of problem is generally solved using Lamport clocks that create in the system a logical order based on causal dependencies and that has to be respected by all users.

But, in a videoconferencing application that computes dynamic media, as audio and video, having strong temporal requirements, the Lamport clocks seem to be quite limited. In fact, in a videoconferencing tool, interactivity is of major importance, and the semantics is closely related to delay and explicit time. Generally, it is acknowledged that for a highly interactive dialogue, delays between users have to be less than 250 ms [6]. Beyond this threshold, audio and video data can be considered too old and of no relevance. Therefore multipoint synchronisation has to design and set up mechanisms allowing the control of end to end delays between users, in order to ensure the suited order between audio and video information, on a temporal basis.

3. Related Approaches

As it has been mentioned in section 1, two main approaches exist in the literature. The first one, and the most used, relies on a timeline basis - this is an absolute time approach - while the second relies on the composition of time intervals - this is a relative time approach.

3.1. *"Timeline"* Approach

This approach, consisting in using timestamps is easy and intuitive. In fact, each audio or video presentation object contains its own presentation date, and the presentation process has just to display it at the right time. This method ensure the intra and inter-streams synchronisation: a 25 images/s video sequence is replayed with a 25 images/s rate with respect of the temporal constraints on each object (intra-stream synchronisation), and the inter-streams synchronisation is also ensured, as each stream synchronises itself on a common temporal reference timeline and are then synchronised the one related to the other.

However, timestamps do not solve the problem of acceptable jitters on information objects. In fact, timestamps are only a date on a time axe, and are then unable to integrate the acceptable variations on the presentation of each object. This approach, then, does not provide a suitable solution to the problem of multimedia synchronisation

3.2. Temporal Intervals Composition

The second approach frequently encountered in literature is based on temporal intervals composition [1]. It means that the presentation time of one object depends on the preceding object presentation completion. Thus, only composing the time durations of all presentation data units can create multimedia synchronisation scenarios. This is the case for instance of the OCPN (Object Composition Petri Net) model [8].

To design guaranteed synchronised multimedia, it is mandatory to consider all problems due to the temporal variability of the computing times (jitters, drift) and to link them to the properties inherent to each multimedia object (for instance to link the jitters of an operating system to the acceptable jitters of each multimedia object). To define these synchronisation properties on the multimedia objects themselves, a model allowing to model the application synchronisation constraints is required. Several studies have already been realised in this domain, and some models have been proposed. In particular, the OCPN model only considers nominal computing times, and does not address the temporal jitters acceptable on most of the multimedia data.

This limitation led [5] to propose a new model, called TSPN (Time Streams Petri Nets) taking into account this temporal variability aspect. By definition, TSPNs use temporal intervals on the arcs leaving places. The temporal intervals are triplets (x^S, n^S, y^S) called validity time intervals, where x^S, n^S and y^S are respectively the minimum, nominal and maximum presentation values.

The inter-streams temporal drifts can be controlled in a very precise way using 9 different inter-streams transition semantics and the position of each time interval on the arcs leaving the places (allowing the computing of each stream apart from the others). Using these transition rules, it is possible to specify synchronisation mechanisms driven by the earliest stream ("or" synchronisation rules), the latest stream ("and" synchronisation rules) or by a given stream ("master" synchronisation rules). For more details, [13] gives a formal definition of the model and the different firing semantics.

Example: Using a TSPN for PNSVS

It is proposed in this part to use the TSPN model to describe the set of the timed synchronisation behaviours that appears in a videoconference application as PNSVS [4].

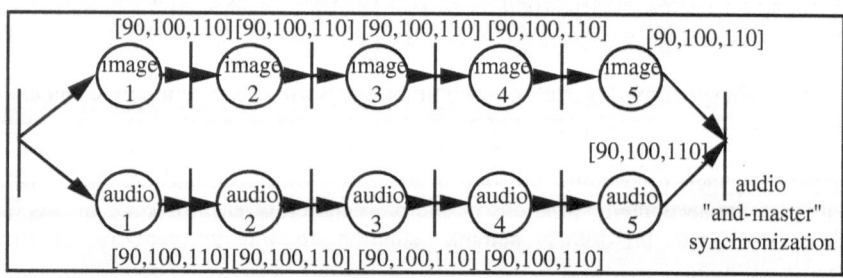

Fig. 1. Videoconference synchronisation constraints: TSPN based modelling

As an example, Fig. 1 describes the requirements for a videoconferencing system whose QoS parameters are:

- A throughput of 10 images per second. Then the nominal presentation time of a video object is 100 ms;
- An acceptable jitters on the audio or video of 10 ms (in absolute value) [6]. It follows that the temporal validity intervals are [90, 100, 110] for one image and one audio packet;

- The audio is the most important media: inter-stream synchronisation is of «and-master» type, the voice being the master. In fact, voice being more important than video, it is defined as the master stream and its temporal constraints have to always be fulfilled; the «and» part of this rule tries as much as possible to wait and respect the constraints on the video stream to reduce its discontinuities;
- The synchronisation quality: the inter-stream drift must be less than 100 ms [6]. 100 ms being the limit under which the temporal gap between audio and video cannot be noticed. Then the inter-stream synchronisation period corresponds to the presentation of 5 images. This is because the maximum drift on 5 objects is 50 ms and the inter-stream drift is then less than 100 ms.

As it has been shown before, as well as in [11] for PNSVS, the temporal intervals composition approach is perfectly suited for point to point multimedia synchronisation. But this approach is not so powerful for multipoint synchronisation. In fact, it is quite difficult with such an approach to synchronise all users and to control the end to end delay as no global clock is used and we do not assume any temporal properties on the delays in the network. In fact, each jitter on an object introduces desynchronisation between users. As with Lamport Clocks, TSPN only allows logical synchronisation (or ordering) between users.

4. An Hybrid Solution for Multipoint Multimedia Synchronisation

As on one side, timestamps are not suited for multimedia synchronisation, and time intervals composition not suited for multipoint synchronisation, it is not possible to use one of these approaches. But on the other side, time intervals composition is perfectly suited to multimedia synchronisation, and timestamps are, until today, the best way to achieve multipoint synchronisation and a global time in distributed systems. It is then proposed in this paper to use an hybrid approach where:

- Multimedia synchronisation is achieved thanks to a time intervals composition mechanism;
- Multipoint synchronisation is achieved thanks to the set up of a global clock in the distributed system.

Concerning multimedia synchronisation in *Confort*, the principle is very similar to the one of PNSVS described in [4], as TSPN and more generally synchronisation mechanisms relying on temporal intervals composition are designed for 1 to N synchronisation schemas. Moreover, this approach has been designed for unpredictable distributed systems where delays, loss level, etc. are changing. In fact, the synchronisation constraints to enforce on each object depend on the presentation duration of the media object, and on the arrival date of the considered object. The synchronisation scenario modelled by the TSPN of Fig. 1 is refined, concerning actual presentation dates, each time a new media object reaches the receiving videoconference entity. In addition, as current general operating systems are asynchronous (as Unix, Windows, etc.), performing temporal synchronisation on media objects in low level layers is useless, as synchronisation features are disturbed when media objects cross communication and operating system layers. The synchronisation task has then to be put at the upper level of the receiving entity, in the application layer. As the synchronisation task is put on the receiving entity, synchronisation approaches based on temporal intervals are then ideally designed for 1 to N synchronisation.

Given the TSPN of Fig.1, the synchronisation task on each receiver has only to implement a synchronisation runtime able to enforce the synchronisation constraints modelled by the TSPN, and to schedule media objects adequately. The principle of such a runtime is described in section 5, which deals with implementation issues.

Concerning multipoint synchronisation and given a global synchronised clock over all the distributed system, it is then very easy to control end to end delay and causality in media object deliveries. In fact, it is sufficient to put in the header of media objects a timestamp (related to the global clock) and to control that the time between the sending and arrival dates is not greater than the maximum end to end delay required. This approach is very similar to the one of RTP (Real time Transport Protocol) that is recommended by IETF for audio / video Transport [2], except that we here assume a global clock. How distributed workstations clocks are synchronised to obtain this global clock is explained in section 5 (related to implementation issues).

Finally, with the temporal intervals composition based approach that allows 1 to N multimedia synchronisation schemas and the synchronisation of all the workstation clocks - and therefore of all the sending entities of the videoconferencing session - an N to N synchronisation schema is achieved.

5. Implementation Issues

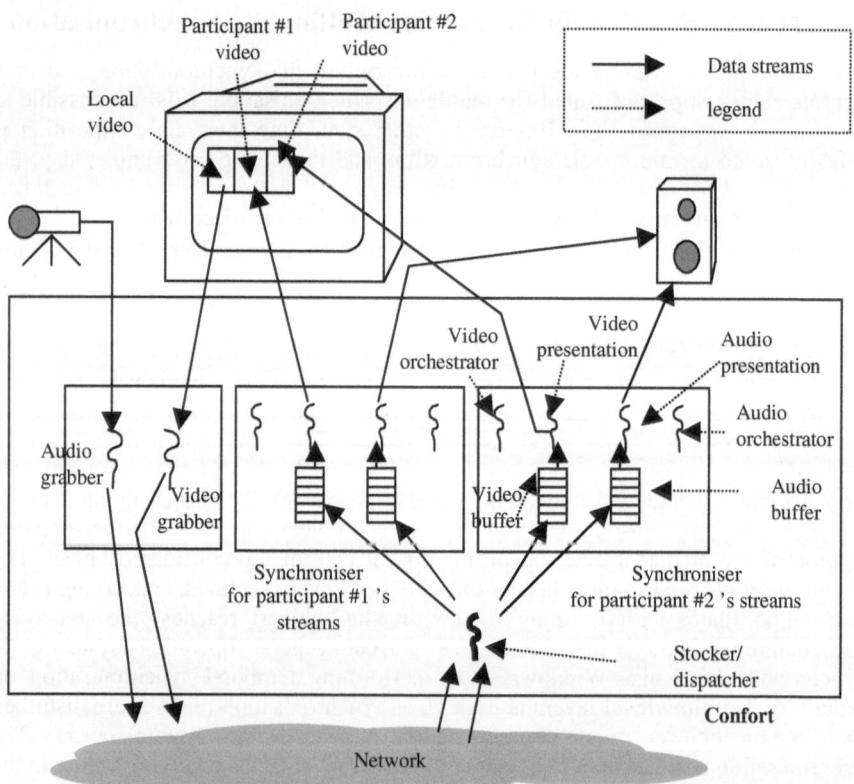

Fig. 2. Confort architecture (3 participants)

This section deals with implementation issues of the hybrid synchronisation task described in the preceding section. Thus, it is shown:

- How multimedia synchronisation can be enforced with a strong respect of audio and video temporal constraints?
- How the global clock can be set-up on a potentially wide area distributed system?
- And how some other functionalities as openness and auto-configurability can be integrated in Confort, in order to make it rapidly and easily usable in any collaborative working environment?

To illustrate all these implementation issues, Fig. 2 depicts the *Confort* software architecture.

5.1. Principle

An important constraint of the Confort development concerns its ability to work on general systems as Windows or Unix, with general hardware. We prefer to avoid dedicated systems and hardware architectures that are not widely available, a videoconferencing tool having to work on as many desktop computers as possible.

However, implementing multimedia synchronisation mechanisms having to enforce temporal constraints requires some real time functionalities of the operating system, to be able to control the processes scheduling. For this reason, Solaris has been selected as it provides a real-time scheduling class, and it was the only one to provide this feature in 1994, when the work on videoconferencing began at LAAS. This real time scheduling class (also called RT) is really interesting as it gives RT priorities to user processes greater than the ones of system tasks, and also a predictable and configurable scheduler [7].

Before addressing the multipoint synchronisation point, some work has been done, at LAAS, to enforce synchronisation in the PNSVS point to point videoconferencing application [10]. And it has been said in section 4 that synchronisation mechanisms are located in the top layer of the receiving machine. Thus, the multimedia synchronisation problem can be solved in Confort as in PNSVS. In fact on each 1 to N multiconnection from one sender to all other participants of the videoconference session (the receivers), we can duplicate the synchronisation automaton of PNSVS on each receiving entity.

Finally, as it is depicted on Fig. 2, each videoconferencing entity (on each machine) consists of:

- A sending task that grabs video and audio objects and multicasts them to all the other videoconference participants;
- For each distant participant, a synchronisation task that synchronises the audio and video streams coming from this distant participant. Then with N participants, each entity will run N-1 synchronisation tasks similar to the one of PNSVS.

As a reminder, the following recalls how the PNSVS intra and inter-streams synchronisation mechanisms are working. In fact, each stream is computed by a dedicated process (implemented thanks to a thread). But before being presented (on the screen for images or loudspeakers for audio), each object has to be computed by a specific hardware (video or audio board) managed by the kernel and with system scheduling class and priority. And there is no guarantee on the time that the computing of each image and audio object will take: this is a typical asynchronous system. It is then required to create a second process (thread), in the RT class, whose function is to control the temporal behaviour of the presentation process. Running with the RT priority, the control process, also called orchestration process, can perfectly control the associated presentation process. It can stop (or kill) it if it overruns the maximum presentation duration, or it can also make it sleep until the minimal presentation time is reached: the intra-stream synchronisation constraints are therefore enforced. Then, as depicted on Fig. 2, each synchronisation task for synchronising audio and video streams consists of 4 threads:

- One audio presentation thread;
- One video presentation thread;
- One audio orchestration thread devoted to the control of the temporal behaviour of the audio presentation thread;
- One video orchestration thread devoted to the control of the temporal behaviour of the video presentation process.

On the other side, inter-streams synchronisation is performed between the orchestration threads, that have to synchronise themselves on a rendezvous point (corresponding to inter-streams transition on the TSPN model), and this with respect to the semantic rule of the inter-streams synchronisation.

5.2. Clocks Synchronisation

Concerning clocks synchronisation, my first work consisted in investigating for existing solutions. Many are related in the literature, and they are always based on temporal information exchanges, over communication networks. But it also appears that none of these solutions do provide any guaranty on the precision they can reach on a general network as the Internet. Even NTP3 [9], that has been designed for synchronising clocks over the Internet, does not provide any guarantee, the synchronisation precision being dependant on the network load and RTT, for example. Nevertheless, on LANs, NTP proved to be really efficient, as LAN provides large bandwidth and low RTT. NTP is then a suited solution for synchronising clocks on a LAN and has therefore been selected. This choice is one among a lot of similar solutions that are as efficient as NTP. The advantage of NTP, compared to other solutions is its availability: many free implementations are available, and this protocol is already widely deployed.

Nevertheless, it remains to solve the problem of synchronising machines on WAN. Up to my knowledge, there is no clock synchronisation protocol really efficient on WAN as the Internet. The selected solution is then the one offered by GPS, which transmit the universal time of reference atomic clocks by satellite links, with a precision of few microseconds. Thus, the solution retained consists in setting one GPS board on each LAN, and then to synchronise all the clocks of all the machines connected to this LAN, with the clock of the GPS board, using NTP. Similar solutions are sometime used for scalable and reliable multicast protocols. Thanks to this approach it is possible to synchronise very precisely the clocks of partners involved in a widely distributed videoconferencing session. Section 6 presents some evaluation measurements of this solution.

5.3. Additional Functionalities of Confort

The other functionalities that have to be integrated in *Confort* are not related to synchronisation. The point here concerns the ability of the videoconferencing tool to be easily integrated in a more general and generic tele-working environment. These kinds of environments may or may not manage the work session membership, and may or may not send control and/or command messages. Depending on the work session control tool of the collaborative environment (maybe a workflow management system), the videoconferencing system has to:

- Be able to receive control and/or command messages from the work session controller, and then to dynamically change the group of participants (by adding or removing some members), change the requested QoS on audio and video streams, etc. This ability is called "openness". For that an interface has been designed, that defines the type of control/command messages, and fields they contain;
- Be able to detect itself membership changes. This ability is called "auto-configurability". Then, if the videoconferencing receiving entity detects that new audio and video streams arrive from a new participant (that were not participating in the session up to now), it automatically allocates the required resources (buffers), starts the synchronisation entity (consisting of 4 threads), and create a video window to display the received images. On the other side, if the receiving entity does not receive any data from one participant for a long time, this participant is removed from the active members, the allocated resources are freed, and the associated synchronisation threads are killed. This principle allows the application to always have a coherent knowledge of the state of participants group membership, even in case of failures or crashes.

The architecture principle to develop such functionalities relies on the use of a stocker/dispatcher thread (Fig. 2). This thread is the master of the overall architecture as, depending on the type of data it receives, it can:

- Store audio and video data in the right buffer, depending on who send it;
- Change what has to be changed if it receives a control/command message (openness);
- Adapt itself the dynamic architecture of the videoconferencing entity if it notices that a new participant is sending data, or at the opposite, if a participant does no more send data (auto-configurability).

6. Results

Confort has been implemented on Sun workstations (Sun Sparc Station or Ultra station), in C++, using the Solaris 2.5.1 or 2.7 operating system. Confort supports Parallax or SunVideo video boards that allow the system to grab, display, compress and uncompress images using the M-JPEG algorithm. Audio is the Sun standard audio system. Workstations running Confort communicate using the UDP transport protocol. Tests have been performed on top of a 10 Mbps Ethernet and a 155 Mbps ATM network.

Confort is a full meshed and full duplex videoconferencing application that can process up to 25 images/s (320 x 240 pixels and 24 bits coded colours).

This part aims to evaluate the performances of the hybrid multipoint synchronisation task developed in *Confort*. First it is required to verify if all synchronisation solutions do respect the temporal presentation requirements, and in particular the quality required for the intra and inter-stream synchronisation. The measurements reported here have been realised in the case of a 10 images/s videoconference application whose synchronisation requirements were modelled by the presentation TSPN given in Fig.1, and with a 3 users videoconferencing session.

Fig. 3. Shows, for the first 100 images received by one of the user from another, the jitters that appear during the presentation of each image. The measured jitters are the difference between the effective presentation duration and the nominal presentation time. This Fig. clearly shows that the maximum jitters value is always fulfilled. Note that the jitters is always negative: this is because in this experiment there is no network problems (no loss and no jitters, the goal being to evaluate the temporal synchronisation mechanisms) and then the needed data are available as soon as the presentation process needs them. Thus, the anticipation mechanism of the temporal intervals composition approach, that stops the presentation of an object as soon as its minimal presentation time has been reached, actually works. The very small variations are due to the real-time scheduler of Solaris 2.

Fig. 4 shows the same measurements but now for the Confort audio stream. As in the video case, intra-stream synchronisation requirements are always fulfilled. Note that now the jitters are always positive. In fact, this is because firing the intra-stream transition is controlled by the audio driver signal. With a time scale expressed in millisecond, the presentation duration of one audio object is always equal to 100 ms; variations are due to the time required by the system to take into account this information.

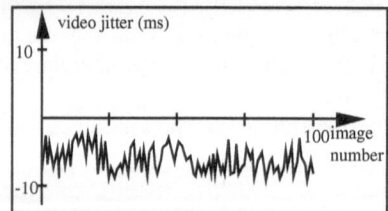

Fig. 3: Measurement of the jitters on *Confort* images

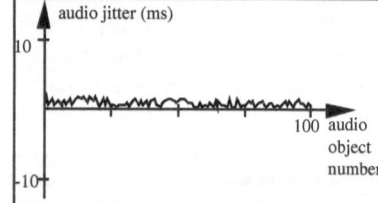

Fig. 4: Measurement of the jitters on *Confort* audio objects

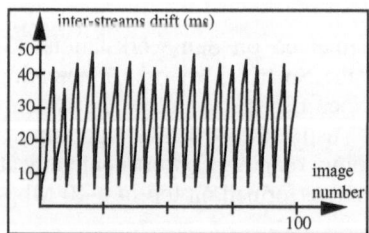

Fig. 5: Measurement of *Confort* inter-streams drift

Finally, Fig. 5 shows the curve representing the inter-stream drift for the audio and video sequences. Fig. 3 shows that the video jitters is always negative and Fig. 4 that the audio jitters is always positive. Thus, the inter-stream drift (the difference between the audio and video presentation dates) is positive. It can be easily seen that this drift increases during each period of 5 objects (what is due to the cumulative effect of the jitters) and it is then reduced to a value near zero at each inter-stream synchronisation. It clearly follows that the inter-stream synchronisation requirements are perfectly fulfilled: the maximum drift never exceeds 50 ms, the maximum allowed value being 100 ms. Note that the maximum value is never reached here because the audio jitters is always very close to zero and thus the inter-stream drift is essentially due to the video jitters.

Therefore, the evaluation of multimedia synchronisation mechanisms proves that they perfectly work. It remains, nevertheless, to evaluate the mechanisms allowing the synchronisation of the clocks of all the workstations involved in the videoconferencing system. These measurements have been performed on a dedicated 155 Mbps ATM local network, on which delays are perfectly known (at a 1 ms scale), and thanks to a network analyser. It appears that the desynchronisation between the three workstations clocks is always less than 2 ms, and this drift between clocks is essentially due to temporal variations in the transport layer, supported by a non real time operating system, and is not due to low level network layers. NTP, then, provides a suited solution for synchronising workstation clocks on a LAN. In fact, at the scale of a videoconferencing application, such a drift between clocks is not significant, and human users cannot notice an end to end delay variation of 2 ms.

Nevertheless, *Confort* has not been tested on general wide area network as the Internet (it has only been tested on a national private network: SAFIR), and then we do not perform any measurement in such case. But, GPS manufacturers ensure that the precision of synchronisation with GPS boards (connected to reference atomic clocks by GEO satellite links) is of few microseconds. It is then of no significance compared to the precision of NTP on a LAN, and compared to the requirement of a videoconferencing application.

7. Conclusion

This paper has presented a new hybrid approach for synchronising multimedia streams in a multipoint videoconferencing application. The multimedia synchronisation relies on a temporal intervals composition basis to take into account the jitters that are acceptable on audio and video objects. Then, the interactivity and the coherence of multimedia information between distributed participants is achieved thanks to a clocks synchronisation mechanism. Taking advantage of a global distributed clock, it is then possible to perfectly control the end to end delay and its variations, and to enforce mulipoint synchronisation. The solution for synchronising distributed clocks is very practical as it relies on dedicated hardware (GPS) and NTP. Nevertheless, it has been proved that the hybrid solution proposed in this paper is really efficient.

The limitation that appears is related to the portability of the Confort videoconferencing tool and the management of its evolutions. In fact, this tool is written in C++ because it was required to access the real time functionalities of Solaris. But, each time the video board or the video drivers version change, it is required to reopen the source code, and to modify the video procedures. To avoid such portability and maintenance problems, it would be interesting to use generic video interfaces as the Java JMF, for instance, that solves all the compatibility and portability problems of video programs. But, today, the mapping between the Java Virtual Machine (JVM) and the real time services of the operating system is not done, the JVM having a really asynchronous behaviour. Nevertheless, the Java approach seems really interesting for portability and version maintenance. That is why next step will consist in studying the mapping between the Java runtime and the operating system kernel services.

8. References

1. J. F. Allen, "Maintaining knowledge about temporal intervals", Communications of the ACM, Vol. 26(11), November 1983
2. Audio Video Transport Group of the IETF, available at http://www.ietf.org/html-charters/avt-charter.html, March 2000
3. G. Blakowski, R. Steinmetz, "A media synchronization survey: reference model, specification, and case studies", IEEE journal on selected areas in communications, Vol.14, No 1, January 1996
4. M. Boyer, P. Owezarski, M. Diaz, "Dynamic QoS renegociation in the PNSVS videoconferencing application", International conference on Interactive Distributed Multimedia Systems (IDMS'98), Oslo, Norway, September 8 - 11th, 1998
5. M. Diaz, P. Sénac, "Time stream Petri nets, a model for multimedia streams synchronization", proceedings of MultiMedia modeling'93, pp 257-273, Singapore, November 1993
6. K. Jeffay, D.L. Stone, F.D. Smith, "Transport and display mechanisms for multimedia conferencing across packet-switched networks", Computer networks and ISDN systems 26 (1994)
7. S. Kleiman, D. Shah, B. Smaalders, "Programming with threads", 534 pages, Prentice Hall 1996
8. T.D.C. Little, A. Ghafoor, "Synchronization and storage models for multimedia objects", IEEE journal on selected areas in communications, Vol. 18, No. 3, pp 413-427, april 1990
9. D.L. Mills, "Network time protocol (version 3): Specification, implementation and Analysis", RFC 1305, March 1992
10. P. Owezarski, M. Diaz, "Models for enforcing multimedia synchronisation in visioconference applications", proceedings of the 3rd MultiMedia Modeling conference - Towards the information superhighway (MMM'96), pp 85-100, World scientific editor, Toulouse, France, November 12-15, 1996
11. P. Owezarski, M. Diaz, C. Chassot, "A Time Efficient Architecture for Multimedia Applications", IEEE Journal on Selected Areas in Communications, special issue on Protocols Architectures for 21st Century Applications, vol. 16, No. 3, April 1998
12. H. Schulzrinne, R. Frederick, V. Jacobson, "RTP: A Transport Protocol for Real-Time Applications", RFC 1889, January 1996
13. P.Sénac, M. Diaz, A. Léger, P. de Saqui-Sannes, "Modeling Logical and Temporal Synchronization in Hypermedia Systems", IEEE journal on selected areas in communications, Vol. 14, No. 1, pp 84-103, January 1996

Utility Based Inter-stream Adaptation of Layered Streams in a Multiple-Flow IP Session

Dimitrios Miras[*1], Richard Jacobs[2], and Vicky Hardman[1]

[1] Department of Computer Science, University College London,
Gower St., London WC1E 6BT, UK
{d.miras, v.hardman}@cs.ucl.ac.uk
[2] Distributed Systems Group, BT Research Labs, Adastral Park,
Ipswich IP5 3RE, UK
richard.j.jacobs@bt.com

Abstract. Future multimedia applications will evolve to content-rich, interactive presentations consisting of an ensemble of concurrent, related to the presentation scenario, flows. Recent research highlights the importance of co-ordinating adaptation decisions among participating flows in order to share common congestion control state. We exploit models that quantify the effects of the dynamics of hierarchically encoded multimedia content on perceived quality and present a mechanism to apportion the session's aggregate bandwidth among its streams that improves the total session quality. Dynamic bandwidth utility curves are introduced to express the variability of multimedia content and represent the level of quality (or satisfaction) an application/user receives under given bandwidth allocations. The relative importance of the participating flows, determined either by the user or the application scenario, is also considered. We discuss our approach and analyse simulation results obtained based on trace-driven simulation.

1 Introduction

Recently, the Internet has seen an explosive growth of real-time traffic. IP QoS mechanisms, particularly the IETF diff-serv and int-serv initiatives, aim to provide the network infrastructure with the necessary mechanisms to support the delivery of streamed multimedia. This, together with the advent of access technologies, like ADSL and cable modems and the continuous expansion of the backbone networks, allow delivery of higher volume and richer content multimedia since access rates of the order of Mbps are feasible. Traditional forms of multimedia collaboration, such as conferencing and audio/video streaming, are to be soon followed by more interactive applications: Internet TV, network games, tele-presence and complex immersive environments.

In future multimedia applications, the user will be presented with a collection of concurrent multimedia flows: several camera views from a single sports event,

---

[*] Author supported by British Telecom Research Labs, UK. Also partially supported by EPSRC project #GR/LO6614 MEDAL

H. Scholten and M. van Sinderen (Eds.): IDMS 2000, LNCS 1905, pp. 77–88, 2000.

multiple video feeds from a virtual classroom. The user will also be able to prioritise incoming flows according to content or willingness to pay.

However, the proliferation of unrestricted real-time traffic over the Internet is threatening network stability [7]; real-time media need to be able to adapt to changing network conditions and share bandwidth fairly with other well-behaved traffic (TCP). In multi-flow applications, a number of related, concurrent flows will exist between a pair of hosts, and probably will experience highly correlated levels of service (delay, loss), if they share the same bottleneck links. We therefore believe that congestion adaptation decisions should consider the session's streams as a group, rather than individually. If concurrent flows exploit the shared state information about service levels, and co-ordinate their reactions, efficient multiplexing and bandwidth sharing can be achieved [8], [2], [3].

Encoded multimedia content (especially video) is characterised by widely variable resource consumption, primarily owing to changes in the spatial and temporal domains of the original sequence. The encoding algorithm, the encoding parameters used, and the image size exacerbate the variability. This means that under a fixed bandwidth allocation, the usability of the flow fluctuates. If the impact on perceived quality can be measured, then an inter-stream resource allocation mechanism can apportion the bandwidth appropriately.

This paper addresses the problem of delivering multiple concurrent multimedia streams from a single source to unicast heterogeneous receivers. We prioritise individual flows according to their time-varying usability, measured using utility curves and quality profiles, and aim to maximise the quality of the session as a whole. Utility curves express the quality a user or application is getting under different resource allocations. We assume that the utility of a stream can be described by a quality index, which maps transmission rate to a value. In this way, we differ from the model where sources adapt transmission rate to congestion without considering the resulting quality of each stream.

This paper is organised as follows: Section 2 introduces the notion of utility curves and quality profiles, and presents relevant work. Section 3 describes our approach for allocating the session bandwidth among participating flows. Section 4 presents simulation results obtained. In Section 5, we identify applications our work can be applied. Section 6 concludes the paper.

2   Utility Curves and Related Work

2.1   Utility Curves

Research in network pricing and resource allocation [16], [18] extensively uses the notion of utility curves (or functions) to investigate methods for the optimal pricing of network resources. Utility functions map the quality (or satisfaction) that a service offers, when allocated a certain amount of resources, to a real number (usually in $[0, 1]$). So, if U is a utility function, then if $x \! / \! y$ (i.e., allocation y is preferred to x), then $U(x) < U(y)$, which means U is an increasing function.

In a networking scenario, utility functions may depend on many parameters: available resources, encoding scheme, encoding parameters (e.g., image size, frame rate, quantiser value), user preferences and the content of the media stream. However, trying to devise a solution based on all of these variables is likely to be intractable. In networked applications, bandwidth is the most important resource, and we therefore limit our consideration to *bandwidth utility functions*.

Bandwidth utility curves can be used to capture the adaptive nature of multimedia applications; these applications can operate over a range of resource availability. Effectively, a bandwidth utility curve is a mapping from the available network resource used by the application/flow to a value that represents the level of satisfaction (or dissatisfaction) of the application/user, at any time. This satisfaction index can be obtained using objective or subjective methods for assessing quality. Objective methods use specific signal metrics to assess quality, such as *PSNR* (Peak Signal-to-Noise Ratio). These metrics, while easy to obtain, often fail to correlate with the perception properties of the human audio-visual system. Quality assessment models based on subjective measurements provide more accurate results, but are more difficult to obtain. A widely used model is *MOS* (Mean Opinion Score) [12], where the perceived quality is usually rated on a 1 to 5 scale.

Utility functions for adaptive flows can be obtained by using MOS score tables, and interpolating between values to produce piecewise linear utility functions (Fig. 1(left)). While a model based on subjective quality assessment is more reliable, it is difficult to generate the required results in real-time. Currently, there is significant effort being put into producing objective quality assessment models that exploit the properties of human perception (see [9], [19], [21] and their references), but, especially for video, they are computationally intensive to run in real-time. These models produce output that most of the times is highly correlated with the results of subjective quality tests.

Min-Max is a special case of a linear utility function, where an application only operates between a minimum and maximum of resources. When the values a resource can take are discrete (or the application is adaptive in discrete steps), then utility curves have a scalar shape. For example, a discrete quality curve is produced when we map the number of received layers from a layered stream to the stream's quality. In Fig. 1(left), some commonly used utility curves are depicted. Exponentially-decay functions (Fig. 1(left)) are introduced in [6], [17] to describe bandwidth utility curves for adaptive applications.

2.2   Dynamic Generation of Bandwidth Utility Curves

Statically defined utility functions do not accurately model the burstiness of encoded video. A mechanism that inspects video on-the-fly is needed. In [10], a framework for dynamic generation of utility curves is presented. A utility generator that uses machine-learning techniques is employed to classify the utility of an object into classes of utility curves. The range of utility curves is associated with scaling profiles that describe scaling actions; these describe the adaptation that occurs for given network conditions. Scaling actions could be, changing the quantiser, dropping frames, changing the colour depth, neglecting coefficients or

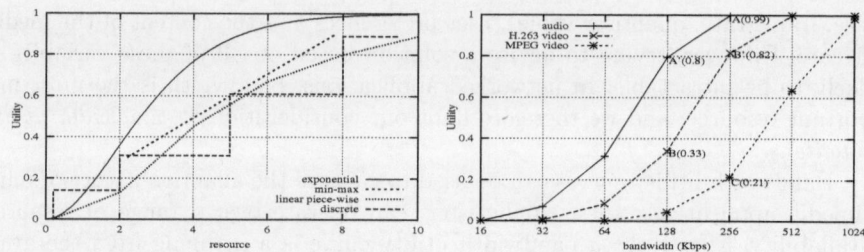

Fig. 1. Common utility curves (left) and hypothetical audio and two video curves (right)

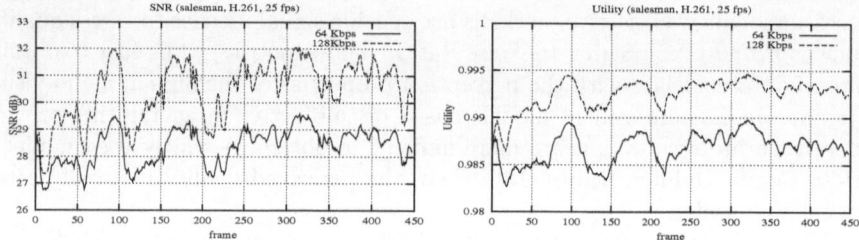

Fig. 2. Correlation of SNR and utility metric used

dynamic rate shaping. The utility profiles and the corresponding scaling profiles are then transmitted; scaling actions can be taken at intermediary or boundary nodes (e.g., at the boundaries between a wired and a wireless network or base station). Similar work is presented in [4], [5], focused on MPEG-4 video.

We use a model similar to [10], and gather utility profiles off-line by assessing pre-encoded streams. The utility metric that is used is based on signal distortion, and is $U(r) = 1 - err^2/signal^2$ (so, utility values are in the $[0,1]$ range), where err^2 and $signal^2$ are the mean square error and mean square signal respectively. The resource r represents the cumulative number of layers that the stream under consideration was encoded with. While a U value is calculated for every video frame, there is no reason to perform the inter-stream allocation on a per-frame basis, as this is a very fast time scale. Fig. 2 depicts the correspondence between this metric and the *signal-to-noise ratio (SNR)* for a video sequence (salesman), and shows their correlation. Such utility values are biased by the way the frame is encoded (i.e., if it was I, P or B-frame) or by very fast (few ms) scene changes. We want to eliminate spikes and very short fluctuations in the utility curves, to avoid misinterpretations when the allocation process runs. In this way we try to track more meaningful longer-term changes in the content, where changing the number of transmitted layers will have the desirable effect. We use a exponentially weighted moving average $U_{avg} = \alpha \cdot U_{avg} + (1 - \alpha) \cdot u_{cur}$ to smooth the utility curve, where u_{cur} is the current frame's utility value, and α is the averaging weight parameter. Fig. 3(left) depicts the instantaneous utility values derived from a video sequence, and its associated moving average approximation.

2.3   Session Bandwidth Sharing and Related Work

SCUBA [1] is a protocol for expressing receivers' interest in streams of a multicast session. Preference for an individual flow, is scalably multicast from each receiver as a weight. Session bandwidth is allocated among the various streams according to the averaged preferences of the group as a whole. The cumulative transmission bit-rate for stream i with weight w_i is then $w_i \cdot B_{session}$. The protocol also determines a layer mapping, which is effectively the assignment of transmission priorities to each layer of every stream. Receivers join layers of higher priority first.

Youssef et al [22], present a model to perform inter-stream adaptation, where multicast senders adjust their transmission rates (or layers sent), according to feedback reports obtained from receivers. A QoS manager and an inter-stream adaptation module are used by the receiver to dynamically select the receiving streams' operating status, according to network parameters, like delay, loss and jitter. Feedback is scalably multicast to the group, and enables senders to detect the highest rate (or layer) any receiver is expecting for the encoded stream.The transmission rate (or number of transmitted layers) can then be adjusted at the sender so that resources are not wasted in sending information (layers) the receivers cannot receive/handle. Bandwidth is also allocated based on the dynamically changing priorities of streams.

Work presented in [2], [3] describes a framework for the construction of end-to-end congestion control to allow an ensemble of flows that are destined to the same receiver to share common congestion information and make collective adaptation decisions. The framework is providing applications with an API to inform of the aggregate fair-share of bandwidth for a collection of flows. A hierarchical round-robin scheduler is used to apportion bandwidth among flows based on weights (either pre-configured, or derived from receiver hints).

Assigning bandwidth based only on preference weights does not guarantee that the receiver is getting best quality; this assumes that the resulting quality each flow offers to an application is proportional to the resource allocated to it. For reasons explained earlier, this is not always the case, so this method may give sub-optimal results. Consider Fig. 1(right), which depicts three instantaneous utility curves, for one audio and two video streams (such utility curves correspond to adaptive applications, [6]). Assume that the three streams have preference weights of 0.4, 0.2 and 0.4 respectively, and that the probed fair-share bandwidth of the session is 640 Kbps. Then a weight-proportional allocation would allocate the points (A, B, C) as shown on Fig. 1(right). However, we can obviously do better. By simply giving 128 Kbps from the audio stream to the H.263 video steam stream (points A', B'), there is a noticeable improvement in the utility of the H.263 video stream, while the audio quality is not substantially degraded. This is because the usability of a flow depends on the nature of the media (audio, video), the media content and the encoding scheme used, so a proportional sharing of the session bandwidth does not always suffice.

3 Weighted Utility-Based Session Bandwidth Allocation

In this section we present our approach to utility-based inter-stream session bandwidth sharing. We assume the transmission of hierarchically encoded streams to cope with receiver heterogeneity [15]. Layered coding features the decomposition and encoding of the original signal into a number of cumulative signals (layers). Popular streaming applications (like *RealPlayer* and *Windows Media Player*) transmit a single stream pre-encoded at a certain bit-rate (for example, 56, 128 Kbps), and the appropriate version of the stream is chosen to match the receiver's capabilities. This implies that many different versions of the same stream need to be stored at the content server side. In contrast, for layered coding, only a multi-layered encoded version of the stream is needed.

We base our mechanism on the existence of a companion congestion manager module (like [2], [3]) to provide information about the aggregate bandwidth availability for the session. This bandwidth may be reserved or be a fair-share over a best-effort network path (TCP-friendly, or any other fairness criterion).

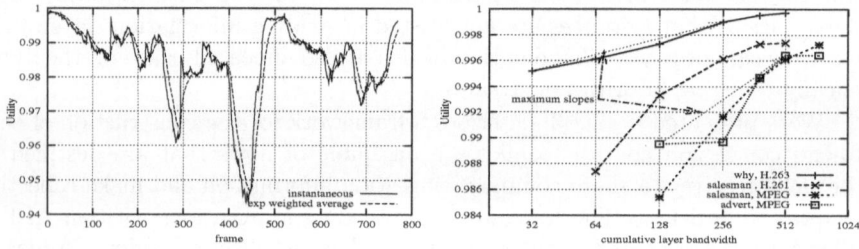

Fig. 3. Smoothing of a utility curve (left), and utility curves and concavity (right)

3.1 Algorithm to Determine Transmitted Layers

Assume that the application consists of N different layered streams. By considering layered flows, there are discrete operating points for each stream. For each stream, a quality profile exists in the form of tuples $(t, l, U(t, l))$, where t is the time index of the utility value (i.e., frame or group of pictures number), l the number of layers the stream is encoded into and U the corresponding utility value. Each stream i is assigned a preference weight w_i, and $\sum_{i=1}^{N} w_i = 1$. Denote by $B_{aggregate}$ the aggregate session bandwidth, provided by the companion congestion manager module. The aim of the allocation is to: maximise total session utility: $\sum_{i=1}^{N} w_i \cdot U_i(c_i)$, subject to: $\sum_{i=1}^{N} R_i(c_i) \leq B_{aggregate}$, where $c_i = 0, 1, \ldots$ denotes the number of allocated layers for stream i, and R_i is the cumulative bandwidth requirement for c_i layers of stream i. This belongs to the general *knapsack problem*.

Rajkumar *et al* [14] apportion a resource to multiple contending applications with one or multiple QoS parameters to maximise the overall utility. The algorithm works optimally if the utility functions are increasing and concave. In the case of non-concavity, a greedy algorithm is presented that performs near-optimal allocation. Our approach is based on this greedy algorithm, as the concavity criterion cannot always be met. This is shown on Fig. 3(right) which depicts a snapshot of real utility curves from the video sequences used in our experiments (Table 1). The algorithm operates as follows:

1. Assign the initial constraints[1]. The available bandwidth then is $B_{avail} = B_{aggregate} - \sum_{i=1}^{N} R_i(c_{min})$, where $R_i(c_{min})$ is the bandwidth requirement corresponding to the minimum constraint.
2. Calculate the weighted slopes, $w_i \cdot \frac{U_i(l+1) - U_i(c_i)}{R_i(l+1) - R_i(c_i)}$, $\forall$ layer l above the currently allocated layer c_i. It is clear that in the case of non strictly concave utility curves, the calculated slopes may include more than one of the discrete operating points of the flow under consideration (shown as dashed lines in Fig. 3(right)).
3. Find the maximum slope(s) among all unallocated layers, under the constraint of available bandwidth to satisfy the corresponding assignment.
4. Repeatedly increase the bandwidth for those streams found with maximum slope until all layers are allocated, or the available bandwidth cannot support any further allocation (exit condition). For each layer allocated, accordingly, subtract its resource requirement from B_{avail}.

3.2   Time-Scale Issues

Users prefer non-fluctuating quality, so it is important in the transmission of layered media that the number of layers does not change often. This means that, the bandwidth-sharing algorithm should only run at sufficiently distanced time periods. On the other hand, reducing the frequency of the allocation may result in low responsiveness of the algorithm, as sampling utility curves at low frequency in not meaningful.

The period of inter-stream allocation also reflects the frequency with which an accompanying congestion manager module is queried to indicate the levels of aggregate network availability for the ensemble of flows. Such network probing interval (hundreds of *ms* or even *secs*) is significantly larger than the (video) inter-frame interval. Usually, in order to reflect meaningful network changes, it should be in the order of a few *RTT*s. Determining an optimal period for inter-stream allocation is not a trivial problem, so we set it as an application parameter, τ. This is the synchronous mode of algorithm operation. We also provide for an asynchronous operation mode, so that significant spatio-temporal changes in the content can be accommodated before the expiry of τ. When such changes occur, an exception is raised, the timer τ is reset, and the algorithm runs. The exception is triggered whenever the difference between the instantaneous

---

[1] Initial constraints may exist in cases where a minimum acceptable level of service is required. This can be expressed by a minimum number of layers that need to be transmitted for the corresponding streams.

utility value and its moving average (§2.2) is bigger than a threshold value. This threshold value may be determined by the application programmer, based on subjective user evaluation. We intend to investigate methods of efficiently evaluating the value of the adaptation period according to quickly changing network parameters (RTT) and the error in the accuracy of utility values in further work.

4   Experimental Results

We ran simulations using *ns-2* (Network Simulator, v.2) [13]. We used three original video sequences with diverse amounts of scene activity which accordingly represent diverse requirements in terms of bandwidth consumption. The first (*Salesman*) is relatively low motion, CIF size (352x288), 450 frames, video-conferencing like sequence; the second (*Why*) is a high motion, QCIF size (176x144), 1500 frames sequence extracted from a TV commercial. The third (*Advert*) is also a high motion, CIF size, 770 frames sequence from another TV commercial. We derived four encoded streams from the original three, encoded up to six discrete bit-rates (layers) using an H.261 and an H.263+ encoder, developed at BT Labs [11], and a publicly available MPEG codec[2]. Table 1 summarises the features of the four chosen streams. In total, twenty-one combinations of (encoder, bit-rate) sequences were produced. A pre-configured preference weight was given to each flow, also shown on Table 1.

These four streams were assessed off-line using the metric described in §2.2. For each encoding point we generated the corresponding utility curves (Fig. 2), on a per-frame basis. We do not scale the utility values obtained to a meaningful range (i.e., 0-5), as this does not benefit to the execution of the algorithm. Those quality values are smoothed on a frame-per-frame basis, as explained in §2.2.

Table 1. Properties of video streams

Sequence	Size	Encoding	Cumulative bandwidth	Preference weight
Why	QCIF	H.263+	32, 64, 128, 256, 384, 512	0.25
Salesman	CIF	H.261	64, 128, 256, 384, 512	0.2
Salesman	CIF	MPEG	128, 256, 384, 512, 740	0.25
Advert	CIF	MPEG	128, 256, 384, 512, 740	0.3

The simulated network topology consisted of a media server concurrently transmitting the four video flows to a single receiver over a network with a single bottleneck link. The bottleneck link capacity was set to 1.5 Mbps. We ran the simulation for 120 *sec*. We set the running average parameter α to 0.9, the allocation period τ to 2 *sec* and switched off the exception operation mode.

---

[2] http://www.mpeg.org/MPEG/video.html##video-software

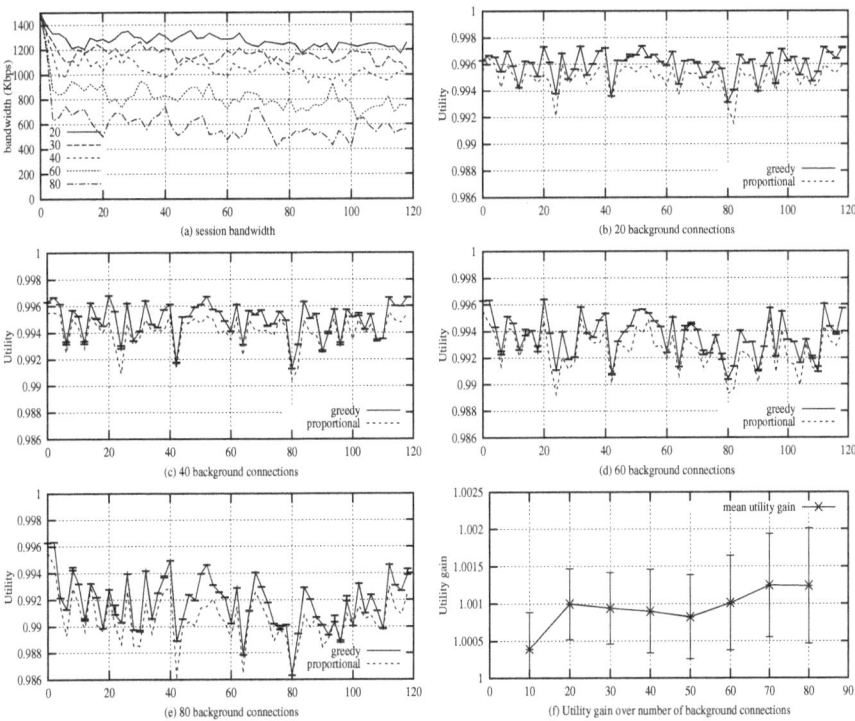

Fig. 4. (a) simulated session bandwidth over time, (b)-(e) improvement in total session utility (over time) for different levels of background traffic, and (f) mean utility gain and standard deviation over number of background connections

Recent work shows evidence of self-similarity in Internet traffic [20]. To simulate the background traffic at the bottleneck link, we subtracted from the bottleneck link the superposition of several ON/OFF UDP sources, whose ON/OFF times are drawn for a pareto distribution with a shape parameter set to 1.5. The mean ON and OFF times were 1 *sec* and 2 *sec* respectively. During ON times the sources transmitted packets of size 1500 bytes at the rate of 32 Kbps. The number of concurrent background connections was varied between 10 and 80. Fig. 4(a) shows the variation of the bandwidth available to the session over the simulation time. During the simulation, all utility profiles were periodically snapshot (on a per-frame basis) to produce their moving average curves. Every $\tau secs$, the corresponding moving average values for each flow and encoding point were chosen, which formed linear piecewise curves like the ones shown in Fig. 3(right). The algorithm was applied on these curves and the number of layers to be transmitted for every flow was determined.

Fig. 4(b)-(e) depict the improvement in total session utility compared to a weight proportional sharing of the session bandwidth. The error bars represent the distance of the greedy solution from the optimal, obtained by solving the knapsack. As shown, the optimal solution coincides with the greedy solution most of the time. We performed measurements to determine whether the amount of background traffic affects the output of the algorithm. Fig. 4(f) depicts the mean and standard deviation of the utility gain ($U_{total,greedy}/U_{total,proportional}$) for different levels of background traffic; it shows that the benefit obtained from a utility based apportion of the bandwidth is not degraded with decreasing levels of bandwidth availability, rather, it remains almost constant or even increasing as the number of background connections increases.

5   Applications

The Internet multimedia communication model will be increasingly interactive, and more content-rich. For example, Internet TV coverage of a football event will have, apart from the main field view, several other video feeds: zoom view, reverse angle, players close-ups, replays, etc. Virtual classrooms, immersive shopping worlds and tele-presence are also applications that involve multiple concurrent and diverse flows of changing importance.

In applications that involve multiple flows, the component flows might be compressed using different algorithms, might have different quality and frame rate requirements, and consequently will have different quality profiles. Application/service providers will wish to get the maximum benefit from their network resources, in terms of total viewers' satisfaction. In conditions of finite network resources, all flows cannot transmit at peak quality and quality trade-offs should be considered at the sender. However, this must be tempered in the light of other considerations, such as server workload sharing, bandwidth reservations and charging, and a resource sharing mechanism that maximises the perceived quality should be enforced. This will improve the usability of the service for both the user (who will get maximum benefit from their spending) and the provider (who will be able to achieve efficient utilisation of resources, while offering maximum possible quality).

In order to determine a receiver's preference weights, notice of mechanisms that imply preference might be used. In a sports game, for example, a viewer that is watching the main field image will use a bigger portion of the screen to display that video, (e.g. SCIF) and this could be taken to mean better quality is desired. At the same time, several smaller size (e.g., QCIF) images from other video feeds could also be positioned on screen. When a user wants to switch the main feed of interest, the currently lower quality flow will become the main one, user's preferences will be re-calculated, and eventually this will be reflected in the way bandwidth (or layers) is allocated by the source. This implies the existence of automated application cues or GUI assisted mechanisms for the setup (on- or off-line) of the user preference profiles. In addition, the mechanism requires that content servers will be able to generate (on-line or off-line) quality profiles for their codecs, content, etc. Advances in the generation of quality profiles that include assessment results highly correlated to human perception [9], [21] will strengthen the potential of mechanisms such as the one proposed here.

6   Conclusions

In this paper we are inspired by two trends in multimedia networking; the evolution of more interactive and content-rich applications made possible by advances in networking and significant market interest and the need for rate policing of real-time traffic and the benefits of making collective adaptation decisions over an ensemble of related flows that share common congestion information. We have presented arguments that illustrate the importance that the time-varying nature of multimedia content may have in an inter-stream bandwidth allocation process, when efficient objective quality assessment models are in place. For this purpose, we utilised a simple, SNR-based metric, and described a mechanism for inter-stream session bandwidth sharing that improves the total session quality.

Although a quality assessment model that features SNR-like measurement as the quality index does not always provide reliable results, we did not intend to describe an efficient objective quality assessment model, but rather to indicate the usefulness of utility-based dynamic inter-media bandwidth sharing. More effective real-time quality assessment models will be soon available with advances in coding techniques and higher availability of processing power. Off-line quality assessment can also be used in the case of pre-recorded, stored media to obtain quality profiles. These quality indices can be dynamically consulted at appropriate time scales (i.e., on a GOP per GOP basis) and adaptation can be applied accordingly.

The simulation results obtained showed an improvement in comparison to a scheme that shares the session bandwidth according to the preference weights of the flows. We note that the mechanism presented is coarse-grained since as it works at the granularity of layers. Future work will investigate how to further enhance the model to cope with fine-grained adaptation by adjusting and varying the transmission rates of the layers, to utilise any excess bandwidth below that needed to allocate extra layers.

Acknowledgement. We would like to thank Orion Hodson, Graham Knight and Antony Steed at UCL for helpful comments and suggestions in this work.

References

1. Amir, E., McCanne, S., and Katz, R. Receiver-driven Bandwidth Adaptation for Light-weight Sessions. In *Proc. of ACM Multimedia '97*, Seattle, WA. Nov 1997.
2. Balakrishnan, H., Rahul, H S., and Seshan, S. An Integrated Congestion Management Architecture for Internet Hosts. In *Proc. of ACM SIGCOMM 99*, Cambridge, MA. Sep. 1999.
3. Balakrishnan, H., and Seshan, S. The Congestion Manager. Internet draft, `draft-balakrishnan-cm-01.txt`, Oct. 1999.
4. Bocheck, P., Nakajima, Y., and Chang, S.-F. Real-time Estimation of Subjective Utility Functions for MPEG-4 Video Objects. In *Proc. of Packet Video '99 (PV99)*. New York, USA. April 26-27, 1999.

5. Bocheck, P., Campbell, A., Chang, S-F., and Liao, R. Utility-based Network Adaptation for MPEG-4 Systems. In *Proc. of NOSSDAV '99*. Basking Ridge, NJ. June 1999.
6. Breslau, L., and Shenker, S. Best-Effort versus Reservations: A Simple Comparative Analysis. In *Proc. of ACM SIGCOMM '98*. Vancouver, Canada. Aug 1998.
7. Floyd, S., and Fall, K. Promoting the Use of End-to-End Congestion Control in the Internet. In *IEEE/ACM Trans. on Networking*, 7(4), pp. 458-472. Aug. 1999.
8. Floyd, S. Congestion Control Principles. Internet draft, `draft-floyd-cong -01.txt`, Jan. 2000.
9. Lambrecht, C., and Verscheure, O. Perceptual Quality Measure Using a Spatio-Temporal Model for the Human Visual System. In *Proc. of IS&T/SPIE*, San Jose. Feb. 1996.
10. Liao, R., Bouklee, P., and Campbell, A. Dynamic Generation of Bandwidth Utility Curves for Utility-based Adaptation. In *Proc. of the Packet Video '99* (PV99). New York, USA. April 26-27, 1999.
11. Nilsson, M., Dalby, D., and O'Donnell, J. Layered Audio-visual Coding for Multicast Distribution on IP Networks. In *6th International Conference on Multimedia Computing and Systems*. June 7-11, 1999.
12. -. Recommendation ITU-R BT.500-7, Methodology for the Subjective Assessment of the Quality of Television Pictures. ITU-R Recommendations, Geneva. Oct. 1995.
13. -. UCB/LBNL/VINT Network Simulator, ns-2[3]. 1998.
14. Rajkumar, R., Lee, C., Lehoczky, J., and Siewiorek, D. A Resource Allocation Model for QoS Management. In *Proc. of the IEEE Real-Time Systems Symposium*. Dec. 1997.
15. Rejaie, R. and Handley, M. Quality Adaptation for Congestion Controlled Video Playback over the Internet. In *Proc. of ACM SIGCOMM 99*. Harvard Univ., Cambridge, MA, USA. Aug. 31-Sep. 3, 1999.
16. Sairamesh, J., Ferguson, D. F., and Yemini, Y. An Approach to Pricing, Optimal Allocation and Quality of Service Provisioning in High-Speed Packet Networks. In *Proc. of IEEE INFOCOM 95*.
17. Shenker, S. Fundamental Design Issues for the Future Internet. In *IEEE JSAC*, 13(7), Sep. 1995.
18. Shenker, S., Clark, D., Estrin, D., and Herzog, S. Pricing in Computer Networks: Reshaping the Research Agenda. In *Proc. of ACM SIGCOMM Computer Communication Review*. Apr. 1996.
19. Wang, S., Sekey, A., and Gersho, A. An Objective Measure for Predicting Subjective Quality of Speech Codecs. In *IEEE JSAC*, vol. 10, pp. 819-829, June 1992.
20. Willinger, W., Taqqu, M. S., Sherman, R., and Wilson, D. V. Self-Similarity Through High-Variability: Statistical Analysis of Ethernet LAN Traffic at the Source Level. In *Proc. of ACM SIGCOMM 95*. Caimbridge, MA, USA.
21. Wolf, S., and Pinson, M. Spatial-temporal Distortion Metrics for In-Service Quality Monitoring of Any Digital Video System. In *SPIE International Symposium on Voice, Video and Data Communications*. Boston, MA. Sep. 1999.
22. Youssef, A., Abdel-Wahab, H., and Maly, K. Controlling Quality of Session in Adaptive Multimedia Multicast Systems. In *Proc. of 1998 International Conference on Network Protocols (ICNP '98)*. Austin, Texas, pp. 160-167. Oct. 1998.

---

[3] `http://www-mash.cs.berkeley.edu/ns/`

An Interaction Control Architecture for Large Chairperson-Controlled Conferences over the Internet

Lukas Ruf, Thomas Walter, and Bernhard Plattner

Computer Engineering and Networks Laboratory
Swiss Federal Institute of Technology, ETH
ETH Zentrum Gloriastr. 35
8092 Zürich Switzerland
{ruf,walter,plattner}@tik.ee.ethz.ch

Abstract. This paper presents a novel approach for interaction control in large, synchronous and loosely coupled but chairperson-controlled conferences. Based on the IP-Multicast protocol which is extended by mechanisms allowing resource reservation for prioritized flows, the proposed architecture supports interaction control among conference members as well as focus control of each conference participant.

Interaction control is performed by using a scalable signaling protocol: a conference may be recursively split into sub-sessions each of which provides an identical functionality independently of the others and establishes a singular session. A session is controlled by a chairperson who manually grants, revokes or rejects an interaction request of a registered session member. The granted interaction is announced to the session participants such that all participants may have their focus automatically set to the audio and video streams providing session participant.

The such provided focus control enables conference participants to individually manage their own audio and video stream perception, i.e. a participant either decides individually to whom the personal attention should be granted or follows strictly the session's focus granted by the chairperson.

The proposed conference control architecture provides the network-management platform to large virtual classrooms for university-like lectures over high-speed Internet-connections.

While focusing the application of the conference control architecture to university-like lectures, it remains generally applicable for any audio- and video-supported chairperson-controlled conference over the Internet.

1 Introduction

The use of information and communication technology in teaching and learning has initiated a transition to remotely sourced university lectures; [10,23,26] are a few examples. In such a scenario, lecturers and students are separated in space, i.e. students can attend lectures from almost everywhere and in particular from home. As already noticed by others it can be foreseen that this tendency will continuously increase [2,3,6]. The reasons are manifold: Professionals performing continuous life-long learning to keep in pace with the steady progress of research might be interested in reducing traveling time when attending a course. Students may prefer attending a lecture from home to extract

H. Scholten and M. van Sinderen (Eds.): IDMS 2000, LNCS 1905, pp. 89–103, 2000.

the relevant information better. Educational institutions might be interested in reducing costs by transposing the ever demanding need for university space to virtual classrooms.

Teaching and learning in a virtual classroom can be seen as a large chairperson-controlled conference. It is a large conference since, in general, the number of participants is huge. It is chairperson-controlled because if a student intends to ask a question then she or he signals her or his intention to the lecturer. The lecturer grants or rejects attention to the asking student, and, thus, acts as a chairperson. In addition to asking questions, students should have the possibility of chatting with each other. Like in a traditional classroom setting this interaction should bypass the control of the lecturer. In order to create a feeling of being in a single virtual classroom we require also that interaction is supported by exchanging audio and video streams [7].

In any large conferences, a form of coordination is required in order to prevent significant information from disappearing in the vast "background noise" created by chatting conference members. A possibility to cope with this problem is to enable a single speaker to provide information while all other conference participants may either focus on the speaker or, alternatively, create smaller discussion-groups. For the coordination among sites we have designed an application-level protocol that supports interaction control among conference members as well as focus control of each conference participant. Similar to traditional lecture situations, a centralized management directs the mainstream of the lecture while enabling and disabling request-reply like questioning. This paper focuses on protocol requirements and specification of a conference control architecture to be used in a synchronous[1] distance education environment for university-lectures. Even more, the protocol can handle sub-conferences between groups of participants. For high-quality audio and video communication, the systems foresees the use of resource reservation. The proposed conference control protocol is generally applicable in any conference-like situation which requires audio and video support and is led by a chairperson.

1.1   Overview of This Paper

In Section 2, we discuss assumptions and requirements of the assumed scenario. The protocol implementation is presented in Section 3. A discussion of related approaches is provided in Section 4. Section 5 provides an overview of the results and concludes the paper.

1.2   Terminology

In this paper we define our terminology as follows:

- Lecture participants are divided into *lecturer* and *students*. The term *lecturer* may be applied to any conference-controlling chairperson.
- The *local classroom* denotes the place where the lecturer is located.
- *Local students* attend a lecture in the local classroom.

---

[1] In synchronous conferences, participants are separated only in space. The asynchronous model further separates the participants in time.

- *Remote students* participate in the lecture over the Internet.
- The term *virtual class* denotes the aggregation of lecturer, local and remote students.
- An abstraction of the location of the virtual class is named *virtual classroom*.
- *Lecture-interaction* denotes a dialogue between students and the lecturer that is percept by the virtual class.
- The lecturer controls lecture-interactions by using a *lecture-controlling computer*.
- A *participating site* denotes a computer which is connected to the virtual classroom.

2   Basic Assumptions and Requirements

In this paper, we assume a distance education environment where students are in a local classroom as well as at remote sites. The virtual classroom is established over the Internet to which all remote students are connected. Each participating site runs a multimedia-capable computer which has attached camera, speakers and microphone. The computer is used for the following purposes: primarily, it receives audio and video data from the virtual classroom. Received audio and video are decoded, decompressed and presented to the corresponding end-devices. During interactions with the lecturer or other participants, the student's computer acts as audio and video source. It compresses, encodes and sends audio and video data to the virtual classroom.

In the near future, available bandwidth to users at home will increase by using broadband cable TV or any mode of the digital subscriber line technology as the underlying link. Differentiated [4] and integrated service qualities are about to be installed and configured in edge-border routers of Internet service providers. High-quality audio and video distribution over the Internet will become reality and, thus, will allow students to actively participate in virtual classes from remote locations.

Observing traditional university-like[2] lectures, the lecturer is the acting person. The audience is listening while viewing teaching aids (e.g. slides) and looking at the lecturer. In a lecture, participants may require a temporary change of the lecture-attention (floor control) by requesting the focus of the lecture to ask a question. The other participants are aware of this interaction.

Separation in space should be bypassed in the virtual classroom so that the differences between a traditional and a virtual class are minimized. A major drawback of space-separated lecture participation is the inability to perform an individual discussion with a desk-neighbor. We therefore introduce the concept of an individual focus. Focus control allows a lecture participant to decide personally to whom it is granted: to the lecture or to the virtual desk-neighbor that wants to discuss an individual aspect (cf. Section 3.8).

Today, distance education platforms lack the possibility of controlling and setting the individual focus to the participants own point of interests. However, this should not be done on the cost of the information flow of the lecture but additionally, i.e. the personal focus can be set to a discussion or chat with other participants.

---

[2] By the term *university-like* we mainly describe the behavior of students: they decide on their own whether they want to attend a lecture, do not want to participate or temporarily leave the classroom for "playing cards". At universities, it is the freedom of thoughts and behavior that creates this extraordinary spirit.

Summarizing, in a synchronous [3] interactive distance lecture environment the following requirements are to be met:

- A virtual classroom must be established to integrate fully the remote participants into the lecture.
- High-quality audio and video (a/v) data must be transmitted bi-directionally to overcome the space-separation.
- Floor control and focus control according to a lecture-like policy must be integrated.
- Interactivity with the lecturer and interactivity between remote participants must be supported.
- Scalability and flow prioritization are preferred over lossless transmission.
- Platform and application independence is required to keep track with the development and deployment of off-the-shelf products.
- A uniform management platform is needed that covers lecture participation and individual audio and video supported discussions among participants.

3 Implementation

Distance education systems as in use today, mainly cover the distribution of audio and video data from a lecturer to remote students [23]. Interaction between student and lecturer is performed by sending e-mails or asking lecture-assistance via textual chatting-features.

Some other distance education scenarios simply offer conference protocols [10] based on the MBONE [15] tools. Again others [5] try to integrate audio and video distribution and whiteboard-applications in a single, hand-tailored application and, therefore, miss the required platform-independence to reach not only experienced users but beginners as well. Even extensions to applications and transmission of data require major applications changes.

Assuming huge numbers of participating students as in traditional lectures, the transition to lectures in the virtual classroom goes with the possibility of a major extension to the number of listening and actively participating students. Thus, a centralized approach to lecture membership- and floor-control like in [5] may impose a major hurdle to the aspect of scalability.

To summarize, currently available distance education platforms have the following drawbacks:

- No audio and video interactivity neither between remote students and the lecturer nor among remote students.
- No distance education-tailored platform supporting the conceptual behaviors of university-like lecturers in a virtual classroom.
- No application and platform independence.
- Restricted scalability.

---

[3] In contrast to tightly coupled ones (refer to [11]) where conference participation and information access is strictly controlled, the loosely coupled session model allows a broader and more opened joining semantics.

Recognizing the drawbacks mentioned above, we propose an architecture for university-like lectures in a virtual classroom to manage *scalable session control, floor control* and thus *interaction control* based on the concept of an *individual focus*.

3.1 Conceptual Overview

Our control scheme follows strictly the procedures of students attending a university-lecture, i.e. the general focus of interest is set to the lecturer. A student that intends to ask a question signals the intention and waits until the lecturer grants the requested attention. Besides the "official" interaction with the lecturer, students may whisper with their virtual desk-neighbors without disturbing the main flow. Lecture-interactions as well as individual discussion-groups of remote students are performed in real-time using audio and video data-streams.

The provided distance education platform is implemented in a strictly layered-approach. The coordination and selection of high-bandwidth data streams are separated from input and output applications by providing a local gateway application, called focus-control tool (see Figure 1). Thereby, we can provide platform and application independence. Any application on any platform that provides the required interfaces with respect to networking infrastructure and data formats of the virtual classroom, can be installed on remote students' systems.

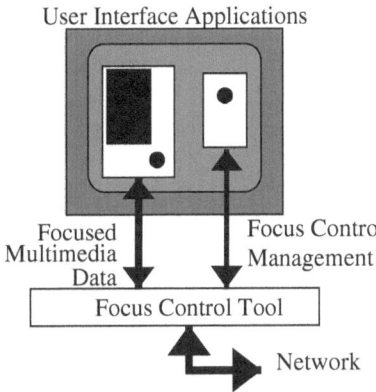

Fig. 1. Focus Control Tool And User Interface

3.2 Abstractions

Our architecture is based on the following abstractions:

- *Session*: A session denotes a communication group.
- *Session Holder*: Every session is controlled by a session holder.
- *Sub-session*: Sessions may be "forked" into sub-sessions.

94   L. Ruf, T. Walter, and B. Plattner

- *Parent Session*: The term parent session denotes the origin of a sub-session.
- *Floor*: The floor of a session denotes the transmitted information.
- *Floor Holder*: The floor holder provides the source of information within a session.
- *Participant*: A participant joined a session.
- *Registered Participant*: Only registered participants may actively participate, i.e. interact with the session.

3.3 Communication Channels

The proposed conference control architecture manages the flow of data through four primary communication channels (see Figure 2):

- *SAC*: The *session announcement channel* provides the session description to join a session; see Section 3.5.
- *FDC*: By *the focused data channel*, audio and video data are received.
- *ICC*: By the *interaction control channel*, floor control is bilaterally coordinated but controlled by the chairperson of a session. While the session holder decides to whom the floor is granted, the decision is announced to future session holders in advance by the interaction control channel.
- *TAC*: The *teaching-aids channel* provides the flow for teaching aids (e.g. slides) and annotations to those.

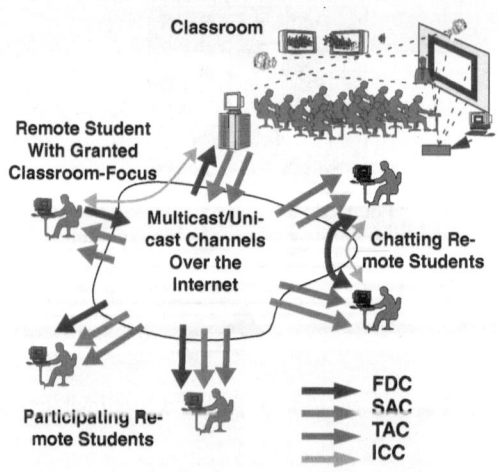

Fig. 2. Channel Overview

Session Announcement Channel: If a newly created session's scope addresses a multicast [16] group, it becomes an open session. Registered participants of the parent session may join this sub-session. By the session announcement channel (SAC), participants may set their focus to the current floor holder whose audio and video is sent.

Focused Data Channel: Using the focused data channel (FDC), audio and video data are transmitted. An FDC exists per participating computer. It is dynamically set (cf. Section 3.7) according to the session announcements received by the SAC (cf. Section 3.3).

Interaction Control Channel: By the interaction control channel (ICC), the dialogue between participants is controlled (cf. Sections 3.8 and 3.4). The ICC is initiated by a participant that intends to interact with the session.

Teaching-aids Channel: The teaching-aids channel (TAC) exists only in the root session which is defined as the lecture itself (cf. Section 3.4). All participants of the virtual classroom continuously receive the teaching aids transmitted by this channel even if they participate in sub-sessions. The TAC remains valid during a teleteaching session.

3.4   Session Establishment

Conceptually, every session is established if a participant sends a *session request* to another participant (a state transition diagramme is shown in Figure 3). If the callee is willing to change his actual focus and to establish a sub-session, a *session grant* message is returned. Otherwise, the session request is *rejected*. In the established sub-session, the callee becomes the session holder, i.e. the chairperson. The callee and the session holder are marked as registered participants.

Both, the session request message as well as the session grant message, are manually initiated. The caller retrieves the session participants register (SPR) via a web-based interface and receives the address of the callee by selecting the appropriate entry. After having selected an entry, a session request message is automatically sent to the callee.

The callee now decides whether he wants to create a private, unicast session or an open, multicast one. In either case, the callee becomes the session holder.

If a multicast session is established, the session holder sends a registration message to the parent session. This message extends his or her entry in the parent SPR by the address of the newly created SAC. By representing this additional information, the callee becomes noticeable as a sub-session holder. The address of the appropriate SAC in the SPR is used to join a sub-session (cf. Section 3.5).

Neither registered nor un-registered session participants provide continuous video or audio streams to the floor. To deal with bandwidth constraints only floor holders are allowed to send their streams. By that, session participants are not continously aware of other attendees. Registered participants are marked within the SPR of the session holder.

A unicast session does not require a registration of the session holder in the parent's SPR; a unicast session in the virtual classroom represents the desk-neighbor whispering in a traditional lecture.

The sequence of messages sent during a session establishment is visualized in Figure 3. In both cases, the initially established channel to send a session request message to the session holder becomes the ICC during the interaction period (cf. Section 3.8).

At the beginning of the lecture, the lecturer and the lecture information itself are registered in the SPR of the web-page which belongs to the lecture.

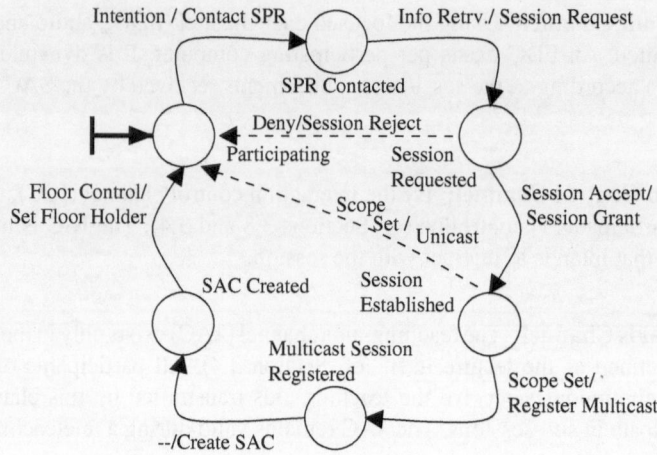

Fig. 3. Message Sequence

3.5   Session Join

Joining a session requires the address of the appropriate SAC. This information is retrieved by traversing down the tree of sub-sessions starting at the root session's SPR or at the SPR of the current session if a sub-session should be joined. If an SPR is contacted for getting information of a sub-session, the requesting participant gets registered in the contacted SPR but not in the targeted sub-session, i.e. her or his participation is logged in the SPR. Clicking on the web-based SPR-representation transmits the information where to connect to the SAC of the targeted sub-session.

The information being transmitted on the SAC describe the sub-session as follows:

- Master session's SAC-address.
- Master session's TAC-address.
- Parent session's SAC-address.
- Floor group-address.
- Session holder.
- Floor holder.
- Validation checksum of the SAC-data.

The information to connect a SAC is used to perform a multicast-join operation into the SAC of the sub-session and, if applicable, into the TAC of the lecture. Receiving the information transmitted on the SAC provides the address of the floor, i.e. the FDC. The received data is used to set the focus of the participant to the current floor holder (see Section 3.7). Joining a sub-session changes the focus of the participant. However, teaching aids are always sent on the TAC.

3.6   Session Close

A session close is either performed individually by the session participant or, for all members of the session, by the session holder. The protocol provides the information to reconnect to the parent session or return to the root session.

If the session holder decides to close the session, this intend is signaled to the session before the SAC is closed. This session close-announcement is sent to allow the participants to have their focus automatically set to the parent session or to the root session otherwise.

If a participant intends to cancel its participation in a sub-session, he or she either returns to the parent or root session: a multicast-leave operation on the current FDC and SAC and reconnects to the appropriate SAC and FDC as described above. If a registration in the current session was once performed, the participation is de-registered if a controlled session close is performed by the session participant. Otherwise, if the participation is canceled by a system crash or abnormal program termination and the exiting participant was registered, the entry in the SPR is kept for the lifetime of the session. Session requests and join procedures (see Sections 3.4 and 3.5) to non-existent registrants are correctly handled by the focus control tool: no new session is established, the participation is kept unchanged.

3.7   Focusing

Focusing on the floor holder is performed by using the information provided on the SAC. The requirements for audio and video transmission impose quality of service constraints on bandwidth and delay. Thus, channel reservation for the receiver site is required as the individual focus-concept relys on information-channels. However, unwanted data may still arrive to the participant's computer. Therefore, the Focus Control Tool (see Section 3.8) provides a local filtering according to the currently set focus. The Resource Reservation Protocol (RSVP [18]) is used to setup quality of service constraints in edge-routers[4] at the participants' Internet service providers (ISPs). For scalability reasons, it is assumed that the backbones interconnecting the ISPs provide enough bandwidth and small delays to fulfill the requirements for live video and audio transmissions. Differentiated services [4] between ISPs will guarantee the required qualities for the virtual classroom. It is further assumed that backbones establish RSVP-tunnels for signaling purposes. As soon as a participant joins a session or recognizes an announcement-change, the current focus is teared down by sending the appropriate RSVP-TEARDOWN message to the routers; afterwards, an RSVP-RESERVE message is sent along the receiving path. RSVP-signaling is performed by the focus control tool (cf. section 3.8) independently of the user application outputting the received data. All this gives the required application independence in virtual classroom sessions.

---

[4] By the term edge-router we denote the routers at the outer border of the Internet, i.e the ISPs, and not the interior routers of backbone networks.

3.8   Interaction Control and Floor Control

Interaction control is performed by using the ICC. A participant that intends to acquire the floor establishes a unicast ICC to the session holder. A session participant becomes registered, if not already done, by sending a floor request. If a new session must be created, the channel established to send a session request to another participant becomes the ICC between the session participants (see Section 3.4).

By the ICC, a floor request is sent to the session holder. The session holder manually decides whether he or she wants to grant or refuse the floor to the requester. Refusing a floor request results in the termination of the ICC (see Section 3.6). If the floor is granted, a notification called *floor grant message*, is sent by the ICC to the requester. Besides the notification sent by the ICC, the change of the floor holder is, of course, announced on the SAC. The interaction policy imitates the known protocols of traditional university lectures: if the session holder intends to explicitly ask a participant, he or she asks the question by audio and video to the session and expects the callee to establish an ICC and follow the procedure as described above.

Until an interaction process is either closed by the session holder or canceled by the participant, the ICC remains established. Closing the ICC denotes the termination of the interaction. Using the ICC, request-reply alike dialogues are possible. It is further assumed that during discussion several ICCs per session are kept alive: the session holder decides "on the fly" to whom he or she wants to grant the floor.

If the session holder intends to ask a participant explicitly, the session holder articulates the question on the floor and addresses the participant by voice – as done nowadays in traditional university-like lectures – and expects that the participant responds by a floor request message as described above.

The Focus-Control Tool. (FCT) provides the gateway functionality as described above (see Figure 1). Session participation, floor and session request as granting both of them are handled by the focus control tool. It provides a clearly defined interface to allow graphical display tools being connected.

The FCT receives the SAC, FDC and TAC. After data streams passed filtering and validation tests, they are locally provided to the applications. This gateway functionality provides the above requested platform and application independence. If the display tools running under UNIX allow data reception for example via UNIX domain-sockets [9], filtered data are locally provided by this interface. If, on the other hand, display tools being started under other operating system only provide traditional IP-interfaces [8], valid data are transparently provided on such interfaces. Applications, originally not multicasting capable, may participate in the virtual classroom: the FCT transparently performs the unicast to multicast and vice versa gateway functionality. Intercepting the SAC-data, resource reservation signaling can be provided by the FCT to applications that are unaware of quality of service possibilities. The amount of transmitted audio and video data is reduced since FCT sends them out to the floor only if the floor is granted to that participant.

3.9   Teaching Aids and Annotations

The floor holder of the root session, the lecture, is allowed to provide annotations to the transmitted teaching aids. Since the floor focusing and filtering of the data streams is provided by the FCT, the teaching aids display- and modification tool could behave as if it is was a single-user application. Annotations to the teaching aids are added on the lecture-controlling computer. Modified information is sent to the participants of the lecture.

3.10   Media Scaling

By the term media scaling the transformation of input to output data regarding the available output-bandwidth is meant. Expecting the functionality and widespread use of active Internet nodes [1] to become real in the near future, we rely on the possibilities offered by node-plugins to perform the required media scaling so that a participating computer always receives the audio and video data in a best-possible quality.

3.11   Failure Handling

The proposed protocol follows the commonly available approach for reservation-failure handling in an Internet environment: If path-reservations fail, best-effort transmission is used per flow. Failure scenarios are identified for lost RSVP-reservation messages and the in-capability of routers on the path to provide the required resources. In case of a not established user-lecturer interaction, it is up to the student to signal the interaction-intent.

Currently, no service redundancy to provide failure recovery is foreseen in case of session controlling computer crashes. In a future generation of the currently provided virtual classroom description, failure recovery procedures must be integrated. A possible solution to provide the fault-tolerance might be the mirroring of the lecture-controlling computer. An in-depth analysis of this topic is subject of further studies.

4   Related Work

Besides the vast literature available on the topic of distance education, only few publications address the thematic of the underlying signaling protocol. Therefore, we concentrate on the ones that provided the most influence during the design of this protocol suite. The signaling protocols of the MBONE tools and the protocol concept of digital lecture boards (DLB), the ITU-T.120 recommendation as well as the MACS environment.

4.1   Session Description Protocol (SDP), Session Initiation Protocol (SIP), Session Announcement Protocol (SAP), Questionboard (QB)

The protocols SAP, SIP and SDP [19], [22], [20] are primarily designed to convey the information required to support the MBONE [15] tools. They are, spoken in general terms, the basics required to establish the session directory SD [21] and to be used by vic [25], vat [24] and wb [27]. By the use of the MBONE tools, loosely coupled conferences are created.

SAP provides the protocol to announce different sessions to be registered in the SD. It describes the information layout of the announced session. The principal purpose of SAP is the transmission of the SDP and SIP data. By SDP, the session is described in the SD. It provides the description of session participants to the SD which imposes limitations on scalability: explosion of messages. SIP, on the other hand, describes a method to invite computer users to participate in an MBONE session. Data transmitted provide information to start automatically the required applications if configured correctly.

The idea how to join a session in our architecture stems from [19]. The definition of the transmitted information on the SAC was influenced by SDP. We focused on the computer processing-performance aspects and therefore reduced the SAC-data to a binary and thus easier to process form while the MBONE signaling protocols are character-based. While SIP provides the invitation to join a session, we changed the direction and provide a more university-lecture oriented join mechanism. The concept of the SPR was influenced by the SD. We provide a web-based static form of the the information to join a session on the parent's session holder computer.

Floor control as provided in the herewith presented interaction control architecture is similar to the concepts evolved by the MBONE questionboard (qb) [14]. While qb provides moderated sessions where the floor is granted by the moderator as well, it limits its application due to the following reasons:

- Qb is designed for the MBONE. By that it relies on the conference bus [15] as designed for the MBONE tools. For proper operation, the MBONE tools are required.
- Qb does not allow the explicit revocation of the granted floor by the moderator.
- Hierarchical (sub-)session management is not provided.

Nevertheless, an aspect not covered by our approach resembles qb's recovery mechanisms by multicasting "hello-alike" host-startup messages.

The loosely coupled conference-join semantic of our conference control architecture is similar to the MBONE tools (IP multicast) while our centralized coordination resembles the original need for this protocol: the distance education environment.

4.2   The Upper Rhine Virtual University (VIROR)

In VIROR [26] the universities of Mannheim, Heidelberg, Freiburg and Karlsruhe in Germany developed a tele-teaching application-platform. They rely on the MBONE-tools to perform the protocol-oriented communication constraints and provide a tools-wrapping application. A drawback in their approach clearly is the dependency of the underlying operating system platform while not separating the management-flow from the user-interfacing information-flow.

In VIROR, evolved from the project Tele-Teaching Uni Mannheim-Heidelberg [5], floor control is implemented in a similar manner as our protocol does, but requires synchronized finite state machines on each participating computer. It further differs in the degree of interactivity between lecture participants; private chat-sessions between virtual desk-neighbors are not provided (see Section 3.8).

4.3   ITU Recommendation T.120 – Data Protocols for Multimedia Conferencing

The ITU-T.120 family of protocols [12] is designed to be used as basis for multimedia conferencing over public switched telephone networks. Similar to our approach, the ITU-T.120 family provides a centralized moderating functionality. In contrast to our interaction control architecture, the ITU T.120 family of protocols requires reliable communication. Multicasting is not intended. By that, this centralized approach provides explicitely, server-controlled connections per remote location. A specialized Multipoint Control Unit is required to provide the virtual interconnections among the single multimedia terminals. So, dynamic scalability is limited.

4.4   MACS – Modular Advanced Collaboration System

The approach taken by the project MACS [13] provides a similar concept as the one provided by our architecture. A floor and session control library performs the appropriate actions. This library provides an interface to allow application programmers to adapt their code. Portability is reached by the use of JAVA as library programming language. Currently, no hierarchical session management does exist.

While MACS provides a similar concept of establishing sessions, scalability limitations could exist in the tight management restrictions imposed by the registration mechanism. Hierarchical session management covered by a single user-interface is not foreseen. No integration of resource reservation protocols or differentiated service mechanisms is foreseen in contrast to our architecture (cf. section 3.7). Portability as provided by the JAVA-based library requires involved applications being adapted while our approach using the gateway functionality (cf. section 3.8) provides platform *and* application independence.

5   Conclusions

In this paper we have proposed a new virtual classroom architecture. The architecture provides interaction and floor control protocols to coordinate large audio and video conferences over high-speed Internet connections. Interactivity rules imitating university-lectures provide a scalable coordination platform for virtual classes. Scalability is reached by the distribution of coordination-competence to session holders, and the minimization of registration overhead. The approach of using native IP-multicast over the Internet supports a large number of receiving conference participants.

High-quality audio and video streams that enable participant to perceive everyone's interaction with the session holder are managed by using RSVP on the outer-border routers and differentiated services in the backbones. Audio-visually supported discussions between the lecturer and remote students as well as those among students establish a virtual classroom. Due to the loosely coupled organization of multimedia sessions and the proposed use of RSVP only on the edge border routers of the Internet, scalability of the provided architecture should be given. Further measurements of the deployed architecture within larger trials are required to prove the scalability.

Management is done by providing a control tool which acts as a gateway. Thus, a platform-independent control and management tool coordinates the reception of data flows from the virtual classroom as the session per se while off-the-shelf products can be used for multimedia input/output purposes. Altogether, our proposed coordination architecture and protocol cover the requirements put forward in Section 2 and deals with the problems mentioned in Section 3.

In this paper, aspects of security and membership control and effective memberships within a lecture (who is allowed to actively or passively participate) are not covered. The problem of multicast-address allocation is out of the scope of this paper. The reader is referred to [17] for further discussions.

The proposed platform will provide the network resource control to the project Easy Teach and Learn$^{(r)}$ [7] of the Communication Systems Group at Swiss Federal Institute of Technology.

Acknowledgments. We would like to express our acknowledgments to the Swiss Federal Institute of Technology (ETH Zurich) and the Hasler Stiftung for funding the project of Easy Teach and Learn.

References

1. Keller, R., et al.: "An Active Router Architecture for Multicast Video Distribution", to appear in Infocom 2000
2. "Meeting Online Can Save Money, Boost Productivity", ComputerWorld, 03/06/2000, http://www.computerworld.com/home/print.nsf/all/000306F3F2
3. "E-Learning: A Catalyst for Competition in Higher Education", http://www.cisp.org/imp/june_99/06_99baer.html
4. Blake, S., et al.: An Architecture for Differentiated Services, IETF, RFC 2475
5. Geyer, W., et al.: The Digital Lecture Board – A Teaching and Learning Tool for Remote Instruction in Higher Education. ED-MEDIA/ED-TELECOM 99, Freiburg, BRD, 1998
6. ETH World, Swiss Federal Institute of Technology, http://www.ethworld.ethz.ch/
7. Walter, T., et al.:Easy Teach & Learn(r): A Web-based Adaptive Middleware for Creating Virtual Classrooms, to appear in HPCN2000
8. Stevens, W.R.,"Unix Network Programming, Network APIs: Sockets and XTI, Volume 1", Prentice Hall, 1998
9. Stevens, W.R.,"Unix Network Programming, Interprocess Communications, Volume 2", Prentice Hall, 1998
10. Interactive Remote Instruction, http://www.cs.odu.edu/ tele/iri/
11. ITU Recommendation H.245: Control Protocol for Multimedia Communication
12. ITU Recommendation T.120: Data Protocols for Multimedia Conferencing, 1996
13. Brand, O., et al.: MACS – Modular Advanced Collaboration System, Praxis der Informationsverarbeitung und Kommunikation, 22 (1999) 4, p. 213-220.
14. Malpani, R., Rowe, L.A.: Floor Control for Large-Scale MBone Seminars, ACM Multimedia'97, Seattle Washington, 1997
15. Macedonia, M. et al.: MBONE provides Audio and Video across the Internet, IEEE Computer, V27,4, April 1994
16. Deering, S.: Host Extensions for IP Multicasting, STD 5, RFC 1112, August 1989
17. Handley, M.: Session Directories and Scalable Internet Multicast Address Allocation. Proc.ACM Sigcomm 98, September 1998, Vancouver, Canada.
18. Zhang, L. et al.: RSVP: A new Resource Reservation Protocol, IEEE Network, Vol. 7, No. 5 (1993)

19. Handley, M., et al.: Session Announcement Protocol. IETF. Internet Draft draft-ietf-mmusic-sap-01.txt (1998)
20. Handley, M., et al.: SDP: Session Description Protocol. Network Working Group, RFC 2327 (1998)
21. Jacobson, V., Session Directory, Lawrence Berkeley Laboratory. Software on-line. ftp://ftp.ee.lbl.gov/conferencing/sd
22. Handley, M., et al.: SIP: Session Initiation Protocol. Network Working Group, RFC 2543 (1999)
23. Stanford Center for Professional Development, http://stanford-online.stanford.edu
24. Jacobson, V., et al.: vat, UNIX Manual Pages, Lawrence Berkeley Laboratory, Berkeley, CA.
25. McCanne, St., et al.: vic: A Flexible Framework for Packet Video, Lawrence Berkeley Laboratory, Berkeley, CA.
26. VIROR, The Upper Rhine Virtual University, http://www.viror.de
27. Jacobson, V., et al.: Using the LBL Network Whiteboard, Lawrence Berkeley Laboratory, Berkeley, CA.

Using DMIF for Abstracting from IP-Telephony Signaling Protocols

R. Ackermann[1], V. Darlagiannis[1], U. Roedig[1], and R. Steinmetz[1,2]

[1] Darmstadt University of Technology
Industrial Process and System Communications (KOM)
[2] German National Research Center for
Information Technology (GMD IPSI)

Abstract. IP Telephony recently finds a lot of attention and will be used in IP based networks and in combination with the existing conventional telephone system. There is a multitude of competing signaling protocol standards, interfaces and implementation approaches. A number of basic functions can be found throughout all of those, though. This includes the addressing of participants using symbolic names, the negotiation of connections and their parameters as well as the enforcement of a dedicated handling of data streams by means of QoS signaling activities. Thus, a generic abstraction hiding underlying protocol specifics is very desirable and useful. The Delivery Multimedia Integration Framework DMIF - as part of the MPEG approach towards distributed multimedia systems - forms a general and comprehensive framework that is applicable to a wide variety of multimedia scenarios.

In this paper we describe a more generalized and abstract view to basic IP Telephony signaling functions and show how these can be hidden below a common DMIF interface. This will allow for the implementation of inter-operable applications and a concentration on communication functionality rather than protocol details. We expect that this will also allow for better exchangeability, interoperability and deployability of emerging signaling extensions.

Keywords: Internet Telephony, Signaling, SIP, H.323, MPEG-4, DMIF

1 Introduction

IP-Telephony applications are considered to have a huge economic potential in the near future. Because companies and service providers start to consider it to be getting ready for carrier-grade usage, it may also speed up the deployment of state-of-the art QoS, security and billing components in local as well as in the backbone networks. Though IP-Telephony might be seen as (just) a specific application today, it is part of an ever emerging scene of more general multimedia applications. Considering the high dynamics and multitude of concurrent approaches in signaling protocols, interfaces and implementations, a consistent and comprehensive framework can speed up development and allows a faster, better and more generic implementation as well as the interoperability, exchangeability and re-use of modular components.

H. Scholten and M. van Sinderen (Eds.): IDMS 2000, LNCS 1905, pp. 104–115, 2000.

2 IP-Telephony Signaling Protocols and APIs

IP-Telephony signaling protocols are used to establish a conversation comparable to a classic telephone call using an IP infrastructure. Typical applications
and scenarios are recently based on different protocol suites. Mainly, there are
two major approaches - the H.323 [6][7] protocol family and the Session Initiation Protocol SIP [4] with a changing distribution and relevance. Though today,
a high percentage of applications and scenarios is still H.323 based (and we will
therefore initially focus on it), it is supposed that in the near future the use of
the SIP protocol may increase [3][13]. Both protocol types will even be usable
together with appropriate gateways [?][14]. Additionally, the close interaction
with the existing Public Switched Telephone Network (PSTN) on the basis of
interacting media gateways plays a very important role. For that domain, protocols like MGCP [2] / H.248 [5] that describe the interaction towards the PSTN
SS#7 [12] and may use both H.323 and SIP for signaling within the IP telephony
world are under recent development and standardization.

2.1 Signaling Using the H.323 Protocol Family

The H.323 protocol suite compromises a variety of communication relationships,
which are handled via dynamically negotiated channels for a number of H.323
protocol components such as RAS, Q.931, H.245. These use Protocol Data Units
that are encoded as described in ASN.1 specifications. Though the H.323 protocol suite has proven to provide the intended communication services especially
for usage in LANs, it is considered to be complex, not easy to extend and having a considerable signaling overhead that can not be neglected in a global
environment.

2.2 Signaling Using the Session Initiation Protocol SIP

The Session Initiation Protocol SIP has initially been used as a protocol for
multicast applications and provides generic control functionality. Its basic operations which are directly related to call setup are registration of participants
and redirection or proxying of control data traffic. This allows to access telephony services through single points of contact, may hide infrastructure aspects
and is also applicable for building hierarchies.

Additionally to its primary function, SIP allows to control call proceeding
and additional services in a very generic, efficient and extensible manner, e.g. by
means of Call Processing Language (CPL) scripts. SIP protocol functionality can
either be provided by centralized components but also at "smart" end system
nodes. Over the last period of intensive work, SIP has emerged towards the core
protocol of a comprehensive framework, addressing additional features such as
QoS support, firewall interaction and call routing as well.

2.3 Application Interfaces - TAPI and JTAPI

The Microsoft Telephony API (TAPI) [11] has been developed in a joined effort
by Microsoft and Intel and is provided as part of Win9x and WinNT. Its targets

are to isolate features of the underlying hardware from the applications by means
of a standard API as well as to specify a Telephony Service Provider Interface
that the underlying services have to meet.

It supports basic Computer Telephony Integration (CTI) applications like
automated dialing but starting with the TAPI version 3.0 under Windows 2000
involves sophisticated features such as IP multicast conferencing, a H.323 stack
and Interactive Voice Response (IVR) functionality as well. Inherently it is limited to the Windows platform though and does not up to now cover SIP, though
protocol descriptions state, that it provides powerful means to incorporate new
protocols or protocol extensions as so called Third Party Service Providers.

The Java Telephony Application Programming Interface (JTAPI) [15] is an
object-oriented interface that allows the development of portable telephony applications in Java. It uses a modular approach that places additional functionality, so-called extensions, on top of a common JTAPI core. The API itself just
describes interfaces which have to be implemented for the underlying hardware
or protocol infrastructure. As a current drawback it must be stated that there
is still only a small though rising number of JTAPI peer class implementations.

Both APIs have current limitations and are inherently targeted at telephony
services. We do not consider JTAPI or TAPI as comprehensive alternatives or
competitors to our approach - they can even be combined with it or provide
services.

3   The Abstraction Framework

3.1   MPEG and DMIF

MPEG-4 [10] is a new multimedia standard that is much more powerful and
comprehensive than the previous MPEG standards. To begin with, it provides
an object-based description of content, which can be both naturally captured
and computer generated. Though the term MPEG-4 is often associated with the
specification of a set of video codecs working with individual visual objects, the
standard is much more comprehensive.

Among others, MPEG-4 defines the Delivery Multimedia Integration Framework (DMIF) [8][9]. DMIF is a framework that abstracts and thereby encapsulates the delivery mechanisms from the applications.

The frameworks API, called DMIF Application Interface (DAI), works with
Universal Resource Locators (URLs), which specify appropriate delivery mechanisms for specific scenarios. URLs can also specify the required network protocol,
which provides a protocol abstraction for the applications. Additional parameters of a connection such as e.g. Quality of Service (QoS) requirements can be
passed as arguments through this generic interface as well. DAI is language and
platform independent. A basic description of its primitives is given in Table 1.

Additionally, DMIF defines an informative DMIF-Network Interface (DNI)
for the network related scenarios. DNI allows the convenient development of
components that can easily adapt their signaling mapping to different protocols.

Table 1. DMIF primitives and functionality

Primitive	Description
DAI_ServiceAttach	Allows the initialization of a service session with a remote peer, specified with a URL.
DAI_ServiceDetach	Allows the termination of a service session.
DAI_AddChannel	Allows the establishment of end-to-end transport channels in the context of a particular service session.
DAI_RemoveChannel	Allows the replacement of existing transport channels.
DAI_UserCommand	Allows the application-to-application exchange of messages.
DAI_SendData	Allows the transmission of media in the established channels.

3.2 Framework Architecture

Before defining the basic objects of our architecture, it is important to identify main use cases, which are typically required by IP Telephony applications. Basically a user should be able to register himself in the "IP-networked world", to enable his locating and identification for other participants. After that, it is possible to receive calls or to originate them to remote users. So, in a necessarily limited scenario there are three main use cases, which are shown in Figure 1, using a UML use case diagram.

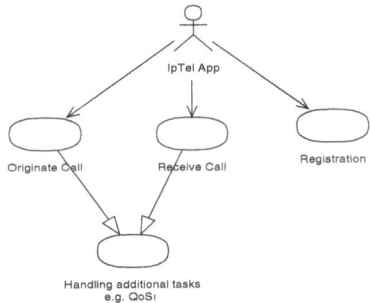

Fig. 1. Basic Use Cases

As an example for an additional service, we refer to the signaling for ensuring the desired QoS.

From the analysis of the use cases we derive the architecture of the proposed framework. It is shown in Figure 2.

In our framework the DMIF Application Interface is implemented with two interfaces, DMIFSession and DMIFApplication. The first provides the set of methods that are offered to the application from the DMIF layer, while the former is the set of callback functions for the DMIF layer to inform the application about events and messages. The DMIF layer is provided to the applications through

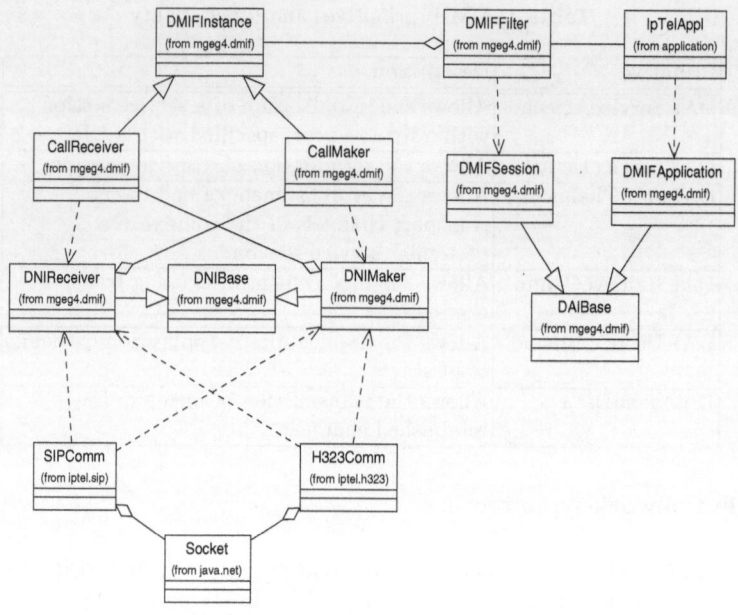

Fig. 2. *Framework Architecture*

the DMIFFilter. Its responsibility is to parse application requests and to activate the appropriate DMIFInstances to handle them.

The DMIFInstances are formed of two different objects: the CallReceiver and the CallMaker. A CallReceiver object initially allows the user to register himself. It is then responsible for the acceptance of incoming calls as well as for their handling. A CallMaker executes the requests for outgoing connections. Both CallReceiver and CallMaker behavior is independent of the used signaling protocol. They communicate with the appropriate signaling object using the DNI interface. Specific implementations of signaling protocols are the SIPComm object, which uses the SIP protocol and the H323Comm object, which uses the H.323 protocol suite.

4   Usage Scenarios and Protocol Mappings

After having identified basic functions we now show how calls can either be received or originated using both H.323 or SIP as the underlying signaling protocol. The protocol details are hidden from the applications which use the same interface and primitives in any of the cases. They - using the appropriate URL identifier - just have to specify which of the available protocols should be used .

4.1   Registering and Receiving Calls Using H.323

In this scenario, the IP Telephony Application (IpTelAppl) uses the H.323 protocol to register the participant with a Gatekeeper and enables him to accept calls. The sequence of protocol steps is shown in Figure 3.

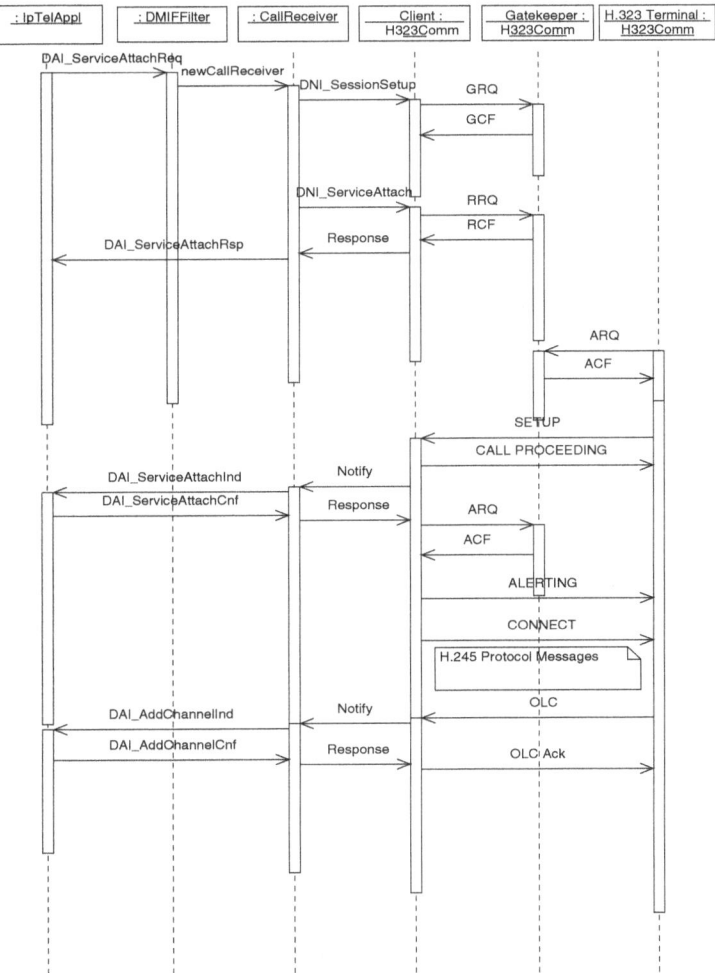

Fig. 3. *Registering and receiving calls - H.323 scenario*

It should be noticed that the DAI interface is used between the IP Telephony application and the DMIFFilter and the DNI between the CallMaker and the H323Comm object. The IpTelAppl attaches to the local Gatekeeper, by using the DAI_ServiceAttachReq primitive, where it can pass the address of it, if it is known. The DMIFFilter, parses the passed URL and activates a CallReceiver object to handle the details of the operation. The CallReceiver may inquire at the local H323Comm object about the location of the Gatekeeper, if its address is not already specified, using the DNLSessionSetup primitive. In this case the local H323Comm object broadcasts a request to locate the Gatekeeper. After the Gatekeeper has been identified, the CallReceiver object requests the H323Comm object to register itself at the local Gatekeeper. The local H323Comm object is completing the request, by exchanging one more pair of messages with the Gatekeeper (RRQ and RCF). After the successful completion of the registration task, a handler to the CallReceiver is returned to the IpTelAppl for further usage.

Suppose that later another H.323 terminal wants to call the IpTelAppl. It can obtain the address of the local H323Comm object from the Gatekeeper. Then, it submits a Setup message to the local H323Comm object to request the establishment of a new session. The IpTelAppl is informed about this request from the CallReceiver object. The IpTelAppl instance can decide to accept the new call and the local H323Comm object requests admission from the local Gatekeeper (ARQ and ACF messages). After that it replies to the remote H.323 terminal that it accepts the connection (CONNECT message).

A number of H.245 messages follow in order to exchange the capabilities of the two terminals. Then, an Open Logical Channel (OLC) message is sent to request for a media channel. The CallReceiver indicates this to the IpTelAppl (DAI_AddChannelInd), which confirms it. Finally, the local H323Comm object sends an acknowledge to the remote H.323 terminal. The last procedure might be repeated for more media channels.

4.2 Registering and Receiving Calls Using SIP

Figure 4 shows the registration of an IpTelAppl in order to receive calls for the SIP case. The IpTelAppl requests user registration with the DAI_ServiceAttach command. The DMIFFilter therefore creates a new DMIF Instance, the Call-Receiver to handle the registration and possible future calls. A CallReceiver requests from the SIPComm object to establish a connection with the SIP location sever. The SIPComm object sends a REGISTER message to the Location Server to store its location information for future incoming calls. The handler returned to the IpTelAppl is used to proceed future interactions.

Later, when a new INVITE message is received from the SIPComm object (Client instance), the IpTelAppl will be informed. It then can either confirm the acceptance of the incoming call or reject the new invitation. In the case of acceptance, a SIP 200 OK response is replied to the caller, and the call is established after the final ACK is received.

4.3 Call Setup Using H.323

In this scenario - shown in Figure 5 - the IpTelAppl wants to setup a call to a remote H.323 terminal. Only the most important and relevant (to the DMIF layer) H.323 messages are shown.

The IpTelAppl calls the DAI_ServiceAttachReq to originate a new call. It passes the URL of the remote participant for symbolically addressing the intended communication partner. The DMIFFilter parses the request and creates a new CallMaker object to handle the details of this operation. It requests the session

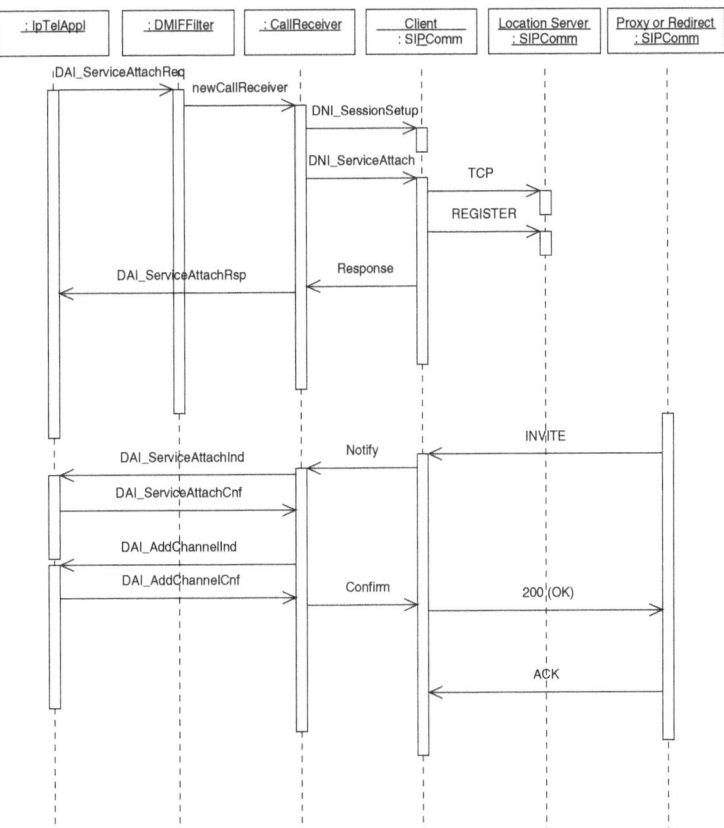

Fig. 4. *Registering and receiving calls - SIP scenario*

setup from the local H323Comm object, which communicates with the Gatekeeper to request for admission to place the call and to address the remote party (ARQ and ACF messages). Then, the CallMaker calls the DNI_SessionAttach primitive to request the establishment of a new session with the remote terminal from the local H323Comm object. A set of messages is exchanged between the two H.323 peers, compromising both Q.931 and H.245 protocol elements. At the end, the IpTelAppl receives a positive response.

Once the connection is established, the application may initiate the setup of additional media channels with the DAI_AddChannelReq primitive. The local H323Comm object is negotiating the channels with the remote H.323 terminal (OLC and OLC Ack messages). This procedure is repeated between the two terminals for every media channel.

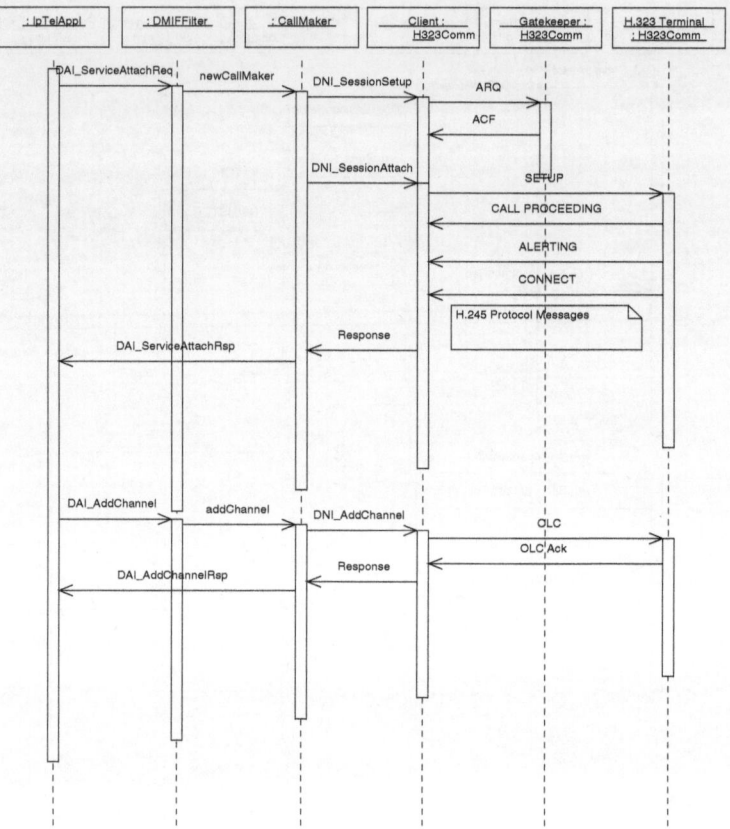

Fig. 5. *Originating Calls - H.323 scenario*

4.4 Call Setup Using SIP

In this scenario - described in Figure 6 - the IP Telephony application is setting up a new call, using the SIP protocol. No specific details of the protocol are required for the application.

It constructs an appropriate URL, which denotes the intended usage of the SIP service. The IpTelAppl requests the DMIFFilter to attach with the requested remote participant. The symbolic address of the remote user is passed as a SIP URL with the appropriate parameters. The DMIFFilter creates a new CallMaker object to handle the signaling task for this new call. The CallMaker interacts with the SIPComm object (Client instance) to establish the call (if TCP is used) with a SIP proxy or redirect server, using the DNI_SessionAttach primitive. The Client has to locate the appropriate SIP proxy or redirect server first to request the remote user. After the (successful) connection with the server a response is given back to the IpTelAppl, with a handler to refer to the same objects for later requests.

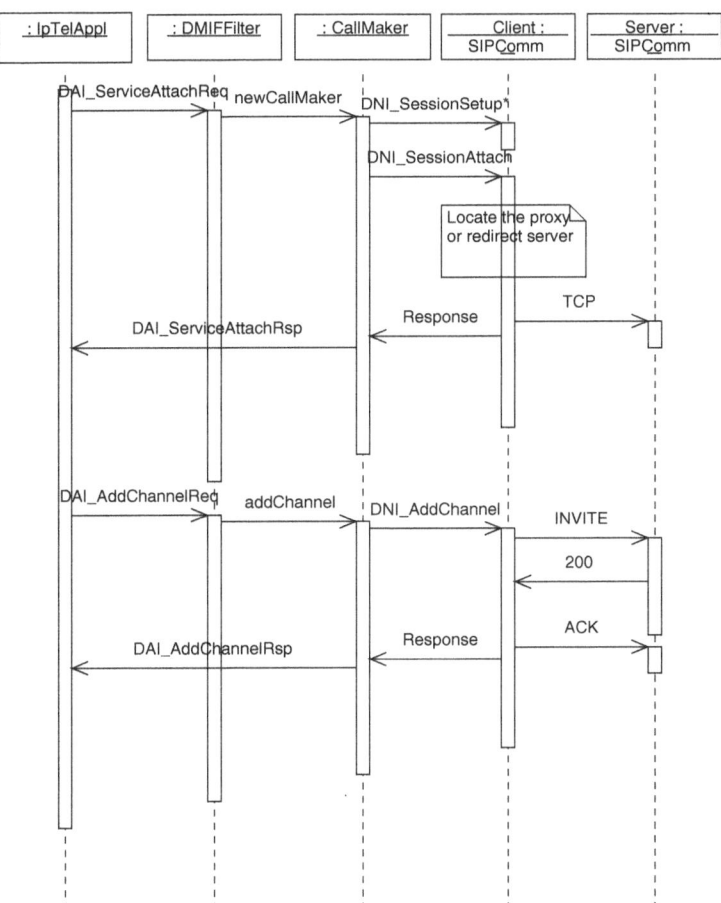

Fig. 6. *Originating Calls - SIP scenario*

In a next step the IpTelAppl may add a voice channel. IpTelAppl calls the DAI_AddChannel primitive from the DMIFFilter to request for a channel with specific QoS, if this is supported from the underlying network. The previously created CallMaker object is identified and is requested to handle the new request. The CallMaker maps the Application QoS to the Network QoS and calls the DNI_AddChannel primitive to ask the SIPComm object to request for a new channel with the remote user. SIPComm, interacts with the SIP proxy and redirect server to locate and invite the remote user, using the INVITE message. In the basic scenario, it will receive a positive SIP response (OK 200), and will complete the invitation with the SIP ACK message.

5 Conclusion and Future Work

In the paper we have proposed a framework based on DMIF and possible mappings for common IP Telephony signaling operations. We are not intending to develope "yet another implementation" for one of the emerging signaling standards, but try to find a generic approach by identifying basic functionalities.

Based on our experiences with implementing a protocol gateway between H.323 and SIP the "conventional way" [1] we assume, that the approach fits well with the recent functionality of established signaling protocols while being flexible enough to also incorporate changes or cope with even new protocols. Though we have concentrated on describing scenarios involving end systems, it is applicable for the development of infrastructure components using a variety of signaling protocols as well.

The basic motivation for choosing DMIF is its standardization and the experience, that generating software in a standardized instead of a per-application or per-protocol way can speed up development and permits more generalized solutions. We consider our framework feasible and intend to implement it using different underlying protocol stack software thus enabling applications to use the described interface and to have means for the evaluation of its performance and implementation as well as runtime-overhead.

References

1. R. Ackermann, V. Darlagiannis and R. Steinmetz: Implementation of a H.323/SIP Gateway. Technical Report TR-2000-02, Darmstadt University of Technology, Industrial Process and System Communications (KOM), Jul. 2000.
2. M. Arango, A. Dugan, I. Elliott, C. Huitema, and S. Pickett: Media Gateway Control Protocol (MGCP). Internet Draft, Internet Engineering Task Force, Feb. 1999. Work in progress.
3. I. Dalgic and H. Fang: Comparison of H.323 and SIP for IP Telephony Signaling. In Proc. of Photonics East, Boston, Massachusetts, Sept. 1999.
4. M. Handley, H. Schulzrinne, E. Schooler, and J. Rosenberg: SIP: Session Initiation Protocol. Request for Comments (Proposed Standard) 2543, Internet Engineering Task Force, Mar. 1999.
5. K. Hoffmann, K. Pulverer, P. Blatherwick, B. Bell, M. Korpi, T. Taylor, R. Bach, and C. Ruppelt: Megaco/H.248 Generic Packages. Internet Draft, Internet Engineering Task Force, Sept. 1900. Work in progress.
6. International Telecommunication Union: Visual telephone systems and equipment for local area networks which provide a non-guaranteed quality of service. Recommendation H.323, Telecommunication Standardization Sector of ITU, Geneva, Switzerland, May 1996.
7. International Telecommunication Union: Packet based multimedia communication systems. Recommendation H.323, Telecommunication Standardization Sector of ITU, Geneva, Switzerland, Feb. 1998.
8. ISO: Information Technology - Coding of audio-visual objects, Part 6: DMIF. ISO/IEC IS 14496-6, 1999. ISO IEC JTC1/SC29.
9. ISO: Information Technology - Coding of audio-visual objects, Part 6: DMIF version 2. ISO/IEC IS 14496-6v2, 2000. ISO IEC JTC1/SC29.

10. R. Koeman: MPEG-4, Multimedia for our time. IEEE Spectrum, 36(2), pages 26-33, February 1999.
11. Microsoft Corporation: The Microsoft Windows Telephony Platform with TAPI 2.1. White Paper. http://www.microsoft.com/ ntserver/commserv/techdetails/prodarch/tapi21wp.asp.
12. T. Russell. Signaling system #7: McGraw-Hill, New York, 1995.
13. H. Schulzrinne and J. Rosenberg: A Comparison of SIP and H.323 for Internet Telephony. In The 8th International Workshop on Network and Operating Systems Support for Digital Audio and Video (NOSSDAV 98), pages 83-86, Cambridge, England, July 1998.
14. K. Singh and H. Schulzrinne: Interworking between SIP/SDP and H.323. Internet Draft, Internet Engineering Task Force, Jan. 2000. Work in progress.
15. Sun Microsystems: Java(tm) Telephony API. Product Description. http://java.sun.com/products/jtapi/.

Short-Range Connectivity with Bluetooth

J.C. Haartsen

Ericsson, Enschede, the Netherlands

Jaap.Haartsen@erh.ericsson.se

Abstract. In the last decades, progress in microelectronics and VLSI technology has fostered the widespread use of computing and communication applications in portable electronic devices. Up until now, information transfer between these devices has been cumbersome relying on cables and infrared. Recently, a new universal radio interface has been developed enabling electronic devices to connect and communicate via short-range radio connections. The BluetoothTM radio technology eliminates the need for wires, cables and connectors between cordless or mobile phones, modems, headsets, PDAs, computers, printers, projectors, and so on, and paves the way for new and completely different devices and applications. Bluetooth is regarded as a complement and an extension to existing wireless technologies, addressing the short-range and inter-device connectivity. The technology enables the design of low-power, small-sized, and low-cost radios that can be embedded in existing portable devices. Eventually, embedded radios will lead towards ubiquitous connectivity.

The Bluetooth radio operates in the unlicensed ISM band at 2.45 GHz. This band has to be shared with other applications that spread radio energy which can disturb the Bluetooth link. Several design features have been implemented to make the Bluetooth air interface as robust as possible against potential interferers. The Bluetooth radio applies frequency hopping with a nominal hopping rate of 1600 hops/s. Instantaneously a narrowband channel of 1 MHz is used to transfer the data packets; every 625ms, this channel is positioned on a different carrier selected according to a pseudo-random hop sequence covering 79 carriers that span the 80 MHz ISM band. The interface supports both synchronous links suited for circuit-switched applications like voice communications and asynchronous links suited for packet-switched applications like data. The voice and data protocols have been optimized to deal with interference present in the unlicensed band. Attention has been paid to low-power modes to maximize battery-life in portable devices.

This paper gives an overview of the Bluetooth radio interface. It addresses key design issues that make Bluetooth a unique air interface that differs from existing wireless technologies. Examples are the support for ad-hoc connectivity, the application of an unlicensed band, the low power consumption, and the enabling of single-chip radios. The paper also presents some user scenarios and future developments

H. Scholten and M. van Sinderen (Eds.): IDMS 2000, LNCS 1905, p. 116, 2000.

A QoS-Control Architecture
for Object Middleware[*]

L. Bergmans[1], A. van Halteren[2], L. Ferreira Pires[1], M. van Sinderen, and M. Aksit[1]

[1]CTIT, University of Twente, The Netherlands
{bergmans, pires, sinderen, aksit}@ctit.utwente.nl
[2] KPN Research, Enschede, The Netherlands
A.T.vanHalteren@kpn.com

Abstract. This paper presents an architecture for QoS-aware middleware platforms. We present a general framework for control, and specialise this framework for QoS provisioning in the middleware context. We identify different alternatives for control, and we elaborate the technical issues related to controlling the internal characteristics of object middleware. We illustrate our QoS control approach by means of a scenario based on CORBA.

1. Introduction

The original motivation for introducing middleware platforms has been to facilitate the development of distributed applications, by providing a collection of general-purpose facilities to the application designers. Currently, commercially available middleware platforms, such as those based on CORBA, are still limited to the support of best-effort Quality of Service (QoS) to applications. This constitutes an obstacle to the use of middleware systems in QoS critical applications, or in case services are offered in the scope of Service Level Agreements with strict QoS constraints. This limitation in the available middleware technology has inspired much of the research that is currently being done on QoS-aware middleware platforms.

Ideally, a middleware platform should be capable of supporting a multitude of different types of applications with (a) different QoS requirements, (b) making use of different types of communication and computing resources, and (c) adapting to changes, e.g., in the application environment and in the available resources. The architectural framework presented in this paper has been developed to be flexible and re-usable. The main benefit of our framework is that it allows us to combine and balance solutions for the control of multiple QoS characteristics.

From the research perspective, a framework-based approach supports the incremental introduction of new solutions such as control algorithms, and allows us to compare different solutions in the same setting. From a middleware developer's perspective, this approach is attractive because it supports incremental development and the construction of product families in which different family members address different sets of QoS characteristics.

---

[*] This work has been carried out within the AMIDST project (http://amidst.ctit.utwente.nl/).

H. Scholten and M. van Sinderen (Eds.): IDMS 2000, LNCS 1905, pp. 117-131, 2000.
© Springer-Verlag Berlin Heidelberg 2000

This paper identifies the problems that have to be solved in order to elaborate our architectural framework, and discusses some techniques that can be used to solve these problems. The use of the architectural framework is illustrated by means of a scenario.

This document is further structured as follows: Section 0 introduces some basic concepts and discusses the role of a QoS-aware middleware when supporting object-based applications; Section 0 introduces our approach and its background, which stems from control theory; Section 0 discusses the technical issues that have to be addressed to realise the proposed framework, presents some requirements for each of these issues and indicates our solutions to fulfil these requirements; Section 0 illustrates the use of our architectural framework with a simple CORBA application using a naming service, and shows how QoS requirements can be enforced by the middleware platform; and Section 0 draws our conclusions.

2. Concepts of QoS-Aware Middleware

This section discusses the concepts that underlie our QoS-aware middleware architecture and identifies the role of a QoS-aware middleware platform in the support of distributed applications.

2.1 Distributed Applications

Distributed applications supported by a QoS-aware middleware consist of a collection of interacting distributed objects. Since in this paper we concentrate on the support of *operations* (invocations of methods), we assume that an object may play the role of either a client or a server on an interface. We also assume the ODP-RM computational model, in which objects may have multiple interfaces [7].

During the development of a distributed application, the interfaces of the application objects have to be specified. In principle this specification should define the attributes and operations of these interfaces. In the case of CORBA, one only specifies the server interface using IDL, and makes use of this specification for creating stubs and skeletons, or to dynamically create requests for operations. However, in general one could specify server and client interfaces, and define rules that can determine whether a server interface is capable of servicing a client interface [14, 1]. Extensions of IDL that allow the definition of both client (required) and server (offered) interfaces have already been proposed in the CORBA component model [13].

When considering QoS-aware middleware, we suppose that the interface specifications are extended with statements on QoS that can be associated with the whole interface or with individual operations and attributes. In the case of a client interface, these statements describe the required QoS, while for a server interface these statements describe the offered QoS. QML [4] and QuO [17] are languages that allow one to specify the QoS associated to interfaces.

2.2 Objects Life-Cycle

After the objects of a distributed application have been implemented, the application is deployed. We consider the general case in which persistent objects and late binding can be supported by the middleware platform. In this case, an object has the following life cycle:

1. Object creation, in which interface references for the server interfaces of an object are created and can be referred to by other objects;
2. Object activation, in which an object starts execution, which implies that all local resources necessary for the object to execute should be properly allocated;
3. Object deactivation, in which local resources allocated to an object may be released, although the interface references may still be valid in case persistent objects are supported;
4. Object destruction, in which the object is deactivated (if it is still active) and its interface references are destroyed.

A QoS-aware middleware platform can use object activation to refine the offered QoS, by restricting the ranges originally described for the offered QoS at design and implementation time. The run-time status of the middleware platform and the communication and computing resources should make it possible to determine this offered QoS more precisely.

2.3 Explicit Binding

Object interfaces have to be bound to each other in order to allow these objects to interact through the middleware. In CORBA, this binding happens implicitly when the client object issues a request (implicit binding).

For QoS-aware middleware platforms, however, implicit binding is not desirable, since the QoS requirements may demand that resource allocation procedures are performed before the request is executed. Unfortunately, we can not predict the speed and reliability of these procedures. In the worst case, we may still have to activate the server object. This means that we can not always guarantee the QoS requirements by using implicit binding. Therefore,in QoS-aware architectures *explicit binding* is necessary, which consists of taking explicit actions at the computational level in order to establish the binding before interacting [7]. Our case for explicit binding for QoS-aware operation support is somewhat similar to the reasoning in [1] for stream interface bindings.

The client object requests the establishment of the binding, giving to the middleware a reference to a server interface. This request also contains the required QoS, which can be retrieved from a QoS specification repository. The middleware platform then searches for the server object. In case the server object has not been activated, the middleware platform activates this object and continues the establishment procedure. Otherwise, the middleware platform compares the offered QoS with the required QoS and uses its internal information to determine an agreed QoS. This process is called *QoS negotiation*. In case the binding establishment has been successful, the client and server objects are informed that a binding has been built. From this moment on these objects can interact through the binding. Figure 1 shows the establishment of a binding using a QoS-aware middleware.

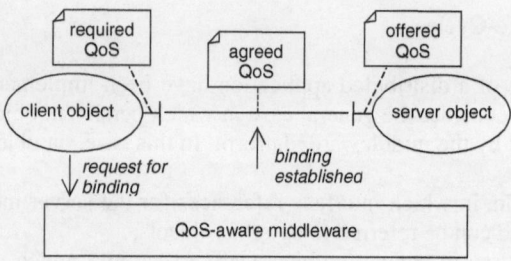

Fig. 1. Binding establishment using a QoS-aware middleware platform

The agreed QoS is determined by considering the required QoS on one hand, and the composite QoS capabilities of the server object (the offered QoS) and the middleware platform on the other hand. The agreed QoS serves as a *contract* between the application objects and the middleware platform, which should be respected during the operational phase when the objects interact through the binding.

The binding establishment may also result in the creation of a binding object. This object binds the client object and the server object, and offers a control interface that allows, for example, the inspection and modification of the agreed QoS. In this paper we ignore the adjustment of the agreed QoS through such an interface, but this is an interesting topic for further work.

Our QoS control approach considers that a binding has been successfully established and that the agreed QoS has to be maintained. The middleware is responsible for that, and is constantly adjusting its internal characteristics and the usage of computing and communication resources in order to achieve it.

3.   Control Framework for QoS Provisioning

This section introduces our approach and subsequently discusses a specialisation of a generic control system model for the purpose of controlling QoS in a middleware context.

3.1 Approach

The design of our QoS-aware middleware architecture is constrained by two conflicting requirements: a) the architecture has to be flexible enough such that it enables us to experiment with different QoS strategies and cope with different kinds of application demands; and b) certain aspects of the architecture have to be fixed so that the robustness and portability of the architecture can be guaranteed.

For this reason we start off with a generic control system model, which we specialise, such that it applies to QoS-control in a middleware context. This specialised model forms our architectural framework, i.e. the fixed part of our architecture. Although some decisions are made with respect to the scope of control, the architectural framework is independent of any specific QoS-control strategy or algorithm. Therefore, different solutions can be compared and evaluated with this framework.

A synthesis-based approach [16] can be used to arrive at a complete QoS-control architecture, e.g. for a specific application or system environment. In this approach, requirements are converted into technical problems. For each technical problem, possible solution techniques are sought. The candidate solution techniques are then compared with each other from the perspective of relevance, robustness, adaptability and performance. Whenever a suitable solution technique is found, the fundamental abstractions of this technique are used to derive the architectural abstractions. This process is repeated until all the problems are considered and solved. Finally, the architectural abstractions are specified and integrated within the overall framework. Since solution domain knowledge changes smoothly, this approach provides us with stable and robust abstractions with rich semantics. The discussion of technical issues in Section 4 partially illustrates this approach.

3.2 Generic Control System

The main objective of QoS-aware middleware is to establish and enforce an agreed QoS that satisfies the demands of applications, given the available resources. We observe this is essentially a controlling problem, and therefore the QoS-control framework should be synthesised from the fundamental abstractions of control systems.

A *control system* [3, 8] consists of a *controlled system* in combination with a *controller*. The interactions between the controlled system and the controller consist of observation and manipulation performed by the controller on the controlled system. The building blocks of the control process are shown in figure 2:

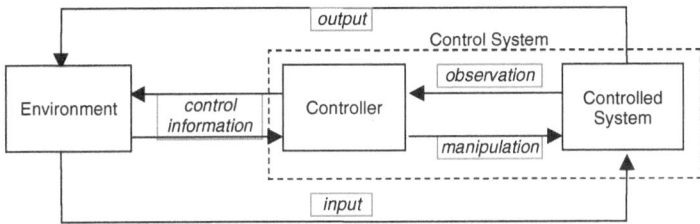

Fig. 2. Building blocks of a control process.

The generic control model abstracts from the type of observation and the type of manipulation that can be employed by the controller on the controlled system. The relationship between the controlled system and the controller can be realised using different strategies. With a *feed-forward control* strategy, manipulation through control actions is determined based on manipulation of the input to the controlled system. A *feed-back control* strategy can be applied for behaviour optimisation. According to this strategy, measurements of the output delivered by the controlled system are compared with a desired behaviour (a *reference*) and the *difference* between them is used by the controller to decide on the control actions to be taken.

3.3 QoS-Control System

In QoS-aware middleware, the 'controlled system' is the middleware functionality responsible for the support of interactions between application objects, while the 'controller' provides QoS control. Here, the environment represents the operational context of the middleware, which consists of application objects with QoS requirements and QoS offers. The middleware platform encapsulates the computing and communication resources at each individual processing node, which may be manipulated in order to maintain the agreed QoS.

Figure 3 shows the specialisation of the generic control model for controlling the QoS provided by a middleware.

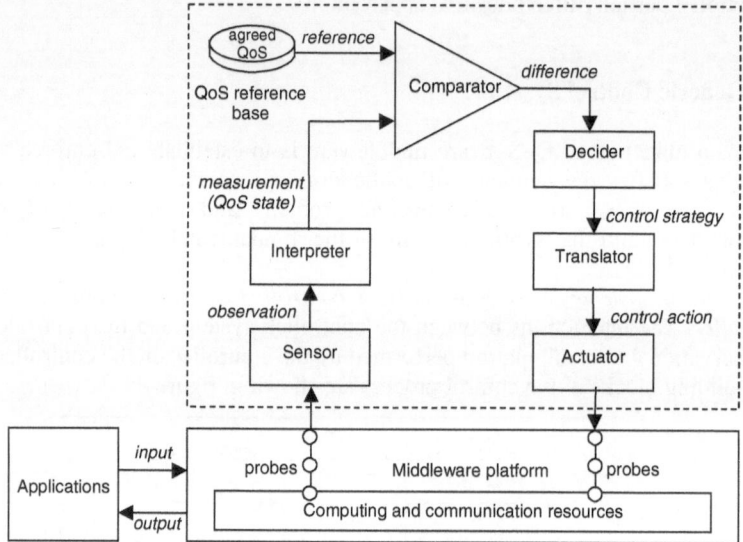

Fig. 3. QoS-control architecture.

In Figure 3 we identify two symmetrical structures, one for handling QoS measurement concerns and another for handling QoS manipulation concerns. A *probe* is a point of observation or manipulation that is available or must be planted in the controlled system, i.e., the middleware platform. Many probes may be planted in the controlled system, for both observations and manipulations.

A *sensor* is a mechanism that uses a probe to obtain observations. Observations can only be useful if they are interpreted in terms of measurements that can be compared with the reference, i.e., they are represented using the same units and have the same semantics. For example, observations can be time moments of the sending of a request and the receiving of the corresponding response. The needed measurement could be the average response time, which implies that the average of the difference between the time moments observed should be calculated in order to generate the measurement. This calculation is performed by an *interpreter*. In general, the interpreter combines observations, which could even come from different sensors, in order to generate measurements.

A *comparator* compares the measurement and an associated reference value (an agreed QoS value), determining the difference. A *decider* gets the difference and applies some algorithm to establish a control strategy, consisting of the objectives to be reached in this execution of the control loop. The control strategy must be translated in a collection of control actions, i.e., manipulations of the controlled system. A *translator* is responsible for translating the control strategy to a collection of control actions. An *actuator* schedules the control actions such that they are carried out using one or more probes. The translator distributes the control actions among the actuators, realising in this way the control strategy.

4. Technical Issues

This section identifies and elaborates a number of the technical issues that have to be addressed in order to realise the QoS control architecture. The first 5 issues correspond to the realisation of the components in our architecture. We explain the requirements, and propose some possible solutions and solution strategies.

4.1 Collecting Observation Values

In order to collect observation values we have to develop probes and sensors. Probes connect the middleware to the control mechanisms and are independent of the actual measurements. Sensors collect the actual measurements and they typically depend on the amount and types of data that are collected.

The fundamental requirement on the probes and sensors is that they must have a minimal impact on the middleware platform. This introduces two issues: (a) how to minimise the impact on the middleware code, and (b) how to map the probes to one or more specific places in the middleware code (cross-cutting problem [9]).

Reflection is a technique in which a system is explicitly represented in terms of a meta-object, allowing one to manipulate the (structure of the) system by manipulating its meta-object. A reflection-based approach suits well to the collection of observation values.

Crosscutting of concerns requires either careful documentation and management of probe insertion points, or entirely new tools and techniques for specifying and implementing crosscut concerns. Recent work in the area of Aspect-Oriented Programming (see, e.g., [9]) addresses these issues.

4.2 Interpretation of Observations

The interpretation process depends on many factors: the involved observation data, the required measurements, and the rules or strategies for interpretation. The number of interpretation rules and their complexity also determine the interpretation process.

The interpretation part should not become a possibly large collection of unstructured ad-hoc code. This implies that a generic model should be developed to define how observations are translated to measurements, such that interpretation code can be reused or generated automatically as much as possible. In case statistical information determines measurements, a lot of input data may be required, such that the amount of storage and processing should be as much as possible reduced.

The interpretation process is essentially a transformation from a set of input values to a set of output values. The variation in input values lies both in sources, types and time, and depends on the sources of the input, i.e., how the middleware has been designed. The resulting output should be independent of the specific implementation details of certain middleware and applications, and it should be suitable for the comparison process. This means that a common *QoS meta model* should be available that determines the types and values of both measurements and the references.

The interpretation of observations can be done through calculations, heuristics (logic rules), stream interpreters or conversions. We need to model these different techniques in a uniform way, with explicit dependency relations to a structured representation of the observations and measurements. Interpretation rules should all be a specialisation of a single abstraction, i.e., the interpreter. Each individual instantiation can be considered as a *micro-interpreter*. For each QoS measure, there should be a clear specification of the interpretation rules in terms of formulas or guidelines. In Figure 4 at the end of this section the extensions to our architecture to meet the needs of interpretation are shown.

4.3   Determination and Representation of the Difference

The comparator compares the measurement with the reference model and determines the difference. This comparison can vary from subtraction in the simple case of one QoS characteristic with a numeric value, to complex calculations possibly using heuristics in the case of multi-faceted QoS characteristics. The main task of the comparator is to deliver an abstraction of the 'problem to be solved' that is as far from the implementation details of the environment as feasible.

The difference produced by the comparator serves to detect (potential) violations of the QoS. Such violations depend on the agreed QoS. Hence, the difference must be obtained by comparing the actual measurements with corresponding references specified by the agreed QoS.

The difference could be represented as a 'distance' vector, where each element of the vector corresponds to a relevant QoS characteristic.

Measurements and references should be described in such a way that they can be compared (see Section 0). For this purpose we use a QoS meta-model, which consists of a collection of concepts that allow one to specify both the measurement and the reference, and the difference. Another benefit of having a QoS meta-model is the ability to build QoS specification repositories. We adopt and adapt QML [4] to specify the QoS meta-model and its instantiations.

Figure 4 illustrates the use of the QoS meta-model in our overall architecture. The agreed QoS is determined before entering the operational phase, through negotiation based on QoS requirements, QoS offers and the capabilities of the middleware platform. In this paper we assume that the agreed QoS is not modified during the operational phase.

4.4 Controlling Algorithm

The difference or distance vector computed by the comparator may –or may not– define a situation that requires controlling (i.e., correcting) actions to be taken. The controlling algorithm is responsible for selecting an appropriate strategy. The strategy to be chosen depends partially on the specific state and configuration of the middleware. Rather than mixing middleware state and configuration information with the measurements and difference, this information must be available independently. For this reason, we introduce a *middleware control model*. This model is an abstraction (model at a meta-level) of the middleware, which specifies what can be parameterised or tuned in the middleware, or which components can be plugged in, deactivated and activated.

The task of the decider is two-fold: firstly to ensure that the agreed QoS can indeed be supported by the middleware platform, and secondly to optimise the overall QoS characteristics, by balancing the different, often contradictory, requirements. In its most general form, controlling is an artificial intelligence task that involves domain knowledge and heuristics about managing and controlling QoS, and the interdependencies between QoS characteristics.

We have not selected a particular solution for the controlling algorithm: our goal is to offer a framework that allows the experimentation with –combinations of– different techniques such as mathematical algorithms, heuristic rules and the use of fuzzy logic as a means of expressing and reasoning about weak but conflicting optimisation [12]. Figure 4 shows the extension of our architecture with an explicit middleware control model.

4.5 Control Strategy and Middleware Manipulation

A control strategy is the output of the controlling algorithm, and it should be an implementation-independent representation of the solution strategy for pursuing certain QoS characteristics. Control strategies are strongly related to the controlling algorithm.

Control actions are abstractions that represent concrete functional behaviour, but are independent of the implementation details of the specific middleware software. Control strategies represent sets of control actions that are to be applied to the middleware in a co-ordinated way. The representation of control strategies must consist of at least the following parts: a) set of control actions; b) a set of probes in the middleware where the control actions can be applied, and c) a co-ordination specification, which could be a script or any other form of executable specification.

There are a few ways to affect the behaviour of a running system like a middleware platform: a) by invoking operations of a local API; b) by modifying the internal state of the system, c) by replacing components of the system with different implementations, and d) by meta-level manipulation of the system itself. A control action can only be a specialisation or instantiation of one of these.

The implementation of control actions through actuators and probes introduces technical issues comparable to the ones discussed in section 0, and therefore they are not discussed further.

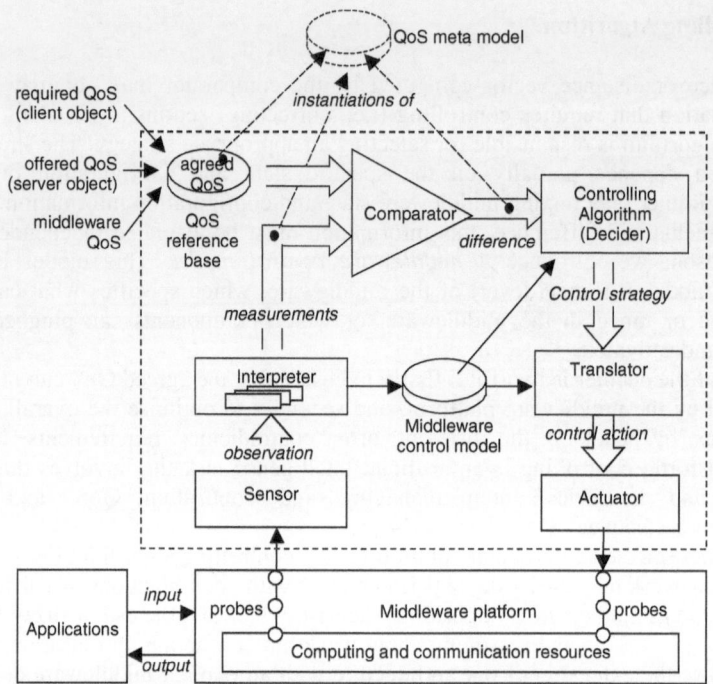

Fig. 4. A detailed version of the architecture incorporating some of the enhancements that are discussed in this section.

4.6 Feasibility of the Overall Control Loop

The performance overhead introduced by our architectural framework has to be carefully considered when using the framework in practical settings. The technical solutions should not make the overall QoS worse than what it would be without them. Several QoS requirements are related to performance (e.g., delays and throughput). Implementations of our architecture may require a lot of additional activities and overhead, which may conflict with the QoS requirements they try to enforce. By adopting a tailorable framework approach, we may choose to build instances of the framework with components ranging from simple, low-overhead components up to complex components. This approach can help coping with the performance overhead by using more efficient versions wherever necessary. In the future, the use of a meta-controller to switch dynamically between different versions may be considered.

Feed-back control loops may make the controlled system oscillate between two undesirable states, depending on the corrective measures and their effects. In some cases, mathematical models based on control theory can help predicting whether the system is stable during operation, allowing one to avoid oscillation. In case mathematical models are not available or are not precise enough, some heuristics may show whether the system is stable or not. Alternatively, additional (meta-level) controllers could be introduced to detect instability and take measures to avoid it, e.g., by actuating on the controlling algorithm. The use of fuzzy logic in the controlling algorithms may also help to avoid that the control loop oscillates during operation.

5. Scenario

This section demonstrates our architectural framework by means of a scenario for a simple CORBA application using a naming service.

5.1 Scenario Set-Up and Use of QML

Figure 5 depicts a QoS provisioning scenario for a CORBA naming service application. The application consists of a client object that intends to invoke a method on a `NamingContext` object through a QoS-aware ORB.

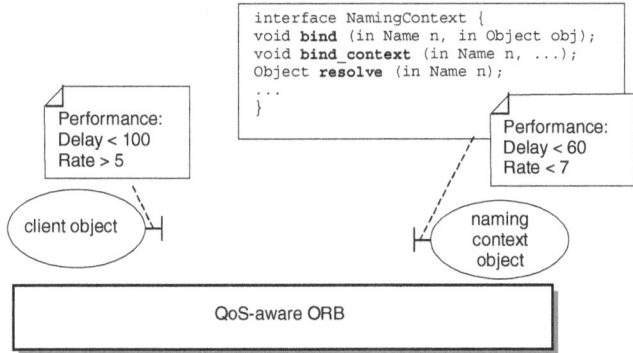

Fig. 5. A naming service application scenario.

The offered QoS of the server object and the required QoS of the client object are depicted in a simplified form:

- the client object requires a *delay* (time necessary to complete a request) smaller than 100, and a supported *rate* (number of requests per time unit) of at least 5;

- the `NamingContext` object offers a delay smaller than 60, and a supported rate up to 7.

We use QML [4] to express the required or offered QoS of an object. The QoS is specified using the QoS dimensions of a QML *contract type*. Figure 6 shows a possible QML contract type that defines the relevant QoS characteristics in this scenario, viz. delay and rate.

```
type PerformanceType = contract {
  delay : decreasing numeric msec;
  rate: increasing numeric req/sec;
};
```

```
PerformanceType contract {
  delay < 60; //maximum delay
  rate  >  7; //minimum rate
};
```

Fig. 6. A QML contract type

Fig. 7. A QML contract

QoS contract types should be defined by the QoS meta-model discussed in Section 4. The QoS that an application object requires or offers is specified by a QML *contract*. A QML contract puts constraints on the dimensions of the corresponding QML contract type. Figure 7 shows a possible instance of the QML contract type of Figure 6. This contract corresponds to the offered QoS of the naming context object in Figure 5; the required QoS of the client object can be defined in a similar way.

5.2 Binding Establishment

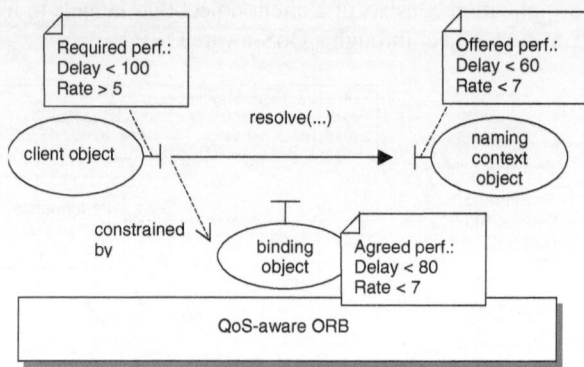

Fig. 8. An established binding with an agreed QoS.

During the binding establishment phase, the client object requests the establishment of a binding with the `NamingContext` object. Whether such a binding is successful depends on whether the application object and the middleware platform together can satisfy the required QoS of the client object. If so, an agreed QoS is established and the middleware should take appropriate actions, such as:

- update of the QoS reference base (see Figures 3 and 4);
- instantiation of sensors that can be used for measurements during the operational phase;
- instantiation of actuators and/or other configuration settings to prepare support for the agreed QoS. For example, when a configurable transport protocol is used, a connection with certain characteristics may be set up.

Figure 8 depicts a possible result of a successful binding establishment phase, where a QoS is agreed and a binding object is created that is aware of the agreed QoS. The offered QoS of the middleware in this case could be an (added) delay less than 20, and a supported rate less than 100. For the QoS characteristics in this scenario, the constraints on the negotiation process that led to the agreed QoS are as follows:

- delay: $QoS_{offered}(server) + QoS_{offered}(middleware) \leq QoS_{agreed} \leq QoS_{required}$
- rate: $QoS_{required} \leq QoS_{agreed} \leq minimum(QoS_{offered}(server), QoS_{offered}(middleware))$

5.3 Operational Phase

During the operational phase, the client object invokes requests and obtains replies from the `NamingContext` object. Invocations may trigger measurements or the installation of timers to measure the actual offered QoS. If the measured QoS approaches certain thresholds related to the agreed QoS stored in the QoS reference base, the control system may take precautions, either proactively or reactively. Proactive control actions are taken when a "danger zone" has been entered, but before a violation has been detected on the agreed QoS; reactive control actions are applied if a violation of the agreed QoS has been detected. Control actions include scheduling of a request on a high-priority thread, transmitting the request over a transport network with priority routing, installing a protocol plug-in that takes advantage of the network QoS by differentiating between the priority of network packets, or optimising delivery to multiple recipients through a multi-cast protocol.

For example, the handling of QoS during the operational phase, where (for reasons of space) we focus on delay, involves the following steps:

- To determine the actual delay, we can measure the time that elapses between the sending of a request and the receipt of a corresponding reply. *Sensors* are responsible for collecting this information from relevant *probes* (e.g., timer, interceptor) planted in the middleware platform.

- The *interpreter* translates the observations received from the sensors into measurements that can be usefully compared to the agreed delay (reference) value that was established for this binding. For example, the delay observations in a certain time period may be used to compute an (average) delay measurement.

- The *comparator* compares the measurement values with the corresponding reference value. The result could be an element in a distance vector:

  $<\Delta Delay, \Delta Rate>$, where $\Delta Delay = Delay_{reference} - Delay_{measured}$

  In a concrete case, we may have a delay 'distance' 20; this means that the actual offered delay has reached 80% of the maximum allowed delay.

- The *controlling algorithm* must decide, based upon the difference values, and other state information, how to deal with the situation. In this case, we assume that over 80% of the maximum delay is the 'danger zone', which requires specific actions to speed up the transport of the request/reply messages. A suitable *control strategy* that may improve the delays, is the activation of a faster/prioritised transport protocol (e.g., RSVP [6]).

- The *translator* uses state and configuration information from the *middleware control model* to determine the availability of the appropriate transport protocol, and the location where it should be plugged-in. The resulting control actions consist of plugging in the protocol on the client side, plugging in the protocol on the server side, and setting the priorities upon this transport plug-in.

- The control actions are performed by the *actuator*, which needs to access the middleware software for plugging in the protocols, and must perform the right priority/delay settings for each of the protocol instantiations. The *probes* used by the actuator can be APIs and/or global variables.

The steps that have been described here are performed repeatedly, either triggered by an internal clock, or by events (timers that expire, requests that are sent or received, etc).

6. Conclusion

In this paper we have presented an architecture to support QoS-aware middleware. We have introduced some assumptions and general concepts for using QoS-aware middleware. The key part, and focus of this paper, is the *QoS-control system*. The QoS controller in this system observes and, if necessary, manipulates the state of the controlled system, i.e. the middleware platform that supports distributed applications. The design of the QoS controller is an architectural framework that is based on models from control theory. This should ensure its stability with respect to evolving requirements, and its applicability to a wide range of controlling techniques.

The QoS-control architecture was discussed in more detail by examining a number of technical issues that must be addressed when realizing the proposed architecture. For each of these issues, we discussed requirements and corresponding solutions or solution approaches. We illustrated our proposal by describing a simple example of an application with QoS requirements, and how this would be dealt with in the proposed architecture.

An initial proof of concept of our approach has been performed in [5]. The prototype, based on the ORBacus implementation of CORBA, measures the QoS by using Portable Interceptors during system operation, and controls QoS at the transport level. Control actions are performed through a pluggable transport protocol that prioritizes IP packets using DiffServ [10] features.

In the paper, we hinted at several topics for interesting future work. These topics address the further development and prototyping of our control architecture, as well as exploring controlling strategies and algorithms that could not be considered so far. In addition, we like to profit from results of related works:

- One of the characteristics of our proposal is that the architecture is largely independent of the specific implementation architectures of middleware systems. The QoS controller is separate from the middleware (and applications) and may interact with these through a number of *probes* (a generic term for interfaces that abstracts from specific implementations). Conceptually (and possibly implementation-wise), this is a reflective model; our QoS controller observes and manipulates the middleware at a meta-level. Several other proposals for reflective middleware have been made, e.g. [2].
- A middleware framework for QoS adaptation has been described in [11]. Both a task control model and a fuzzy control model have been used in this framework to formalise and calculate the control actions necessary to keep the application QoS between bounds. This framework shares many design concerns with our framework, although it has been targeted to the control of applications.
- OMG currently develops Real-time CORBA facilities in the scope of the CORBA 3.0 standard [15]. These facilities allow one to manipulate some middleware characteristics that influence the QoS, such as, e.g., the properties of protocols underlying the ORB and the threading and priority polices applied to the handling of requests by server objects. These facilities are defined in terms of interfaces that have to be implemented in the middleware platform, generalising in this way the control capabilities of the platform.

7. References

[1] G. Blair, J.-B. Stefani. *Open distributed processing and multimedia.* Addison-Wesley, 1998.

[2] G. Blair, C. Coulson, P. Robin, M. Papathomas. An architecture for next generation middleware. In: Procs. of *Middleware 2000*, 191-206. Springer-Verlag, London, 1998.

[3] A.C.J. de Leeuw. *Organisaties: management, analyse, ontwerp en verandering -een systeemvisie.* Assen/Maastricht, Van Gorcum, 1990.

[4] S. Frølund, J. Koistinen. *Quality-of-service specification in distributed object systems.* Techn. report HPL-98-159. HP Labs, Palo Alto, USA, 1998.

[5] M. Harkema. *A QoS Provisioning Service for CORBA.* MSc thesis, Univ. of Groningen, Groningen, The Netherlands, 1999.

[6] IETF. *Resource ReserVation Protocol (RSVP). Version 1 Functional Specification.* IETF RFC 2205, Sept. 1997.

[7] ITU-T I ISO/IEC. *ODP Reference Model Part 1. Overview* (ITU-T X.901 I ISO/IEC 10746-1). May 1995.

[8] W.J.M. Kickert, J.P. van Gigch. A metasystem approach to organisational decision-making. In: J.P. van Gigch (ed.), *Decision making about decision making: meta-models and metasystems*, 37-55, Abacus Press, 1987.

[9] G. Kiczales, J. Lamping, A. Mendhekar, C. Maeda, C. Lopes, J.-M. Loingtier, J. Irwin. Aspect-Oriented Programming. In: Procs. of *ECOOP '97*, Springer-Verlag LNCS 1241, June 1997.

[10] K. Killki. *Differentiated services for the Internet.* Macmillan Computer Publishing, 1999.

[11] Baochun Li, Klara Nahrstedt. A control-based middleware framework for quality of service adaptations. *IEEE J. on Sel. Areas in Comms.* Vol. 17, No. 9, 1632-1650, Sept. 1999.

[12] E.H. Mamdani and S. Assilian. An experiment in linguistic synthesis with a fuzzy logic controller. *Intl. J. of Man-Machine Studies* 7, 1-13, 1975.

[13] OMG. CORBA components. OMG TC Document orbos/99-02-05. March 1999.

[14] C. Szyperski. *Component software.* Addison-Wesley, 1998.

[15] D.C. Schmidt, F. Kuhns. An overview of the real-time CORBA specification. To appear in: *IEEE Computer* special issue on Object-oriented real-time distributed computing, June 2000.

[16] B. Tekinerdogan. *Synthesis-based software architecture design.* PhD thesis. Univ. of Twente, Enschede, The Netherlands, 2000.

[17] R. Vanegas, J.A. Zinky, J.P. Loyall, D. Karr, R.E. Schantz, D.E. Bakken. QuO's runtime support for quality of service in distributed objects. In: Procs. of *Middleware 2000*, 207-222. Springer-Verlag, London, 1998.

An Architecture for a Scalable Video-on-Demand Server Network with Quality-of-Service Guarantees[*]

L.-O. Burchard and R. Lüling

Paderborn Center for Parallel Computing
University of Paderborn, Germany
{baron,rl}@uni-paderborn.de

Abstract. In this paper, we describe an architecture for a scalable distributed server management system which is capable of managing media assets stored on large-scale media server networks. In addition to the functionality for a decentralized management of the whole server network, the server management system allows to make bandwidth reservations even in advance for streaming in real-time and transmitting media assets. This increases the QoS for end-users since bandwidth reservations for streaming applications can be planned and established a long time before the actual streaming process takes place. Moreover, this approach allows a tight estimation of the duration required for copying media assets among different servers of the network. In order to minimize the required amount of network bandwidth for copying media assets and to reduce the total number of copying processes, the management software provides the functionality to place media assets on the servers of the network in a way that assets of particular interest for a certain client are located on the client's local server. In this paper, the architecture of the server management system itself is presented together with the approach used to provide QoS guarantees for end-users. The overall system is using the differentiated service model. Besides the technical concept we also present results of the integration of the system into the corporate network of a german company.

1 Introduction

In large scale media streaming applications, a central server for the storage and delivery of broadband media streams keeps the administrative effort low but has the obvious disadvantage of a limited scalability, concerning both network and storage resources. Especially when MPEG-2 encoded video streams are used, network and storage requirements pose severe constraints on the hardware infrastructure. Compared to a central server system, a network consisting of several

[*] This work was supported by a grant from Siemens AG, Munich in the framework of the project HiQoS - High Performance Multimedia Services with Quality of Service Guarantees

H. Scholten and M. van Sinderen (Eds.): IDMS 2000, LNCS 1905, pp. 132–143, 2000.

media servers has the advantage that it can be scaled to a large extend and allows to support a large number of widely distributed clients but requires additional effort for the content management. Thus, an important aim is to develop mechanisms for the automatic management of broadband media content in a network of distributed media servers which can be connected with each other using the available network infrastructure such as Internet or satellite links.

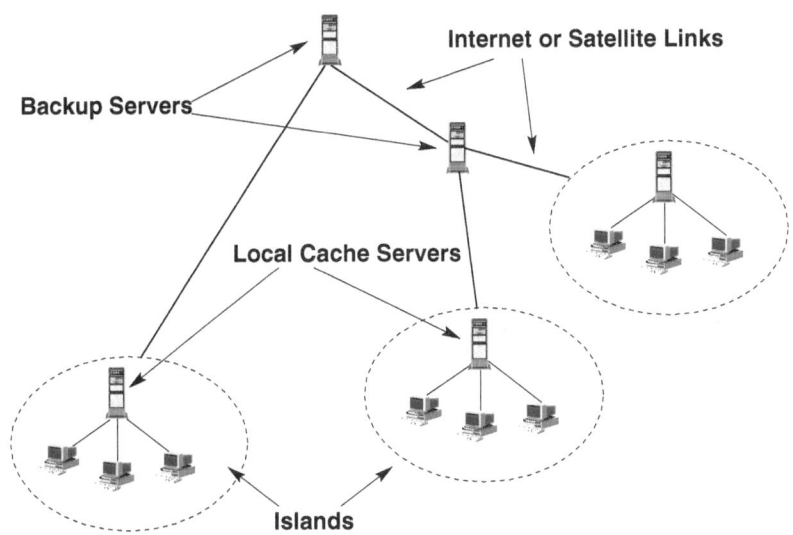

Fig. 1. Structure of a Media Server Network

The management software described in this paper organizes a media server network such that one or more servers are used as caches for a single *island* (i.e. an environment which provides the necessary network infrastructure for streaming high-quality media assets). The media assets are mapped onto the available servers in a way, that the caches store the media assets most likely being accessed from within such an island. Thus, each client has easy and cheap access to any media asset stored on a media server that is located within the same island as the client. The mapping is based on knowledge about clients preferences, e.g. subscriptions of content categories or user profiles. Changes in users behavior and new media assets require a remapping of the media assets, i.e. file transfers between the different servers. Clients inside islands also have access to the assets stored on the servers located in remote islands or somewhere in the backbone network (see Figure 1). However, streaming from a remote server might not be possible due to insufficient bandwidth availability or network connections that are available only for a certain time such as satellite links. Therefore, in these cases assets must be transferred to the client's local media server.

For transferring media assets from one server to another, timing constraints must be considered, i.e. deadlines for the delivery of the media assets to the destination. For example, in a teleteaching environment the media assets required during lectures must be available when the lectures start. This means, in case a client requests a media asset from a remote server, the server management system has to provide an estimation of the time of availability at the client's local server. In order to deal with unreliable network connections and changes of the available bandwidth, the durations can be largely overestimated. However, this leads to unrealistic figures concerning the estimated times and to an inefficient network utilization. In contrast to this, the approach described in this document is to use network resource reservations also for file transfers thus allowing to plan transfers of media assets in advance and to tightly estimate the durations.

In this document, we present the *Distributed Server Management System (DSMS)*, a management software for large scale media server networks. The system provides a mechanism for resource reservations on the network for both streaming and non-streaming traffic based on differentiated services. This allows to plan the transfers of media assets and streaming traffic in advance and to provide reliable information about the duration of file transfers. In addition to that, the DSMS allows the decentralized administration of the media servers in the network, i.e. media assets can be installed on each server independently. The DSMS keeps track of the media assets stored on the servers and 'publishes' them to the whole network.

The rest of the paper is organized as follows: after discussing related work, we provide an overview about the current application environment of the DSMS in a german company. Following that, the system architecture and the advance reservation mechanisms are presented. Finally, we give some conclusions and briefly discuss future work.

1.1   Related Work

In order to keep the requirement low for copying large media files between different servers, it is necessary to map the media assets onto the server network in a way that the need for transferring media files across the network is minimized, which means that assets of particular interest for a client are located close to that client. The methods and algorithms for such an optimization are based on the results described in [7,13]. In these papers, the mapping problem is addressed using simulated annealing heuristics. The work presented in those papers is not restricted to the business TV scenario described in this document, but also discusses techniques for TV Anytime server networks [6], i.e. systems that store broadcast media assets such as free TV programmes and allow clients to access those assets at any time. In contrast to the business TV environment, in the TV Anytime scenario (which is another field of application for the DSMS) the requirement for efficiently utilizing the available network and storage capacities of the server network is much more important, due to the enormous amount of media assets.

The foundations for using advance reservations on networks, i.e. reservations of resources before they are actually required, have been discussed in several papers. In general, those publications concentrate on bandwidth reservations for streaming, i.e. video-conferencing and video-on-demand applications. To the authors' knowledge, the requirement for establishing QoS even for the transfer of large media files has not been considered, yet.

In [5], a model for implementing advance reservations on networks is described. Several issues of the design of an advance reservation mechanisms are presented, e.g. different *channels* or partitions for immediate and advance reservations. The general requirements for the allocation not only of network resources but also computing resources in server and client systems are addressed in [12], discussing the overall scenario in detail.

Several publications discuss extensions of the RSVP [2] protocol in order to enable advance reservation mechanisms. In [10] some basic requirements for implementing advance reservations are discussed, using RSVP as an example for resource reservation protocols. However, a general model is not presented and an actual implementation is not addressed. The focus of [3] is not on a general model, but on the implementation of advance reservation mechanisms on top of existing protocols such as RSVP and ATM. In [4,11], several issues concerning the provision of advance reservation mechanisms are discussed such as admission control or extensions to the existing RSVP protocol. These papers propose an agent-based architecture for advance reservations where each routing domain in a network contains an agent responsible for admission control on behalf of the routers in the network, similar to the approach chosen in this document. However, the considerations in these papers only take streaming applications into account, the transfer of large data files is not considered.

The use of differentiated services (DiffServ, [1]) for advance reservations, as described in this document, has not been addressed so far.

In the following Sections, we present the application environment in the corporate network and its requirements, followed by the description of the DSMS and the approach to provide the required quality-of-service for streaming and transmitting media assets in the server network.

2   Application Environment

The DSMS is used to manage networks of media servers used for high-quality streaming applications, i.e. using MPEG-1 or MPEG-2 encoded streams, in the area of teleteaching, TV Anytime, or business TV applications as described in this document. In the following, the current application of the DSMS in the corporate network of Pixelpark AG, Germany, which was implemented in the framework of the project HiQoS ("High Performance Multimedia Services with Quality-of-Service Guarantees") will be described together with the requirements for QoS guarantees on networks in this environment.

Generally, the server network consists of different *islands*, each containing at least one media server and a number of clients which stream media assets from the local media server. In this environment, media servers capable of delivering MPEG-1 and MPEG-2 encoded streams are used. Within each island (in this particular application, islands correspond to branch offices of the company) the available bandwidth is usually considered to be sufficient for a small number of streaming processes. However, especially in case of MPEG-2 encoded streams with at least 4 MBit/s, resource reservations can be required even within a single island in order to guarantee the jitter-free display of the streams. These reservations can be made using the hardware support for differentiated services which is available in the corporate network[1].

The different islands which build the server network are connected with each other using Internet links which are obviously insufficient for streaming high-quality media assets in real time. Besides using physical networks, as in this scenario, the DSMS provides the opportunity to use satellite connections between the servers, which then lead to hierarchical networks in which the single islands are updated from a central server system [8].

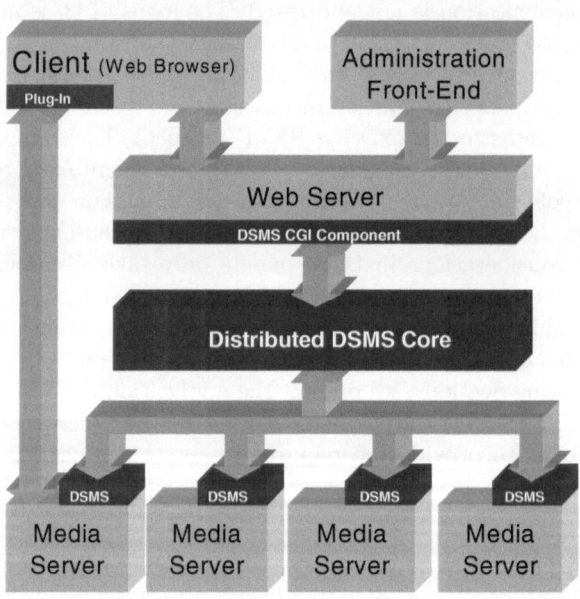

Fig. 2. The Architecture of a Media Server Network managed by the DSMS

For the clients, the access to the media streams stored on the whole server network is provided using a web application that connects to the server management system in order to locate media streams in the server network. In case the requested media asset is not available from the local server, the DSMS determines the estimated time for the transmission to the client's local server. Then, the

[1] The usage of IntServ was not supported by the network hardware in this environment

client can choose whether or not to 'order' the transmission of the media stream to the local server. In case, the requested stream is available with different qualities, the DSMS provides a list with the available streams (together with their quality, in this case bit rate and encoding format were considered to be sufficient to decide which stream meets the user's requirements) among which the client can choose. In some cases, a lower quality stream which can be provided within a short amount of time might be sufficient, whereas in other cases the stream with highest available quality is requested, e.g. for a presentation.

Obviously, missing the estimated deadline for the arrival of a media stream will have severe consequences, e.g. in case the scheduled deadline for an important presentation in which a media stream is required cannot be met. The importance of meeting the deadlines is a major motivation for using the hardware support for different service classes on the network not only for real-time streaming but also for the transmission of media files.

The media streams can be viewed at the end-systems using web browsers with plug-ins capable of decoding and displaying the media assets. In the scenario considered in this document, the communication between clients and servers is based on the RTSP and RTP protocols. The general architecture of the business TV application is depicted in Figure 2 which shows a simplified model of the components of the DSMS and how they are integrated in the application environment. Some of the components are located on each server whereas the core components run as a central service for a larger *domain*, i.e. a larger set of islands which is described more detailed in Section 3.2.

3   The DSMS

In this Section, the functionality and architecture of the DSMS is described.

3.1   Functionality

The DSMS organizes a server network in a decentralized way. Media assets can be installed and removed from each of the servers in the network. Once a new asset is installed on a single server, the DSMS 'publishes' this asset to the rest of the network thus providing access for each client. After that, the new media asset can be distributed in the server network.

Besides the opportunity of dealing with multiple copies of the same media asset introduced due to the requirement for distributing media assets, the DSMS allows a single media asset to be present on the servers of the network with different bit rates or encoding formats. In order to distinguish between the content and the physical copies, each media asset is addressed using a logical and a physical address. The logical address describes a collection of assets with the same content but different properties. The physical address uniquely defines a physical media stream located on one of the servers. The DSMS automatically creates the physical addresses during the storage process and adapts them during the migration. Each client only uses the logical address to request a media asset without having to know the physical locations of the different copies.

An important functionality used to improve the overall QoS for users of the DSMS is to optimize the placement of media assets on the network. A key factor for the perceived QoS of the system is the minimization of asset transmissions, i.e. the aim is to place assets of interest for users on nearby servers, which can be achieved using the algorithms and methods described in [7,13]. In order to minimize the requirement for dynamically remapping the assets onto the server network during run-time, the DSMS provides the functionality for clients to subscribe to *content categories*, which are stored together with each media asset. These categories describe collections of assets with similar content, e.g. news broadcasts. Users can subscribe to content categories which means that the DSMS automatically copies newly installed media assets to the local servers where the corresponding categories were subscribed. In case, a user requests assets of a category not subscribed by one of the users in this user's island, the DSMS copies these assets to the client's local server.

3.2   System Architecture

The DSMS groups a number of islands of the server network as *administrative domains* being controlled by modules which are central instances for a single domain. In addition to that, DSMS modules running on each server node are responsible for the administration of media assets and the communication between different servers (see Figure 3).

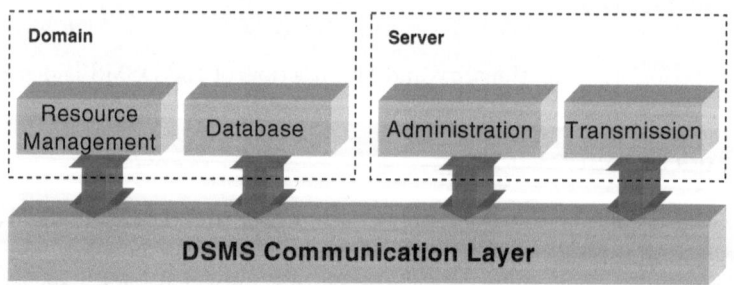

Fig. 3. Modules of the DSMS

The modules running once per domain are the resource management (RM) and a database component. The DSMS uses the RM component for the administration of the advance reservation mechanism. It is responsible for scheduling the migration and transmission of media files and allows the reservation of bandwidth for streaming media assets in advance. The RM manages the available bandwidth resources and grants or refuses access to the network for streaming traffic. The transmission of media files is also scheduled and initiated by this component. The logical and physical addresses are kept in the database component of the DSMS, also running as a central service for a single domain. In order to build a large-scale server network, the central components of different domains can be connected.

The purpose of the modules running on each server node is the administration of media assets on the respective node and the physical transmission of media assets among different servers. The transmission is scheduled and started the domain central RM module which contacts both transmission modules on the server nodes involved in order to initiate the transfer of an asset.

The open architecture of the DSMS allows to use the software not only for managing media servers but also other network services which require to transmit large amounts of data using different types of networks.

4   Advance Reservation for Media Streaming and File Transmissions

According to the DiffServ model, service classes can be used for different prioritizations of network traffic categories. In the application environment described in Section 2, the hardware support for different service classes could be used to implement an advance reservation mechanism, designed but not limited to be applied in the environment of the media server network.

Mapping the media assets onto the server network as mentioned in section 3.1 cannot completely eliminate the requirement for transmitting assets across the network. For example, this occurs each time changes of the user behavior require to remap the assets onto the server network. In case an end-user requests an asset from a remote server it is required to provide the time of availability of each copy of this particular asset to the end-user. This process can be seen similar to booking a hotel room: once a room reservation was made, the guest expects that the room is available at the time of his arrival. In the same way, the request for a media asset requires the asset to be available at the end-user's local server at a given time.

The support for advance reservations of bandwidth for streaming network traffic is an important aspect especially for video-on-demand systems. Unlike other approaches, using RSVP for the resource reservations which either lacks the capability to reserve resources in advance or requires to modify or extend the protocol, in the DSMS implementation the streaming traffic is prioritized using the DiffServ model.

The approach described in this paper is to reserve a fixed amount of the available bandwidth (corresponding to the highest priority DiffServ service class) for streaming and non-streaming media traffic. This service class is managed by the DSMS. Other network traffic, not related to the DSMS, is treated as best-effort traffic (see Figure 4). This allows to reliably estimate transmission times of media assets and therefore to meet deadlines for the transmission of media data and to provide the required QoS for the streaming traffic.

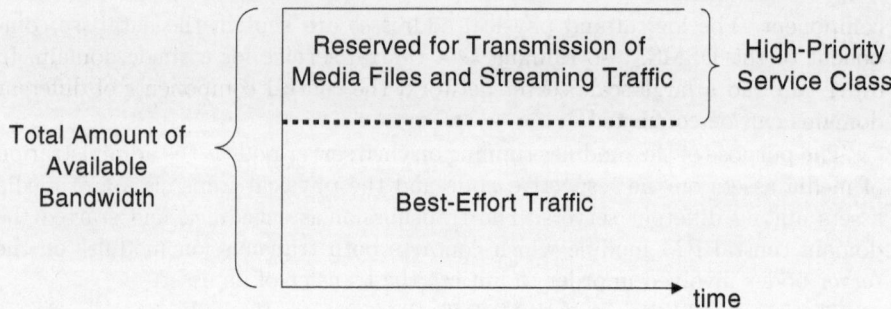

Fig. 4. Usage of a Service Class for Streaming and Non-Streaming Media Traffic

The resource management component (RM) of the DSMS, described in Section 3.2, serves as a *bandwidth broker* [9] which is responsible for allocating network resources, initiating the transmissions and keeping track of the available bandwidth in the DiffServ class reserved for media streaming and transmissions. Each instance of the resource management module schedules the communication within a single administrative domain which can contain more than one island. Currently, the scheduler is implemented according to the first-fit principle: each file transmission is scheduled as soon as possible. In order to schedule transmissions across different domains, the resource management modules can interact with each other and negotiate the actual transmission time and available bandwidth. Due to the bandwidth limitation, streaming traffic across domain borders is not considered in current scenarios but can be included in future implementations.

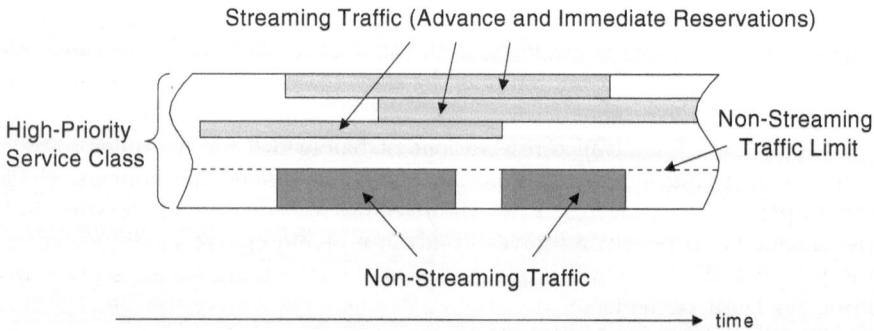

Fig. 5. The Different Traffic Types in the Highest Priority Service Class

The two types of network traffic (streaming and non-streaming) are handled as follows: advance reservations of bandwidth for streaming traffic can be made using a corresponding interface of the RM component. Such a reservation requires to specify the beginning and the end of the streaming session. In case the duration is exceeded, the traffic will no longer benefit from the prioritization. Before actually streaming media assets, clients can make use of the advance reservation mechanism, however this is not required. In case a user wishes to see a media stream from the local server and no advance reservations were made, the RM will grant or refuse access to this asset depending on the current network load in the highest service class thus avoiding conflicts between different transmissions.

In the scenario described in Section 2, it is sufficient to schedule the file transfers not initiated by client requests (i.e. transfers required due to the remapping of media assets or results of subscriptions) during night time when the traffic in the company's network is low. When client requests require the transmissions of media files at day time, only a small amount of the available bandwidth in the high priority service class is used for those transmissions in order to assure sufficient network resources for streaming traffic. The amount of bandwidth used for file transfers can be statically configured and adapted to the requirements of the application (see Figure 5).

In the current implementation, due to administrative reasons both streaming and transmission processes share the same service class. Since the transmission of media files does not pose severe constraints such as latency to the underlying network, a lower priority service class could be used for this type of traffic. However, using best-effort service for those transmissions is not sufficient since the application (i.e. the users) requires the estimated deadlines to be met and these estimations must be tightly calculated in order to achieve an efficient network utilization.

Such a functionality, based upon the hardware support of the DiffServ-enabled corporate network has the advantage of an easy administration and in contrast to other efforts for implementing advance reservations on networks does not require changes to existing protocols or protocol stacks.

The access to the highest service class is restricted using the router's built-in traffic filtering mechanisms. This assures that only traffic from one of the media servers benefits from the prioritization mechanism.

5   Conclusion and Future Work

In this document, a management system for large-scale media server networks is presented. An important feature of the management software is the provision of QoS for both streaming and non-streaming traffic, i.e. transmissions of large media assets.

The DSMS allows to automate the distribution of the media streams within the server network thus keeping the administrative effort low. Media assets can be stored on each of the servers in the network and the distribution of assets is handled by the management system. It is not required to manually transmit media assets since they are automatically copied only to the servers where they are required. This functionality is supported by subscriptions to content categories and in case an asset is not available at a client's local server, the management software allows a user to choose among all the physical copies of this particular asset and to select the stream with most appropriate properties, such as the bit rate or the time required for the transmission to the local server.

In order to efficiently use the available network resources and to provide sufficient bandwidth for streaming and non-streaming traffic within the server network, DiffServ enabled hardware can be used. This allows to schedule the non-streaming network traffic in a way, that provides an efficient network utilization while meeting the estimated deadlines. In addition to that, immediate and advance reservations for streaming traffic can be made, providing the required QoS for the video-on-demand service. In contrast to existing approaches for establishing advance reservation mechanisms, our solution can be easily applied without changes to existing protocols, soft- or hardware.

In future versions of the DSMS, more sophisticated scheduling algorithms for the file transfers will be implemented, e.g. media files could be split and transferred in several pieces in order to more efficiently utilize the available network resources. Moreover, the usage of alternative routes could be implemented in order to avoid bottlenecks on certain network links.

Currently, the DSMS is used in a business TV application. Other applications of the DSMS in the future are TV Anytime systems (i.e. media server networks that work like digital VCRs, providing time-independent on-demand access to TV programmes) where the aspects of mapping media assets onto the server network, in particular the decision which programmes to record, plays a more important role due to the enormous amount of media assets that can theoretically be recorded.

Although specially designed to support a media server network, the generalization of the architecture described in this document seems to be a promising approach to implement advance reservations for any type of network traffic. In such an environment, aspects such as billing and brokerage which were not in the focus of this particular implementation will be introduced to the system. Among the features to examine and to implement will be a dynamic pricing algorithm that allows to adjust bandwidth prices to the demand and vice-versa, i.e. transmission of media files will only be carried out if the gain of transferring an asset is higher than the price for the required bandwidth.

References

1. S. Blake, D. Black, M. Carlson, E. Davies, Z. Wang, and W. Weiss. An Architecture for Differentiated Services. http://www.ietf.org/rfc/rfc2475.txt, 1998. RFC 2475.
2. R. Braden. Resource ReSerVation Protocol - Functional Specification. http://www.ietf.org/rfc/rfc2205.txt, 1997. RFC 2205.
3. F. Breiter, S. Kühn, E. Robles, and A. Schill. The Usage of Advance Reservation Mechanisms in Distributed Multimedia Applications. *Computer Networks and ISDN Systems*, 30(16-18):1627-1635, September 1998.

4. M. Degermark, T. Koehler, S. Pink, and O. Schelen. Advance reservations for predicted service. *Lecture Notes in Computer Science*, 1018:3-??, 1995.
5. D. Ferrari, A. Gupta, and G. Ventre. Distributed Advance Reservation of Real-Time Connections. *Lecture Notes in Computer Science*, 1018, 1995.
6. R. Lüling. Hierarchical Video-on-Demand Servers for TV-Anytime Services. In *Proceedings of the International Conference on Computer Communications and Networks*. IEEE, 1999.
7. R. Lüling. Static and Dynamic Mapping of Media Assets on a Network of Distributed Multimedia Information Servers. In *Proceedings of the International Conference on Distributed Computing Systems*, 1999.
8. K. Morisse, F. Cortes, and R. Lüling. Broadband Multimedia Information Service for European Parliaments. In *Proceedings of the IEEE International Conference on Multimedia Computing and Systems Volume II*, 1999.
9. K. Nicols, V. Jacobson, and L. Zhang. A Two-bit Differentiated Services Architecture fot the Internet. http://www.ietf.org/rfc/rfc2638.txt, 1999. RFC 2638.
10. W. Reinhardt. Advance Resource Reservation and its Impact on Reservation Protocols. In *Proceedings of Broadband Island '95, Dublin, Ireland*, 1995.
11. O. Schelen and S. Pink. An agent-based architecture for advance reservations. In IEEE Computer Society. Technical Committee on Computer Communications, editor, *Proceedings, 22nd annual Conference on Local Computer Networks: LCN '97: November 2-5, 1997, Minneapolis, Minnesota*, volume 22, pages 451-459. IEEE Computer Society Press, 1997.
12. L. Wolf, L. Delgrossi, R. Steinmetz, S. Schaller, and H. Wittig. Issues of Reserving Resources in Advance. In *Proceedings of the Fifth International Workshop on Network and Operating Systems Support for Digital Audio and Video*, 1995.
13. Z. Xiaobo, R. Lüling, and X. Li. Heuristic Solutions for a Mapping Problem in a TV-Anytime Server Network. In *IPDPS Workshop on Parallel and Distributed Computing in Image Processing, Video Processing, and Multimedia (PDIVM2000)*, 2000. To appear.

Augmented Reliable Multicast CORBA Event Service (ARMS): A QoS-Adaptive Middleware

João Orvalho[1] and Fernando Boavida[2]

Communications and Telematic Group
CISUC – Centre for Informatics and Systems of the University of Coimbra
Polo II, 3030 COIMBRA – PORTUGAL
Tel.: +351-39-790000, Fax: +351-39-701266
E-mail: {orvalho, boavida@dei.uc.pt}

Abstract. The CORBA Event Service specification lacks important features in terms of quality of service (QoS) characteristics required by multimedia information. The main objective of the work described in this paper is to augment the standard CORBA Event Service specification with a set of extensions, to turn it into an adaptable QoS middleware multimedia framework. To meet this, some extensions to the CORBA Event Service already developed with the aim of providing multicasting and reliability features have been enhanced in order to allow the close interaction with multicast transport protocols and with QoS monitoring mechanisms. The result was a QoS-aware middleware platform that actively adapts the quality of service required by the applications to the one that is provided by the underlying communication channel. The main quality of service features addressed by the platform – and discussed in the paper – are the support of sessions with different reliability levels, the provision of congestion control mechanisms and the capability to suppress jitter.

1. Introduction

A continuous distributed interactive medium is a medium that can change its state in response to user operations as well to the passage of time [1]. A broad variety of applications use this kind of media, such as multi-user virtual reality (VR), distributed simulations, networked games and computer-supported co-operative work (CSCW) applications. Systems dealing with multimedia events combine aspects of distributed real-time computing with the need for low latency, reduced jitter, high throughput, multi-sender/multi-receiver communications over wide area systems and different levels of reliability [2]. These event-driven systems also require efficient and scalable communications components.

The work being described in this paper consists of a middleware platform for continuous distributed interactive media applications based on some extensions to the CORBA Event Service. The platform - named Augmented Reliable Multicast

---

[1] College of Education, Polytechnic Institute of Coimbra
[2] Informatics Engineering Department of the University of Coimbra

H. Scholten and M. van Sinderen (Eds.): IDMS 2000, LNCS 1905, pp. 144-157, 2000.
© Springer-Verlag Berlin Heidelberg 2000

CORBA Event Service (ARMS) – provides an end-to-end communication framework with QoS-matching capabilities, having in mind VR application requirements. VR applications have specification features concerning interactivity, reliability, continuity, coherence and strict time constraints that render difficult their traffic characterisation in terms of quality of service (QoS) requirements.

ARMS is focused on VR environments based on VRML models, which deal with two different types of information: events and states. Events are time critical and described by small amounts of information, as opposed to states that are generally non-time-critical and require large amounts of information for their description. Therefore, there is the need for different levels of reliability when exchanging event and state information: minimal reliability (with loss detection) for events, and full reliability for states. In addition, ARMS assumes that there is a common time reference, which requires the clocks of all participants to be synchronised by means of NTP or GPS clocks.

In continuous distributed interactive media there are operations that need to be executed at a specific point in time and in a correct order for consistency to be achieved. [1] investigates this problem and proposes a solution by deliberately increasing the response time in order to decrease the number of short-term inconsistencies, leading to the concept of local lag. Instead of immediately executing an operation issued by a local user, the operation is delayed for a certain amount of time before it is executed. The determination of this value is a typical issue of application adaptability.

VR applications are associated with three important functions: scalability, interaction and consistency [2]. The QoS characteristics that influence the previous functions are reliability, losses and delay jitter. Additionally, those functions are influenced by application adaptability factors like frequency of events, synchronisation delay, number of participants, consistency and playout time (display frequency) [2].

The ARMS platform addresses the above-mentioned QoS issues and explores the adaptation between the quality of service required by applications and the one that is provided by the underlying communication system. This paper presents the main approaches taken by ARMS in order to provide QoS adaptability. Section 2 provides a general description of the ARMS architecture. As the work presented in this paper corresponds to an enhancement of a previous, non-QoS-adaptive platform, section 3 presents the characteristics of this previous platform. Section 4 provides details concerning the approach taken by ARMS in terms of reliability, congestion control and jitter. Section 5 describes additional features of the ARMS platform, namely the IIOP/IP multicasting gateway service and the federation of event channels. Section 6 identifies related work. The conclusions and guidelines for further work are presented in section 7.

2. ARMS General Characteristics

Portability over heterogeneous environments is a typical requirement of distributed multimedia systems. Although this portability could be provided by middleware such as CORBA, there exists a widespread belief in the virtual reality community that the quality of service offered by CORBA is not suitable for next-generation large distributed interactive media [3]. Also, another obvious deficiency of CORBA is its lack of support for interaction styles other than request/reply [4]. However, due to their nature, the CORBA Event Service [5] and the CORBA Notification Service [6] have some potential be used for continuous distributed interactive media applications.

Previous research has been focused on limitations of the CORBA Event Service, namely multicasting, reliability and bulk data handling [7,8]. The work presented in this paper extends it to support adaptive QoS middleware functionalities.

ARMS offers a set of QoS-related mechanisms for reliability guarantee, congestion control and jitter control. The QoS management process is supported by object-based monitoring and adaptation functions (Figure 1). Monitoring is the process of observing the utilisation of resources and/or QoS characteristics in the system. ARMS has specific objects for loss and jitter monitoring.

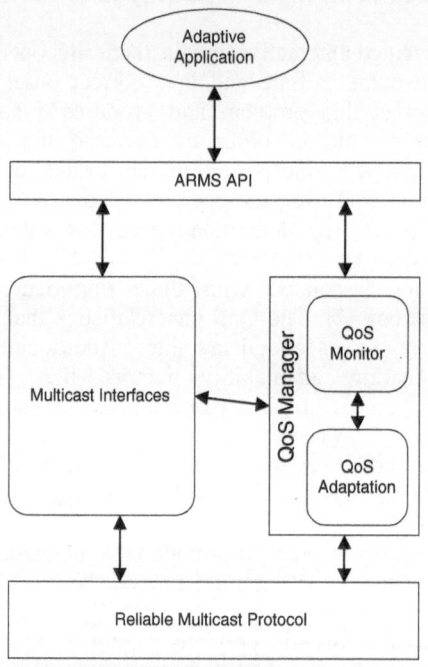

Fig. 1. ARMS architecture

Adaptation mechanisms generally rely on resource control, reconfiguration or change of service [9]. ARMS uses a resource control paradigm, providing adaptation mechanisms for network congestion, losses and jitter. Placing adaptation capabilities in the middleware gives applications the ability to concentrate on specific functionalities, to enforce different adaptation policies and to interact with other components in the system in order to ensure fairness and other global properties [10].

3. Synopsis of ARMS Version 1

The first generation of ARMS offered a set of extensions to the CORBA Event Service specification [7], namely mechanisms for IP multicast communication, reliability and fragmenting/reassembling of large events (Figure 2).

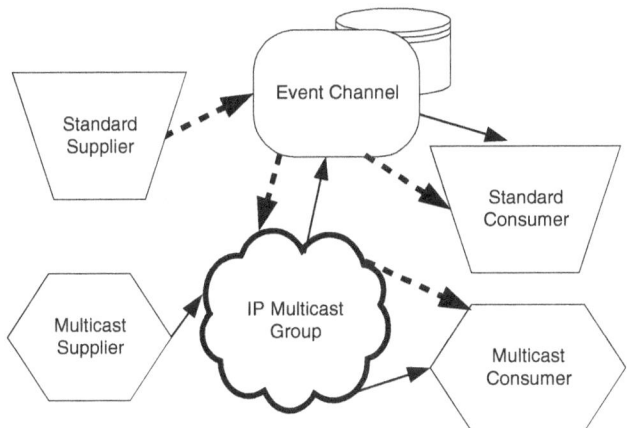

Fig. 2. ARMSv1 high-level view

This service is a pure Java implementation (JDK 1.2 compatible) [11], and so it benefits from all the strengths of Java. These include portability, security, and robustness. It is, also, ORB-agnostic. It is written to the standard IDL-to-Java mapping, so it should work with any Java ORB that supports the standard mapping. As in the original specifications made by OMG CORBA Event Service, suppliers and consumers are decoupled, that is, they don't know each other's identities; thus, the standard Event Service may still be provided by this extended service.

The approach on which the developed work was based is the one called *Push* model. The objective is to allow the consumer to be sent the event data as soon as it is produced. The canonical Push Model allows Suppliers of events to initiate the transfer of event data to Consumers.

The reliable multicast extension can be seen as an alternative way to get the event across to the consumer. This assumption forces the new service to keep all the standard interfaces with the same functionality (the same methods) that are defined by OMG, allowing a choice to be made by the supplier/consumer between IIOP and Reliable IP Multicast. The service implements two kinds if IP multicast interfaces: IP Multicast-Any and IP Multicast-Streams. The IP Multicast-Any deals with *Any* Values while IP Multicast-Streams deals directly with byte-stream values, avoiding the overhead caused by marshalling and de-marshalling of the proprietary *Any*. The reliable multicast solution is based on the Light-weight Reliable Multicast Protocol (LRMP) [12], which deals with IP Multicasting and provides the necessary reliability and better scalability (Figure 3).

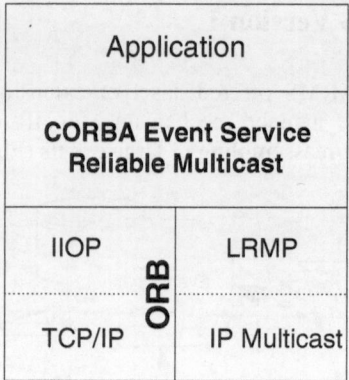

Fig. 3. ARMSv1 architecture

4. ARMS Version 2

The quality of service provided to multimedia sessions is determined, in general, by packet losses and delay. The second generation of ARMS acts as a transport level with QoS monitors that provide information on key aspects of QoS provision such as losses and jitter. Adaptation mechanisms for network congestion (congestion control), losses (reliability sessions) and jitter (jitter filter) have an important influence on the performance of virtual environment applications.

4.1 Reliability Sessions

ARMSv2 supports several reliability levels: Reliable, Reliable-Limited Loss, Loss Allowed, Unreliable with loss notification and Unreliable.

The first is a typical strong reliability session. Reliable-Limited Loss is a session where limited losses are permitted for some types of packets and loss notifications will be triggered when this happens. It provides guarantees for congestion control at sender side and sequence control at the receiver side. Loss Allowed is a session where losses are allowed and accepted for all types of packets but provides loss notification, congestion control at sender side and sequence control at the receiver side. Unreliable with loss notification is an unreliable session, where data are not subject to congestion control at the sender side. However, the ARMS upper level maintains a queue for sequence number control, which allows loss notification. Lastly, Unreliable is a purely unreliable session. Table 1 summarises the characteristics of the various types of sessions offered by ARMSv2.

Table 1. Types of ARMSv2 sessions

	Permitted Losses	Loss Notification	Congestion Control	Sequence Control
Reliable	No	Yes	Yes	Yes
Reliable-Limited Loss	Yes, for some data types	Yes	Yes	Yes
Loss Allowed	Yes	Yes	Yes	Yes
Unreliable with loss notification	Yes	Yes	No	Yes
Unreliable	Yes	No	No	No

4.2 Congestion Control

Currently, different approaches are being discussed to solve QoS and congestion control problems [13]. Complex distributed applications residing in heterogeneous end-to-end computing environments must be flexible and adapt to QoS variations in their end-to-end execution [10]. That is, applications must adapt the bandwidth share they are using to the network congestion state [13]. Usually, there are two distinct levels at which adaptation may take place – the system level (e.g. operating systems and network protocols) and the application level – with different objectives. In order to balance the objectives of these approaches, the ARMS middleware closely interacts both with application needs and with multicast protocols, monitoring network parameters and operating systems resources.

In terms of network QoS control mechanisms, ARMS directly monitors the reliable multicast communication protocol, LRMP, adapting the sender transmission rate to the network congestion state. The adaptation is based on information carried by NACK packets and on local congestion information [8] as the sender window size [12]. Based on congestion information gathered from lower communication objects, ARMS and applications adjust the upper-level sending rate.

QoS mechanisms that are based on adapting the sender transmission rate to the network congestion state don't work well in large multicast groups and heterogeneous environments, because poor performance receivers would impose a low transmission quality. To avoid this, several proposals have been made for hierarchical data distribution [14,15,16]. Nevertheless, in virtual reality environments, data layering approaches are not appropriated for most data, especially for time critical data such as VRML events. Delay is the most important QoS factor for this type of data. Layered data mechanisms solve heterogeneity problems but cause additional delay at the receivers [13]. Thus, to avoid this, ARMS adapts the minimum rate to ensure the fairness of the adaptive congestion control. Receivers should leave the session when the loss rate is very high and the data rate in not reduced by the sender. Nevertheless, layered multicast [14] can be useful and will be explored in subsequent stages of the work.

4.3 Jitter

Variance in end-to-end delay is called delay jitter or, simply, jitter. Critical information such as the case of audio, video and continuous distributed interactive media (distributed simulations, multi-user virtual reality) should be played back continuously, which means that there must be some form of jitter compensation.

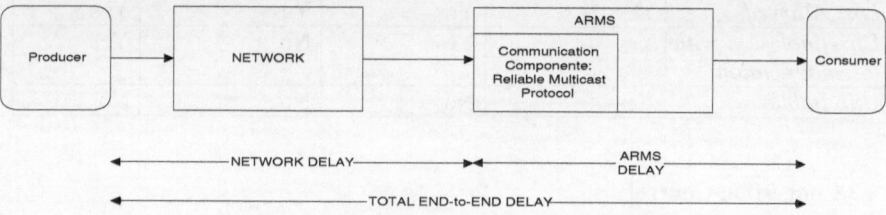

Fig. 4. Components of end-to-end delay

In addition to components for error control, ARMS provides mechanisms for monitoring and controlling jitter (Figures 4 and 5), so that the original temporal relationships can be recovered. ARMS lower level reliable multicast protocol queue sends ordered packets (reliable data) to an upper level Jitter Filter Queue. This Jitter Filter Queue is used to absorb delay variations exhibited by arriving packets. So, this compensation is done by introducing an additional and variable delay, followed by the delivering of packets to the application level, as opposed to a compensation made at the application level [17].

ARMS allows applications to contract this jitter compensation [the threshold synchronisation] for certain types of data, namely unreliable data (pure or with loss notification) and reliable data with limited loss. Packets that arrive after a given threshold are considered to be too late. These can simply be dropped or, alternatively, be marked as late packets and passed to the application. Applications can ignore these packets or can react by requesting that a special NACK be sent by ARMS.

Jitter Algorithms

ARMS implements the Filter Jitter with two different algorithms. The choice of which algorithm to use is made at configuration time. The first algorithm is described in the following paragraphs.

To recover the original timing properties, the Jitter Filter buffers (Figure 5) the packets at the sink until time $T + D$, where T is a source timestamp and D is the bounded maximum end-to-end delay. When networks are unable to guarantee a maximum end-to-end delay bound, the receiver continuously updates an estimate of the maximum delay in order to calculate the buffering time. One of algorithms that can be used to calculate D is based on the RTP specification [18] to estimate the statistical variance of RTP data packet inter-arrival time, measured in timestamp units and expressed as unsigned integer. At a given instant, D is the maximum of all jitter values calculated up to that instant according to the formula:

$$J_i = J_{i-1} + (|Dif(i-1, i)| - J_{i-1})/16$$

where the inter-arrival jitter J is defined to be the mean deviation of the difference Dif in packet spacing at the receiver compared to the sender for a pair of packets [18]. This is equivalent to the difference in the "relative transit time" for the two packets. So, Dif may be calculated as

$$\text{Dif}(i,j) = (R_j - R_i) - (S_j - S_i) = (R_j - S_j) - (R_i - S_i)$$

S_i is the reliable multicast protocol timestamp for packet i and R_i is the time of arrival in reliable multicast protocol timestamp units for packet i, the same for packet j. Jitter is calculated continuously as each packet is received, for each source. Factor 1/16 was chosen to reduce measurement noise while converging reasonably quickly [18,19]. The code below implements the algorithm, where the estimated jitter can be kept as an integer.

```
protected void updateJitter(int timestamp) {
        int elapsed = NTP.ntp32(lastTimeForData - timestamp);
        elapsed = NTP.fixedPoint32ToMillis(elapsed);
        int d;
        if (transit != 0)
          d = elapsed - transit;
        else
          d = 0;
        transit = elapsed;
        if (d < 0)
          d = -d;
        jitter += d - ((jitter + 8) >> 4);
    }
}
```

The second jitter algorithm is based on the work of [20], where statistical analysis of per-packet delay is used to estimate the maximum delay:

$$D = d + r * s$$

where d is the average delay, s is the standard deviation and r is a filter coefficient [20]. The algorithm continuously estimates the average delay and standard deviation, and is based on the 'low pass filtering algorithm' used in TCP for the estimation of the acknowledgement delay time [20]. So, for each packet, the transmission delay, t, is calculated as the difference between the reception time and the emission timestamp. The average delay and standard deviation are the calculated as $d = d_{old} + a(t - d)$ and $s = s_{old} + b(|t - d| - s)$, respectively. The constants a and b ($a,b < 1$) are smoothing coefficients, with the typical values 1/8 and 1/16, respectively.

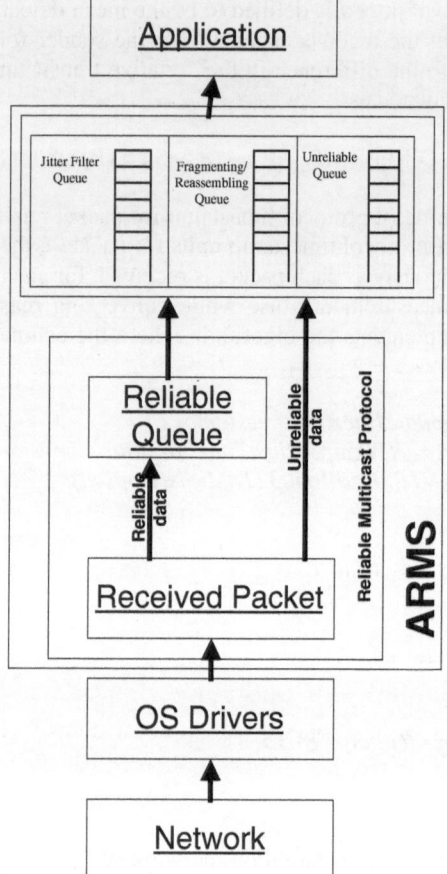

Fig. 5. Diagram of Data Flow

5. Additional Features of the ARMS Platform

In addition to providing QoS adaptation, the ARMS platform has some optimisation features that contribute to its transparency and scalability. With regard to transparency, the platform provides a gateway service between IIOP and IP Multicasting. Scalability is supported by the federation of Event Channels.

5.1 Event Channel IIOP/IP Multicasting Gateway Service

ARMS guarantees interoperability between event suppliers and consumers in a way that is independent of the used communication facilities – standard IIOP or IP multicast. To achieve this, the Event Channel provides a transparent IIOP gateway. As can easily be understood, the interfaces that are provided by OMG must be maintained by the proposed service in order to deal with the *Any* type of data. In this way, interoperability between the proposed service and most of the commercially available ORBs can be achieved. In order to deal with all the values supported by the *Any* type, a new object – Any – was created as an extension of the OMG class, to override the ORBs implementation.

The values to be passed to this service, in order to be sent via multicast, shall be created with the proprietary interfaces of the service, that provide a set of new methods to deal with the resulting buffers and fill the LRMP data packet with the data contained in those buffers. So, by creating a method for extracting the sequence of bytes generated by the *Any* marshalling operation, the result is a stream of bytes that mean nothing to LRMP. Once they get across to the consumer LRMP object, it reassembles them to form an *Any*, and delivers them to the consumer as a valid *Any* that shall be extracted by the application interfaces.

To provide the interchange of Multicast events and Standard events, two scenarios are considered:

- when the supplier is a *MulticastPushSupplier* and the consumer is a normal Consumer;
- when the supplier is a normal Supplier and consumer is a *MulticastPushConsumer*.

For the first scenario, and knowing that the model to be dealt with is the Push Model, the event is sent via LRMP, i.e. is sent to a well known multicast group, and is pushed to Multicast Consumers by one LRMP object that receives events directly. As the Event Channel holds one *ProxyMulticastPushConsumer* for all Suppliers, this proxy acts as a normal consumer to which events are pushed. Each time this proxy is pushed a new event, it invokes the receive method on the Event Channel. This invokes a receive method on the *ConsumerAdmin*, that holds references to all *ProxyPushSuppliers* (standard) and shall invoke the receive method on all those proxies, that will then invoke the push method on the consumer they are attached to. The *ConsumerAdmin* does nothing on the *ProxyMulticastPushSupplier*, or else events would be sent twice to the multicast group.

For the second scenario, the supplier calls the push method on its *ProxyPushConsumer*, that will call receive on the Event Channel, to get the event to standard consumers, and will also ask for any existent *ProxyMulticastPushSuppliers*. If there are any, the proxy shall call the receive method on that proxy. In this way, the event is sent to the multicast group, and all *MulticastPushConsumers* receive it. So, the Event Channel is responsible for doing all the necessary IIOP gateway work.

In either scenario, the gateway operation implies changes to the Any Object that is being forwarded. The Event Channel also takes care of this detail, at the cost of an additional overhead caused by the de-marshalling of the proprietary Any and subsequent marshalling as an ORB *Any*, and vice-versa.

5.2 Federation of Event Channels

Distributed transparency of the Event Channel can lead to a less effective configuration. There are scenarios where consumers and suppliers reside in the same process, host or network and the Event Channel is remote. In these cases, there is a waste of network resources and unnecessary increase of latency. ARMS Event

Channel object uses a configuration facility to federate Event Channels, allowing an Event Channel to be a consumer of another, remote Event Channel. Federation of Event Channels leads to the conservation of bandwidth because only a single event will be sent to all the remote users. Additionally, the average latency is reduced because part of the traffic becomes local. Figure 6 illustrates the use of the federation of event channels when communication is made via standard IIOP. In Figure 7 federation of event channels is used in conjunction with the IIOP/IP Multicast Gateway Service.

The combination of IIOP/IP Multicasting Gateway Service with Event Channel federation contributes to the enhancement of the QoS characteristics of the platform, namely in terms of transparency, latency and scalability.

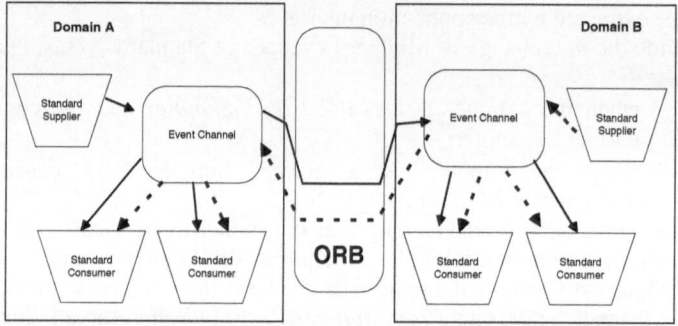

Fig. 6. Federated Event Channel configuration: standard IIOP

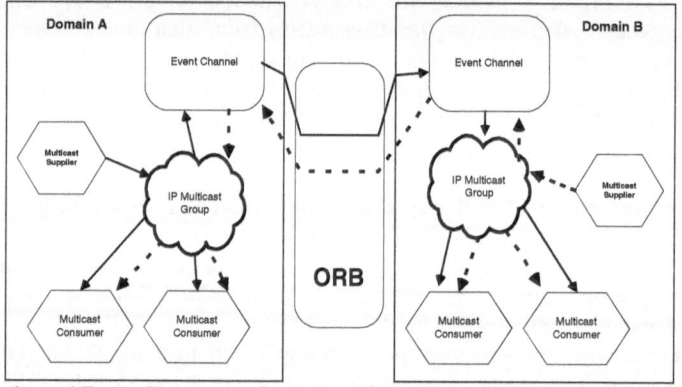

Fig. 7. Federated Event Channel configuration with IIOP/IP Multicasting Gateway Service

6. Related Work

Several approaches exist that try to explore QoS issues in distributed object environments.

OrbixTalk [21,22], which is a commercial IONA implementation of the CORBA *Event Service* specification normalised by OMG [5] written in C++, uses IP multicasting and a reliability mechanism based on *negative acknowledgements*, in order to provide delivery confirmation of each information object. Because the CORBA *Event* Service specification does not address issues for real-time applications, the QoS behaviour is not acceptable for many application domains.

OMG Notification Service [6] is a superset of the CORBA *Event Service* specification. It adds some interfaces, which deal with filtering, security, federation and QoS. However, this specification does not address implementation issues.

TAO's Real Time Events Service [3] it's a powerful implementation of CORBA *Event Service and Notification Service* specifications in C++. Nevertheless, it lacks some QoS characterisitcs, such as reliability. However, OMG CORBA Messaging specification [23] defines several levels of reliability for one-way calls, which are being added to the TAO features to complement this service. This group has been developing considerable work on real-time extensions to CORBA, which enable end-to-end QoS specification and enforcement.

Other CORBA projects, such as QuO [24], implement models for distributed object application, defining, controlling, monitoring and adapting to changes in QoS parameters. The QuO project proposes various extensions to standard CORBA components and services, in order to support adaptation, delegation and renegotiation services to shield QoS variations. The development has a great focus on remote method invocation to remote objects.

ARMS is complementary to the above-mentioned approaches, since it is based on the integration of CORBA middleware and underlying services – such as multicasting – while the referred approaches are concentrated on the CORBA object model.

7. Conclusions and Guidelines for Further Work

Quality of service requirements of continuous distributed interactive media applications encompasses several aspects that are not readily available in standard distributed object platforms. The need for different levels of reliability, congestion control mechanisms and jitter suppression strategies is apparent in applications such as virtual reality, CSCW and distributed interactive simulations.

Middleware platforms can play an important role in quality of service provision, offering flexible mechanisms that adapt the quality of service provided by the underlying communication channel to the quality of service required by applications, according to an established service contract.

This paper presented the main implementation options of the ARMS middleware platform, which builds on a set of extensions to the CORBA Event Service, providing native multicast communication, various reliability levels, congestion control and jitter suppression, with the aim to achieve QoS adaptability. The platform has been implemented at the Laboratory of Communications and Telematics of CISUC, where it is operational and currently being subject to extensive evaluation.

The first set of tests made to the platform was aimed at the verification of the platform operational status. The basic mechanisms for reliability, congestion control and jitter proved to be operational. The following tests will try to quantify the usefulness and effectiveness of these mechanisms. These will provide valuable information concerning the adequacy of placing QoS adaptation mechanisms in the middleware, as opposed to strategies that place them at application level.

Additional lines of research will explore the use of Forward Error Correction (FEC) mechanisms in order to provide some degree of reliability to time critical information, and the use of event filtering based on IP multicast groups in order to improve scalability. This latter functionality will be developed in a centralised service, which will map different event types to different IP multicast groups.

In heterogeneous environments, layered multicast can be useful. This will also be explored in subsequent stages of the ARMS platform, namely by dynamically assigning receivers to multicast groups with different QoS levels, according to the required quality of service.

Acknowledgement

This work was partially financed by the Portuguese Foundation for Science and Technology, FCT.

References

[1]   "Consistency in Continuous Distributed Interactive media", Martin Mauve, Technical Report TR-9-99, Reihe Informatik, Department for Mathematics and Computer Science, University of Mannheim, November 1999.

[2]   "Quality of Service Management in Distributed Interactive Virtual Environment", Dimitrios Makrakis, Abdelhakim Hafid, Farid Nait-Abdesselem, Anastasios Kasiolas, Lijia Qin, Progress Report of DIVE project.
http://www.mcrlab.uottawa.ca/research/QoS_DIVE_Report.html

[3]   "Applying a Scalable CORBA Event Service to Large-scale Distributed Interactive Simulations", Carlos O'Ryan, David L. Levine, Douglas C. Schmidt, J. Russel Noseworthy, Proceedings of the "5Th Workshop on Object-oriented Real-time Dependable Systems", Montery, CA, November 1999.

[4]   "Supporting Multimedia in Distributed Object Environments", D. G. Waddington and D. Hutchinson, *Internal report number MPG-99-17,* 1999.

[5]   "CORBA services: Common Object Services Specification", Object Management Group-OMG, Document formal/97-07-04, ed., July 1997.

[6]   "Notification Service Specification", Object Management Group-OMG, Document telecom/99-07-01, ed., July 1999.

[7]   "A Platform for the Study of Reliable Multicasting Extensions to CORBA Event Service", João Orvalho, Luis Figuiredo, T. Andrade and Fernando Boavida, Proceedings of IDMS'99, Toulouse, France, October 12-15, 1999.

[8]   "Evaluating Light-weight Reliable Multicast Protocol Extensions to the CORBA Event Service", João Orvalho, Luís Figueiredo and Fernando Boavida, , in the Proceedings of EDOC'99 – 3rd International Enterprise Distributed Object Computing Conference, University of Mannheim, Germany, September 27-30, 1999.

[9]   "A General Model for QoS Adaptation", D. G. Waddington and D. Hutchinson, Proceedings of Sixth International Workshop on Quality of Service (IWQoS'98), pag.275-277, Napa, California, USA, May 18-20 1998.

[10] "A Control-Based Middleware Framework for Quality-of-Service Adaptations", Baochum Li and Klara Nahrstedt, IEEE Journal On Selected Areas In Communications, Vol. 17, N° 9, September 1999.

[11] JavaSoft, www.javasoft.com.

[12] Tie Liao: "Light-weight Reliable Multicast Protocol Specification", Internet-Draft: draft-liao-lrmp-00.txt, 13 October 1998.

[13] "The Loss-Delay Based Adjustment Algorithm: A TCP-Friendly Adaptation Scheme", Dorgham Sisalem and Henning Schulzrinne, In Proceedings of Network and Operating Systems Support for Digital Audio and Video (NOSSDAV'98), Cambridge, UK, 1998.

[14] "TCP-like Congestion Control for Layered Multicast Data Transfer", L. Vicisano, L. Rizzo, and J. Crowcroft, In Proceedings of INFOCOM'98, San Francisco, USA, March 1998.

[15] "Receiver-driven layered multicast", S. McCanne, V. Jacobson, and M. Vertteli, In Proceedings of SIGCOMM Symposium on Communication Architecture and Implementation of High Performance Communication Systems (HPCS'97), Chalkidiki, Greece, June 1997.

[16] "Thin Streams: An Architecture for multicasting layered video", L. Wu, R. Sharma, and B. Smith, In Proceedings of Network and Operating Systems Support for Digital Audio and Video (NOSSDAV'97), St. Louis, USA, May 1997.

[17] "Delay and Synchronization Control Middleware to Support Real-Time Multimedia Services over Wireless PCS Networks", H. Liu and M. El Zarki, IEEE Journal On Selected Areas In Communications, Vol. 17, N° 9, September 1999.

[18] "RTP: A transport Protocol for Real-Time Applications", revision of RFC 1889, Internet-Draft (draft-ietf-avt-rtp-new-04.ps), Schulzrinne, Casner, Frederic, Jacobson, June 25 1999.

[19] "Speech Transport for Packet Telephony and Voice over IP", Maurice R. Baker, In Proceedings of SPIE Conference on Internet II: Quality of Service and Future Directions", Boston, Massachusetts, September 1999.

[20] "QoS Adaptive Transports: Delivering Scalable Media to the Desk Top", Andrew T. Campbell and Geoff Coulson, *IEEE Network*, Vol. 11 No. 2., pg. 18-27, March/April 1997.

[21] OrbixTalk da Iona, www.iona.com/

[22] "Reliable CORBA Event Channels"X. Défago, P. Felber, R. Guerraoui, EPFL, Computer Science Department, Technical Report 97/229, May 1997.

[23] "CORBA Messaging Specification", Object Management Group-OMG, Document orbos/98-05-05, ed., May 1998.

[24] "QuO's Runtime Support for Quality of Service in Distributed Objects", Vanegas R, Zinky JA, Loyall JP, Karr DA, Schantz RE, Bakken DE.. Proceedings of the IFIP International Conference on Distributed Systems Platforms and Open Distributed Processing (Middleware'98), The Lake District, England, 15-18 September 1998.

Middleware Support for Media Streaming Establishment Driven by User-Oriented QoS Requirements

Cristian Hesselman[1], Ing Widya[2], Aart van Halteren[3], and Bart Nieuwenhuis[2]

[1] Telematica Instituut, Enschede, The Netherlands
[2] CTIT, University of Twente, Enschede, The Netherlands
[3] KPN Research, Enschede, The Netherlands
hesselman@telin.nl, widya@cs.utwente.nl,
A.T.vanHalteren@kpn.com, L.J.M.Nieuwenhuis@cs.utwente.nl

Abstract. The requirements for the QoS of distributed applications are traditionally expressed in terms of network oriented or systems oriented parameters. In general, the users of these services are not interested or capable of specifying the QoS of their services in such technical terms. In this paper, we propose modeling and engineering concepts for the mapping of end user QoS onto system and network QoS. We introduce QoS agents in structured object middleware that relate end-user QoS specifications to multimedia stream bindings. In fact, the middleware layer supports QoS classes, i.e., a set of QoS characteristics. The end user QoS requirements, generally a set of non-orthogonal specifications, must be supported using the available middleware QoS classes. We also describe the experimental environment that will be used to refine the QoS mapping mechanisms.

1. Introduction

Object middleware, such as CORBA, DCOM and Java RMI, is gaining rapid acceptance as a means to quickly and cost effectively develop a wide range of applications for various areas of industry. The main purpose of these middleware systems is to provide a software infrastructure for interacting application components.

Interactions between software components can be divided into two categories: discrete or continuous. Discrete interactions are typically RPC or message-oriented and are generally used to invoke a computational service. Continuous interactions are generally used for exchanging (multi-) media data and are usually called streams. Middleware platforms often model a stream as a binding object [Gay, Blair]. In this paper we focus on the support of middleware platforms to establish a user-oriented QoS for such binding objects.

Requirements for QoS are traditionally expressed in terms of system-oriented or network-oriented parameters. In this paper we propose a slightly different approach. We start from user-oriented requirements for QoS and identify where and how the middleware can translate them to system or network-level requirements for QoS. We define the architectural concepts that we think are necessary for understanding and decomposing the complexity of stream binding establishment. We further identify

H. Scholten and M. van Sinderen (Eds.): IDMS 2000, LNCS 1905, pp. 158-171, 2000.

how current object middleware systems can be extended to support media stream binding.

Section 2 describes a framework for QoS-aware middleware. Section 3 identifies the interfaces for QoS specification and section 4 describes how user-oriented QoS specifications can be mapped to network and system level QoS. Section 5 presents how our framework can be applied to an implementation of the CORBA A/V Streams specification. Section 6 summarizes with our conclusions.

2.   A QoS Framework for Streaming

In this section, we propose a framework for QoS-aware middleware-based distributed systems that combines objects, layers and planes. We have used the framework to structure the QoS-related functions of a distributed system. This framework is currently being validated in our testbed.

In Section 2.1, we present the object model we use. In Section 2.2 we discuss how the object model fits into our framework from a high level point of view. In Section 2.3 we consider the framework in more detail, in particular in terms of its layers and planes.

2.1 Object Model

Fig. 1 shows the kind of objects that we use to structure our system [Blair]. Each object encapsulates state and behavior and can expose operational and streaming interfaces to other objects.

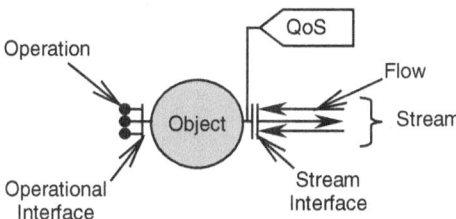

Fig. 1. An object with operational and streaming interfaces.

Operational interfaces allow client objects to invoke computational services onto the object that exposes the interface (the server object). Streaming interfaces allow objects to exchange one ore more flows of continuous information (e.g. audio or video information). Flows are unidirectional and may terminate at one or more sink interfaces. A stream may consist of flows that travel in opposite directions.

Objects may further consist of several other objects at a lower level of abstraction. The higher level object is called a compound object in such cases.

2.2 QoS-Aware Middleware Platform

We propose a middleware platform that is capable to associate a QoS with an object's streaming interface and to control this QoS through the same or another object's computational interface. We propose to extend existing middleware concepts by a QoS framework. We elaborate this framework in particular for middleware that supports multimedia applications using video and audio streams.

The application objects, e.g., cameras, microphones, speakers, files, are endpoints of audio and video stream. The application layer furthermore contains agent objects that act as stand-ins for the end-user (cf. [Jmf]). Agents invoke the services of our middleware platform to bind endpoint objects and to control the QoS of the stream that the binding carries. Agent objects also control the QoS of local endpoint objects.

A binding object allows endpoint objects in the application layer to exchange a stream of multimedia information. A binding object represents a composition of the resources that are involved in tying these application level objects together. This includes media processing resources like codecs, multiplexers, transcoders and packetizers, as well as transport-level resources such as sockets, routers and bridges.

Example
Consider a medical application that allows surgeons to view video clips stored in a database. We assume that a request to connect to the video database originates from the surgeon's machine. Fig. 2 shows the object constellation once the surgeon is viewing a video clip.

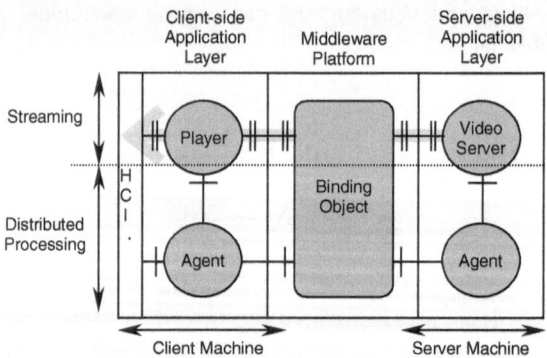

Fig. 2. Binding object interconnecting two application objects.

The video server object in Fig. 2 represents the database. It produces an audio-video stream that flows to the player object via the binding object. The player represents the presentation resources on the surgeon's machine (a display and a speaker in this case) and consumes the stream that the binding produces.

The two agent objects use the operational interfaces of the player, the video server and the binding object to control the QoS of the streams that these objects produce. The client-side agent object may for instance use the binding's operational interface to request the establishment of a certain QoS for the binding or to change the QoS of an existing binding. Similarly, the agent can call the player's operational interface to locally reduce the volume of the video clip's audio part.

The operational interfaces of the binding object have two special properties. First of all, their operations are *application oriented*. This means that the binding object allows the agents to deal with QoS in application-level terms. For example, the binding object allows the client-side agent to specify that it wants the player object to be bound to the video server object using a binding object with an High Definition TV QoS level. As a result, the agent objects do not have to be able to translate this abstract notion of QoS into low-level QoS aspects such as bandwidth, delay, and so on.

The second property is that the binding object's operational interfaces are geared toward the context in which the application is used. We call this the application's *usage context*. This notion is based on our belief that different applications require a different QoS depending on the domain in which they are used, for what purpose the application is used and by whom. For example, if a surgeon uses the system of Fig. 2 to discuss a video clip with one of his fellow surgeons, the client-side agent object will generally request a higher QoS from the binding object than when the surgeon discusses the same clip with one of his patients. We will have more to say on the application-orientation of the binding and our notion of a usage context in Section 3.

2.3 The Platform's Internals

The binding object of Fig. 2 may encapsulate a large number and a variety of resources It may furthermore need to perform complex QoS control activities. We tackle this problem by decomposing the platform into two horizontal layers and three vertical planes. The vertical planes are a data transfer plane, a QoS control plane and a QoS management plane (cf. [Aurre]). Across these planes we distinguish between a middleware (software) layer and the Distributed Resource Platform (DRP) layer (i.e. computing and network resources).

Data Transfer
The data transfer plane contains the objects that are concerned with forwarding the data units of a multimedia stream.

An object in the data transfer plane of the middleware layer encapsulates resources that perform transport-independent as well as transport dependent stream processing. Examples of the former are encoders, transcoders and multiplexers; an example of the latter is an RTP packetizer that can adapt an MPEG-1 encoded stream for transmission over UDP.

The objects in the data transfer plane of the DRP encapsulate the distributed resources (e.g. IP routers, bridges, etc.) that provide end-to-end connectivity.

QoS Control and Management
The objects in the QoS control and management planes of the middleware layer and the DRP govern the QoS of the stream that flows through the data transfer plane.

Each binding object encapsulates a set of objects in the QoS control plane of the middleware layer and the DRP. These objects govern the QoS of the binding's streaming interfaces during its lifetime. We propose that objects in the QoS control plane are responsible for *establishing* a QoS for a binding. The establishment of QoS typically involves the negotiation of an acceptable QoS followed by the reservation and initialization of objects in the data transfer plane to effectuate the negotiated QoS. Other activities that can be found in the QoS control plane involve [ISOQoS, D3.1.2] predicting a binding's current and near-future QoS, keeping a binding's current QoS in line with the negotiated QoS, and releasing a binding and its resources.

The objects in the QoS management plane of the middleware layer and the DRP are not part of a particular binding. Rather, they can be considered part of every binding because their activities transcend the lifetime of individual bindings. Objects in the QoS management plane for instance take care of fault management and statistics collection.

Example

As an example, assume that the binding object of Fig. 2 uses an MPEG-1 encoder to compress the audio-video stream. Also assume that the binding relies on an RSVP reserved UDP transport channel to convey the encoded stream from the server to the client. Fig. 3 shows how the example of Fig. 2 maps onto the layers and the planes of our QoS framework. Observe that it zooms in on the middleware portion of Fig. 2. Fig. 3 does not show the application layer and the objects that it hosts.

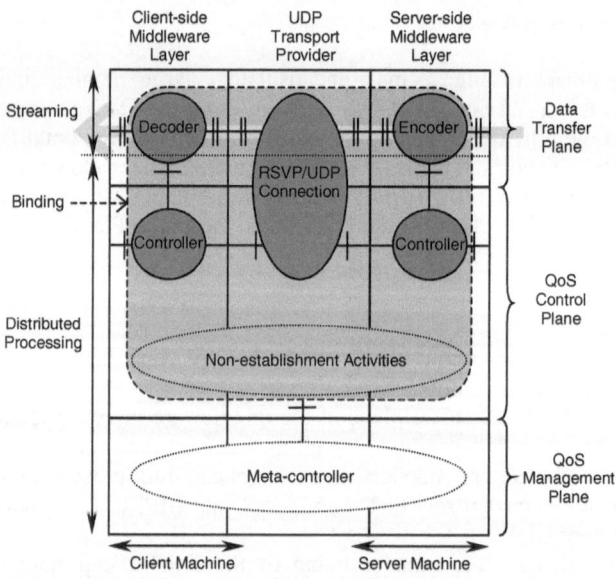

Fig. 3. Two-dimensional version of the QoS framework.

The encoder and decoder are middleware level objects of the data transfer plane. The encoder for instance encapsulates the MPEG-1 encoder and an RTP packetizer. The RSVP/UDP connectivity object is part of the data transfer plane of the DRP and encapsulates the resources that connect the encoder to the decoder.

The controller objects are part of the middleware layer's QoS control plane. They use the operational interfaces that the encoder, decoder and connection objects expose to control the QoS of the streams that these objects produce. The controllers may for instance first negotiate an encoding and a packetizer to use. The controller on the server-side can then invoke the operational interface of the selected encoder, for example to set its output bitrate parameter. After that, the controller can use the operational interface of the connection object to request a QoS that supports the encoder's output. Observe that the notion of QoS at the connection object's operational interface is of a lower level than that at the binding object's operational interface. The former typically expresses its notion of QoS in terms of bandwidth, delay, jitter, and so on, whereas the latter uses application-oriented notions such as "HDTV QoS".

Observe that the objects in the distributed processing part of Fig. 2 and Fig. 3 build on an ORB infrastructure therefore they can invoke each other's operational interfaces even when they reside on different machines. For the sake of clarity, we have not drawn these interactions.

Also observe that Fig. 3 represents non-establishment activities of the QoS control plane (e.g. checking the current QoS against the negotiated QoS) as a single object. The activities of this object are outside the scope of this paper and we have therefore not shown its interactions with the other objects in Fig. 3. For similar reasons, we have also not shown how the object that represents the management activities (the meta-controller object in Fig. 3) relates to the objects in the binding.

3. QoS at Interfaces

In the previous section, we presented our QoS framework. In this section, we elaborate on the specification of QoS at interfaces and the mapping of these specifications to QoS support of the middleware. We define a QoS for a streaming interface by specifying the desired QoS at an operational interface

We use the QoS specification language QML [Frølund] to illustrate our approach.

3.1 Control-Plane QoS Interface Model

In our framework (see Section 2), QoS represents the quality aspects of the interactions between the middleware platform and one or more application objects that act as the endpoints of a stream. QoS realizes the quality aspects of the interactions between these application objects and thereby improves the usability or the utility of the applications that use the platform.

The concepts that we use to specify QoS are based on the user-provider model [ISOQoS, D3.1.2] (see Fig. 6) in which a provider (our middleware platform) mediates the interactions between its users (our application objects).

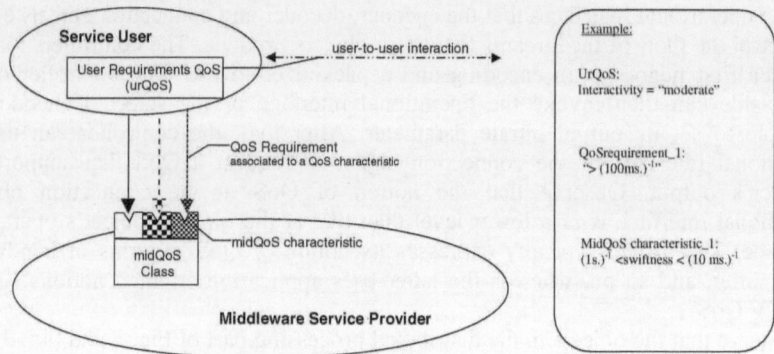

Fig. 4. Control-plane QoS Interface Model

From a top-down perspective, the quality needs of the interactions of application objects drive the QoS that is required at streaming interfaces. These quality needs are called *user requirements* [ISOQoS] with respect to QoS (urQoS). In principle, urQoS is independent of the service provider that mediates the interaction.

The provider may advertise its QoS capabilities in terms of the values that it supports for a certain set of *QoS characteristics* (midQoS characteristics in Fig. 4). QoS characteristics are the quantifiable aspects of QoS such as accuracy, freshness or reliability. The provider's capabilities are in principle also defined independently from the users. Aggregations of QoS characteristics and their associated values are called *QoS classes*. A provider may support several QoS classes. Each class may be optimized for a certain category of services. Examples are QoS classes for real-time services, messaging services, or multimedia services. For reasons of flexibility, a provider may also offer QoS classes of different abstraction levels. QoS classes may for instance be network-oriented or application oriented.

To specify a QoS at a provider's interface, we need to translate urQoS (including the associated values needed) to an instance of a QoS class that the provider offers. To facilitate this, we use the concept of a *QoS requirement* [ISOQoS]. A QoS requirement consists of required values of a QoS characteristic and the qualifiers for these values (e.g. maximum, mean, or minimum).

3.2 User Requirement

The urQoS of Fig. 5 typically depends on what we call the *application context of use*. We have analyzed a number of reports (see for instance [Ewos]) that describe the properties of multimedia data interchange in the medical domain. The reports cover user scenarios as well as the media characteristics and alternative stacks of protocols (including the compatible parameter options) that are suitable for these scenarios. The reports are suitable for our work because communication experts as well as medical specialists have contributed to them.

Our analysis leads to the following elements as determining an application's usage context:

- the application domain: examples are the medical domain of surgery and the game-hall domain;
- the role of the user: examples are imaging specialist, surgeon, game-hall subscriber or guest user;
- the purpose for which the application is used: for example, a surgeon may use a moving image retrieval application to check his diagnostic hypothesis or to view images during surgery. Latency requirements of image retrieval may for instance be different for both cases.

We have furthermore identified the following urQoS dimensions: availability, fidelity, integrity, interactivity, and regulatory. Since these dimensions are not orthogonal the translation of urQoS dimensions to provider-oriented QoS characteristics (midQoS) is not straightforward.

Table 1. User-oriented QoS dimensions

QoS dimension	Quality aspect
Availability	Present or ready for immediate use
Fidelity	Good (i.e. correct) enough in respect of the application purpose
Integrity	Delivering the whole truth in respect of the source
Interactivity	Being responsive
Regulatory	Conformant with rules, the law or established usage

Example of urQoS specification

This example illustrates urQoS dimensions and values that are relevant for a surgeon who retrieves moving images from a medical imaging repository. The surgeon uses a Video-on-Demand (VoD) application to retrieve the images for one of the following purposes:

- to validate a diagnostic hypothesis;
- to have a closer look at the images to prepare a surgical treatment, for example in case the hypothesis has been confirmed;
- as reference material during surgery.

We may specify a template QoS category, i.e. a group of related urQoS dimensions, suitable for the surgeons' VoD application by the following QML contract type:

```
type Surgeon_VoD_QoScategorytype = contract
{//VoD appl. contains list, browse, and retrieve methods
  availability: set {"list","browse",
                "diagnostic phase","surgery phase"};
  integrity: increasing set {"lossy", //for listing
             "functional lossless",   //compression <25:1
             "perceptual lossless"};  //compression <12:1
             //increasing= higher values are better
  //remark: fidelity is modelled via integrity
  interactivity: increasing set{"low","normal","high"};
  regulatory: set {"not relevant","CR standard",
                "Xray standard"};
};
```

The QML contract that specifies the urQoS dimensions and their possible values for a diagnostic hypothesis validation may then look as follows:

```
Diagnostic_validation_QoScategory =
                    Surgeon_VoD_QoScategorytype contract {
    availability == {"list","browse","diagnostic_phase"};
    integrity == {"lossy","functional lossless"};
    interactivity == {"low","normal"};
    regulatory == {"CR standard","Xray standard"};
};
```

For the preparation of a surgery, we may use a more stringent value for integrity, "perceptual lossless" for instance. Similarly, during the surgical treatment we may instantiate the VoD application with the more stringent value "surgery phase" for availability and the value "high" for interactivity.

3.3   QoS Classes and Characteristics

In the user-provider model of Fig. 5, the provider offers services as well as a set of associated QoS capabilities. We use QoS classes to facilitate these capability offers. Similar approaches can be found in ATM networks [Alles] and QoS-aware systems (e.g. [Lazar]).

A QoS class defines a set of QoS characteristics that the provider supports. In this paper, the definition of a QoS class also encompasses the (ranges of) values of its QoS characteristics.

Example:
The following QML contract defines the dimensions of a QoS class for an RSVP-based transport service:

```
type RSVPbased_QoSclasstype = contract {
        delay: decreasing numeric msec;
        rate:  increasing numeric Mb/s
};
```

An instance of the above QoS class type that supports maximum delay guarantees is:

```
RSVPguaranteed_QoSclass =
        RSVPbased_QoSclasstype contract {
        delay < 100;  //bounded delay in a guaranteed service
        rate < 0.064; //ISDN is e.g. a bottleneck link
};
```

The QoS characteristics of a QoS class are generally not completely independent of each other. A provider can furthermore define alternative sets of QoS characteristics and offer them to the users in the form of similar QoS classes. For example, a provider may offer a QoS class that uses the inverse of the delay QoS characteristic, i.e. the swiftness characteristic. A provider may also combine characteristics in a new characteristic (or perform other types of transformations). Instead of using image-height and image-width as characteristics, the provider may for example use image-size characteristic with permissible values like "CIF", "QCIF", etc.

3.4 QoS Interface Specification

In this section, we illustrate an approach to link the user requirements in respect of QoS to the provider-oriented QoS classes in accordance with the QoS model (Fig. 4).

QoS specification scenario at control-plane
In the example QoS class of Section 3.3, the transport provider guarantees that delays will not exceed 100 ms. This means that the provider has to be able to estimate the communication delay between the client and the VoD server before it offers the QoS class.

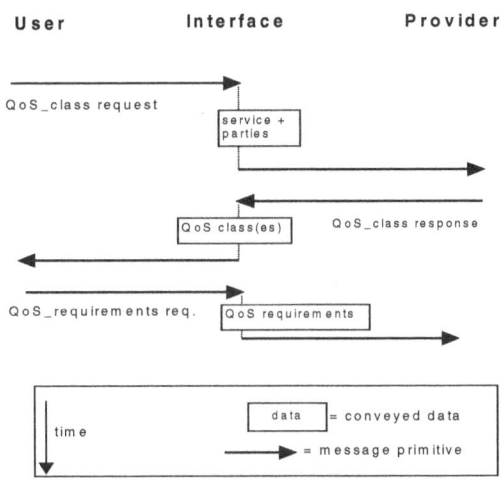

Fig. 5. QoS interface interaction scenario

The scenario of Fig. 5 illustrates how the user needs for QoS can be linked to a provider's QoS capabilities using the QoS model of Fig. 4. The user first queries the provider for the QoS classes that it supports. The user invokes a control-plane operational interface for this purpose. The provider then determines the set of most appropriate QoS classes for this user. What constitutes most appropriate may for instance depend on the delays between the participants of the communications session. The provider can for example consult the information base in the QoS management plane to make an estimate of this delay. After having received the QoS classes, the user can select a suitable one and determine the values of the QoS characteristics that best match its requirements (i.e., urQoS). The user then responds with QoS requirements that represent the required qualifiers and values of the QoS characteristics (see also Fig. 5).

QoS requirement derivation
In the user-provider based QoS model, the user (typically in the form of application objects) must translate the urQoS dimensions and values to the midQoS characteristics.

An important factor in this translation is application domain knowledge. Of particular importance is the application domain information that has been elaborated and documented by communication experts. Examples are international standards like the analyzed CEN/TC251 and EWOS/EG MED medical reports, e.g. [Ewos]. Other ICT standards (e.g. CCIR.601) and results of ergonomic studies are also useful for QoS translations and mappings.

The urQoS dimension availability typically maps to the accessibility, reliability, swiftness and urgency midQoS characteristics. Integrity maps to accuracy, freshness, linkage unity and also swiftness. Table 2 illustrates the translation of availability values relevant in the medical VoD examples to accessibility. Table 3 illustrates the translation of integrity values relevant in the medical VoD examples to accuracy and linkage unity.

Table 2. Example availability urQoS to accessibility midQoS translation

AVAILABILITY	ACCESSIBILITY
"diagnostic phase"	"offline normal" or "offline high"
"surgery phase"	"online normal" or "online high"

Table 3. Example integrity urQoS translations

INTEGRITY	ACCURACY	LINKAGE UNITY
"functional lossless"	"CR degraded" or "Xray degarded"	"sync normal"
"perceptual lossless"	"CR" or "Xray"	"sync high"

4. QoS Characteristic Mapping

This section briefly describes the mapping of middleware QoS characteristics (e.g. accessibility, accuracy or swiftness) to the QoS characteristics of the underlying DRP. Fig. 6 shows the QoS mapping relations when we recursively apply the model of Fig. 4 to the interface between the middleware and the DRP.

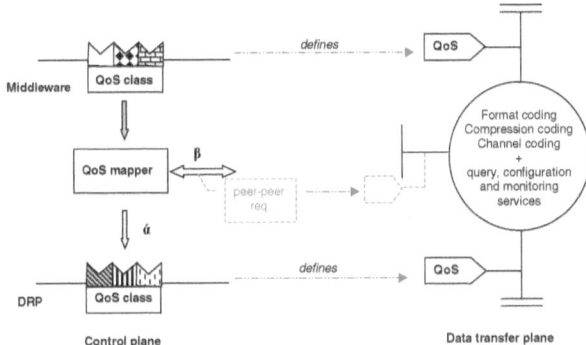

Fig. 6. Control-plane QoS characteristic mapping
and its impact on the data transfer plane

The QoS mapper of Fig. 6 has to take the influence of the objects in the middleware's data transfer plane into account when it determines the QoS requirements for the DRP (ά). For example, if the accuracy midQoS characteristic value permits compression of video images, the video coding object in the data transfer plane typically introduces additional end-to-end delay.

A QoS mapping may also involve peer-to-peer negotiation (β). For instance, the two controller objects of Fig. 2 may need to agree on the encoding format and the encoding parameters to use. The DRP takes part in this negotiation in the role of the provider. This is because the DRP for instance has to provide a connection that has sufficient capacity to carry the output of the encoder. QoS characteristic mapping thus generally involves a *multiparty negotiation.*

In some cases, we can (also) use QoS mapping tables to convert the midQoS characteristics and associated QoS requirements to QoS characteristics that are supported by the DRP. For example, linkage unitity "sync normal" for pointing device-to-video synchronization may map to a maximum delay difference in the range of -580 to +820 ms at the DRP-level [Steinmetz].

5. Implementation

We are currently building a testbed to validate our framework. We have developed a video on demand demo based on the light profile of the CORBA Management and Control of Audio/Video Streams RFP [AVStreams]. The demo uses the Java Media Framework [Jmf] for streaming communications. The server and the client are interconnected by an RSVP-enabled router. The client allows the end-user to select a movie and conveys the user's selection to the server. The server responds by streaming an MPEG-1 video to the client using RTP/UDP.

Fig. 7 shows an overview of the CORBA A/V Streams RFP in terms of the objects that we use (see Section 2.1). The objects shown in gray together represent a stream. The str_ctrl object acts as a control point for application objects, which can be used to

request the establishment of a multimedia stream with a certain QoS. The str_ctrl
object would therefore be present in a controller object of Fig. 3.

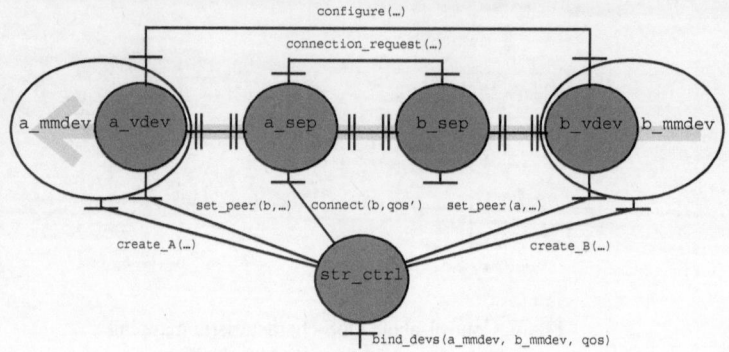

Fig. 7. Overview of CORBA A/V Streams objects.

The a_mmdev and b_mmdev objects encapsulate multimedia devices on the client
and server machines of Fig. 3, respectively. They act as a factory for the *_vdev and
*_sep objects. The a_vdev and b_vdev objects encapsulate the resources at the
endpoint of the stream (e.g. a camera or a speaker) as well as the transport-
independent media processing resources (e.g. an MPEG-1 encoder). They for instance
engage in a peer-to-peer QoS negotiation process to determine the encoder and its
settings to use. The *_mmdev and *_vdev objects would for this reason be divided
over the application and middleware level objects of Fig. 3.

The a_sep and b_sep objects encapsulate the transport specific media processing
resources such as an RTP packetizer and an RSVP entity. These objects would
therefore be distributed over the middleware-level objects of Fig. 3. Observe that the
streaming interface that interconnects a_sep and b_sep is the transport object of Fig.
3.

6. Conclusions

The architectural model for middleware support for QoS provisioning presented in
this paper is a combination of an object model, a layered model, parts of the ISO QoS
model and a model consisting of planes. The model is suitable for structuring the
middleware internals for establishing media streams. A key feature of our model is
the identification of QoS control objects (postioned in a control plane) that are
responsible for establishing a (multi) media stream based on a QoS agreement.

Our approach is application oriented rather than system or network oriented. This
is facilitated by the notion of a usage context and by the fact that the applications and
our middleware-based system talk to each other in application oriented QoS terms.

We have validated parts of our architecture. To support this claim, we have shown
that the CORBA A/V Streams specification fits into our architecture and we have

validated this through an initial implementation. This proves that the composition of our architecture in terms of layers, planes, objects and interfaces is a viable one.

We furthermore show that our notions of QoS specification can be expressed in QML and that the issue of QoS mapping can be realized in a heuristic manner by defining static mappings (e.g. by using mapping tables) between QoS specifications at the different layers.

Acknowledgements

This work was partially sponsored by the Telematica Instituut within the project AMIDST (http://amidst.ctit.utwente.nl). We would like to thank the members of the QoS workpackage for their contribution to this work. We particularly want to thank Dave de Vries for his input.

7. References

[Amidst]   M. van Sinderen, "Amidst Project Description", http://amidst.ctit.utwente.nl
[Alles]   A. Alles, "ATM Internetworking", Engineering InterOp, Las Vegas, 1995
[Aurre]   C. Aurrecoechea, A.T. Campbell and L. Haw, "A Survey of QoS Architectures", Multimedia Systems Journal, Special Issue on QoS Architecture, May 1998;
[AVStreams]   "Control & Management of Audio/Video Streams", document number: formal/98-06-05, June 1998
[Blair]   G. Blair and J-B. Stefani, *"Open Distributed Processing and Multimedia"*, Addison-Wesley, 1998;
[Cen]   CEN/TC251 N98-034, "Quality of Service Requirements for Healthcare Information Interchange", draft CEN report, Feb. 1998;
[D3.1.2]   A. van Halteren et al., "QoS Architecture", Amidst project deliverable D3.1.2,1999, http://amidst.ctit.utwente.nl/workpackages/wp3/index.html;
[Ewos]   EWOS/ETG068, "Multimedia Medical Data Interchange", European Workshop for Open Systems (EWOS), Brussels, 1996, and earlier versions;
[Frølund]   S. Frølund and J. Koistinen, "Quality-of-service Specification in Distributed Object Systems, Distributed Systems Engineering, 5, 1998, pp. 179 – 202;
[Gay]   V. Gay, P. Leydekkers, R. Huis in't Veld, *Specification of Audio and Video interaction based on the reference model of ODP*, Computer Networks & ISDN systems, special issue on RM-ODP, January 1995.
[ISOQoS]   ISO/IEC, "Information Technology – Quality of Service – Framework", ISO/IEC JTC1/SC21 N13236, Geneva, 1997;
[Jmf]   Java Media Framework API Guide, Nov 1999, http://www.sun.com
[Lazar]   A. Lazar, L. H. Ngoh and A. Sahai, "Multimedia Networking Abstractions with Quality of Service Guarantees", Proceedings of the SPIE Conference on Multimedia, Computing and Networking, San Jose, CA, Feb 1995
[Steinmetz]   R. Steinmetz, "Human Perception of Media Synchronization", IBM European Networking Center, IBM – Technical report no 43 9310, 1993;
[Verma]   D. Verma, *"Supporting Service Level Agreements on IP Networks"*, Macmillan Technical Publishing, Indianapolis, 1999;

Interaction of Video on Demand Systems with Human-Like Avatars and Hypermedia

Norbert Braun[1] and Matthias Finke[1]

[1] Zentrum für Graphische Datenverarbeitung,
Rundeturmstraße 6,
D-64283 Darmstadt, Germany
{Norbert.Braun, Matthias.Finke}@zgdv.de

Abstract. When it comes to digital video, users expect more than simple VCR functionality. This paper presents a system that provides both direct manipulation and conversational interaction for digital video. The paper shows the seamless integration of different annotation techniques (synchronous and asynchronous) combined with different presentation and interaction metaphors. Direct manipulation technique is implemented with graphical and acoustical video hyperlinks. These hyperlink techniques are discussed in detail with the focus on their temporal nature driven by the duration of the accessibility of objects within a digital video. An avatar system is used to give the user an additional conversational access to information that is related to the video. This access is independent of the playback time of the video content. These techniques are implemented. Several system components like video server, video presentation and video authoring are shown. The user interaction capabilities based on the techniques mentioned above are presented.

1 Introduction

Beside virtual reality and animation, digital videos play a major role within multimedia applications and, therefore, in video on demand systems. The reason can be found, on the one hand, in the nature of video as a direct view of the real world, not a simulation. On the other hand, there are large datasets of preproduced video, created over years of TV and movie production. A major benefit of multimedia applications lies in the user interaction. The interaction concepts of digital video have to be taken to the advanced interactivity level of animation and virtual reality. This will be a key feature of video on demand systems and the essential benefit of this kind of web-based system in relation to interactive television.

The actual research on interactive digital video is strongly influenced by web technology and multimedia productions. Beside traditional interaction methods like VCR functionality (change video, start, stop, fast forward, etc.) or the electronic programming guide (epg) of digital television, the direct manipulation interaction of the hypermedia metaphor is transferred to video. This method of annotation is divided into

H. Scholten and M. van Sinderen (Eds.): IDMS 2000, LNCS 1905, pp. 172-186, 2000.
© Springer-Verlag Berlin Heidelberg 2000

intramedia annotation (annotating content of one media channel within the same media channel) and intermedia annotation (annotating content of one media channel within another media channel).

Intermedia hypermedia annotation is used by Digital Renaissance [Martel96] in the form of text hyperlinks for the synchronous annotation of the graphic and acoustic content of digital video. Intramedia annotation of the graphic content of digital video is done by VOSAIC [Camp95], Real [Hefta97] and ZGDV [Gerf98]. Intramedia annotation for the audio content of digital video was proposed by [DeRoure98] and [Braun98]. DeRoure applied hypermedia structures to audio; Braun showed an acoustic hypermedia annotation of the audio content of digital video. An overview on the topic of hypervideo is given in [Saw96].

These hypermedia approaches for digital video disregard the temporal structure of the medium. Digital video happens over time, meaning it has a start, a duration, and an end. The same is true for hypermedia annotations of digital video - they have a start, a duration, and an end. This structure has to be shown to the user, as proposed by [Braun99].

Multimedia presentations are often dealing with interaction facilities inside the content as described by [Böhm93]. They are done with doors, touch-sensitive buttons, etc., within virtual reality or via interaction possibilities that are added onto the content, like menus or navigational bars – for interaction onto content and for application navigation. A major metaphor for these applications is conversational interaction with human-like avatars; see [Shne95]. This conversational interaction is heavily used for assistance systems where users are not able to do direct manipulations on objects, but delegate their work to an agent; see [Mandel97].

In order to develop an interaction component on a digital video on demand system, it is of interest to combine the direct manipulation approach of hypermedia and the conversational interaction as done with human-like avatars. Therefore, we present a digital video on demand system that includes two interactive elements:
- hypermedia annotation of the graphic and acoustic content of the digital video
- conversational interaction

The hypermedia annotation is used for synchronous annotation of the digital video's content – and, therefore, for interaction on the objects within the graphics and acoustics of the video. In fact, the hypermedia annotation of the video is a form of direct manipulation – only possible during the temporal appearance of the video objects.

The avatar system is used for a conversational interaction on the video's objects without a limitation of the object's appearance time and not limited to the objects of the digital video either. Hence, the conversational avatar system can be used for synchronous and asynchronous access to information on the video content. Figure 1 shows how an avatar is used within a conversation to show additional information to the user through both gesture and speech.

Figure 1: Conversational Video Interaction

The advantage of the combination of both interaction metaphors for digital video lies in overcoming the video's temporal nature:

- time-dependent interaction on content that is actual visible within the video's presentation
- time-independent interaction on content regardless of the content's playback time; for example, on content that was presented or is yet to be presented by the video

In the following paragraphs, we introduce a hypermedia system and an avatar-based conversational user interface for video interaction. Then, we define the video on demand system and its user interaction component. Lastly, we provide a conclusion and comment on our future work.

2   Hypermedia for Digital Video

Hypermedia systems are widespread due to the explosive growth of the World Wide Web. Hypermedia was developed for discrete media like text or picture, accessible to the user as long as he does not explicitly change the content. The hypermedia annotation is, however, accessible as long as the content is accessible. Due to its appearance in time, continuous media have some special abilities: content is presented within a time interval; it has a start, duration and an end. The content is, therefore, not accessible during the entire duration of a presentation. Due to the temporal basis of video content, most presentation systems have some kind of time index, like a time line with a cursor in it that shows the actual position within the video. This time index could also be shown for the annotation of the hypermedia system.

The hypermedia annotation of digital video can be done in an intermedia way. The term 'intermedia' designates the technique of having different media channels for annotation and content. This kind of digital video interaction is done within the system of Digital Renaissance [Martel96]. The system shows additional text hyperlinks, synchronous to the video's graphics and acoustics. Due to the change of media channel, it is hard for the user to understand which object within the graphics or acoustics is the annotated one. The name of such a textual hyperlink indicates the connection between object and hyperlink. This technique has two disadvantages:

- The syntactical operation of an annotation is dependent on the semantics of a text.
- The user has to focus on the video content and observe the textual hyperlinks simultaneously.

The term 'intramedia' annotation of digital video designates the presentation and annotation of video content within the same media channel. Therefore, it has to be divided into the annotation of the video graphics and the annotation of the video acoustics. In both cases, the annotation can be presented explicitly or implicitly.

- The explicit presentation of a hyperlink is shown to the user during the hyperlink's entire active time.
- The implicit annotation shows the existence of an annotation only upon user demand, meaning the user has to show his interest in an object's annotation to get a message indicating the existence of any additional information on the object.

A typical problem of intramedia hypermedia annotation is the indication of the temporal duration of such a continuous media hyperlink. The following information is of particular interest to the user in order to prevent undue stress when using this technique:

There should be

- a starting point of the annotation that indicates the beginning of the hyperlink.
- a stopping point of the annotation that indicates the end of the hyperlink.
- a temporal progression of the annotation that allows the user to imagine how much time is left until the end of the hyperlink.

In order to make interaction user-friendly, these points have to be worked out for both the hypermedia annotation of the video's graphics, as well as that of the video's acoustics.

Video

An example of hypermedia annotation of the video graphics in an implicit way is illustrated in the system of Real [Hefta97] called the Real Player. The Real Player shows an annotation by a change of the mouse cursor. The system of VOSAIC [Camp95] shows the annotation of the video's graphics explicitly. The annotation is marked as a polygon around an annotated object.

The MovieGoer of ZGDV [Finke2000] is adjustable – the user can choose whether to have an explicit or implicit annotation. If the annotation is implicit, all objects of the video content are at least annotated by a default hyperlink. If the annotation is explicit, the annotated objects are marked by a polygon. The graphical video hyperlinks are activated by the user's mouse click on the annotated object within the video's display. Figure 2 shows how additional information is presented after the user has clicked on a video object.

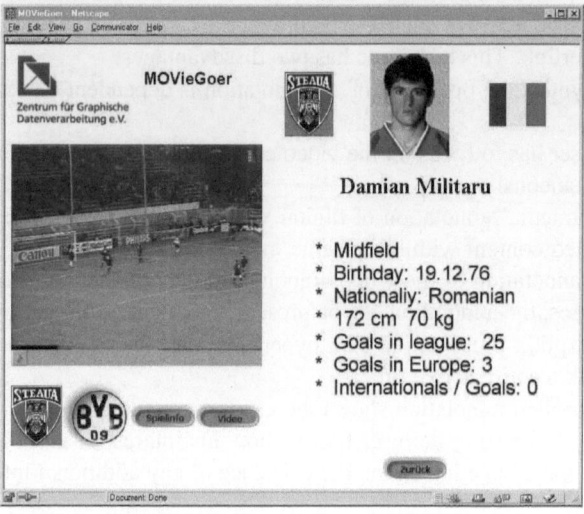

Figure 2: Interactive MovieGoer System

The DIVA System of ZGDV [Braun99] uses explicit annotation technology to show the user the temporal length of the hyperlink. This is done via a change of the colour of the hyperlink's graphic. The hyperlink is started with the polygon shown in the colour A, ending the hyperlink with the polygon coloured in the colour B, and changing the colour of the polygon from the bottom to the top of the polygon from colour A to colour B during the hyperlink's duration. Figure 3 (left hand) shows how this technique is used. The rearmost picture is annotated with a near white polygon while the black colour is growing dominant to the annotation of the fore picture. With this technique, the user can imagine right from the start of the hyperlink how long it will take until the hyperlink ends; see [Braun99].

Audio
Sonic hyperlinks [Braun98] developed by the DIVA System of ZGDV are used as intramedia annotations of audio. A sonic hyperlink is defined as an acoustic annotation of acoustic information. Usually, this is some soft sound playing in parallel to the acoustic content of a digital video. User reaction on this hyperlink is acoustic, too, meaning he can react via his voice by predefined vocal commands. This way, the user has the opportunity to interact with the acoustic video content in a verbal way. The sonic hyperlink's nature as a sound annotation makes it an explicit annotation of content.

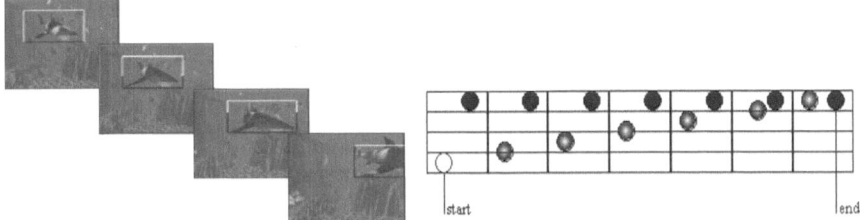

Figure 3: Showing the Duration of: a graphical Hyperlink (left) and a Sonic Hyperlink (right)

The annotation of the acoustics can be done by additional sound or by the variation of the original acoustic content of the digital video. Variations on the original sound of the video lead to user confusion [Braun99]. Therefore, an additional sound is preferred. To indicate the duration of the sonic hyperlink, a two-tone solution proves to be most fitting [Braun99]. Using two tones, A and B, at the beginning of the sonic hyperlink, changing the frequency of A in little steps towards B, and ending the hyperlink when the changed tone A meets the sound of tone B indicates the start, the end and the remaining time on every time position within the sonic hyperlink's time interval. Figure 3 (right hand) shows how the tones change throughout the duration of the hyperlink – until they finally become the same tone.

User interaction on sonic hyperlinks should work by performed voice in order to use one single media channel for the content, the annotation and the user reaction. The user's visual channel is not affected by the interaction on the digital video's acoustic content. Figure 4 shows how a user interacts by using sonic hyperlinks and his voice.´

Figure 4: User Interaction by Voice

Middleware Support for Media Streaming Establishment Driven by User-Oriented QoS Requirements

Cristian Hesselman[1], Ing Widya[2], Aart van Halteren[3], and Bart Nieuwenhuis[2]

[1]Telematica Instituut, Enschede, The Netherlands
[2]CTIT, University of Twente, Enschede, The Netherlands
[3] KPN Research, Enschede, The Netherlands
hesselman@telin.nl, widya@cs.utwente.nl,
A.T.vanHalteren@kpn.com, L.J.M.Nieuwenhuis@cs.utwente.nl

Abstract. The requirements for the QoS of distributed applications are traditionally expressed in terms of network oriented or systems oriented parameters. In general, the users of these services are not interested or capable of specifying the QoS of their services in such technical terms. In this paper, we propose modeling and engineering concepts for the mapping of end user QoS onto system and network QoS. We introduce QoS agents in structured object middleware that relate end-user QoS specifications to multimedia stream bindings. In fact, the middleware layer supports QoS classes, i.e., a set of QoS characteristics. The end user QoS requirements, generally a set of non-orthogonal specifications, must be supported using the available middleware QoS classes. We also describe the experimental environment that will be used to refine the QoS mapping mechanisms.

1. Introduction

Object middleware, such as CORBA, DCOM and Java RMI, is gaining rapid acceptance as a means to quickly and cost effectively develop a wide range of applications for various areas of industry. The main purpose of these middleware systems is to provide a software infrastructure for interacting application components.

Interactions between software components can be divided into two categories: discrete or continuous. Discrete interactions are typically RPC or message-oriented and are generally used to invoke a computational service. Continuous interactions are generally used for exchanging (multi-) media data and are usually called streams. Middleware platforms often model a stream as a binding object [Gay, Blair]. In this paper we focus on the support of middleware platforms to establish a user-oriented QoS for such binding objects.

Requirements for QoS are traditionally expressed in terms of system-oriented or network-oriented parameters. In this paper we propose a slightly different approach. We start from user-oriented requirements for QoS and identify where and how the middleware can translate them to system or network-level requirements for QoS. We define the architectural concepts that we think are necessary for understanding and decomposing the complexity of stream binding establishment. We further identify

H. Scholten and M. van Sinderen (Eds.): IDMS 2000, LNCS 1905, pp. 158-171, 2000.

- Feedback to user actions is the third component of the avatar system model. The feedback is given on user interaction with the hypermedia annotation of the video and on user navigation onto the video.

The three parts together determine avatar behaviour during the presentation of the video.

From the user's perspective, the avatar is either acting as a *conferencier*, guiding the user through the entire video presentation, or he is acting as a companion or colleague, a fellow member of the video audience and, therefore, a bystander to the video presentation. The following paragraphs will discuss the three components of the avatar system in relation to the video presentation.

Synchronous Video Contribution

This component is determined by the video's presentation over time. While the video is active, the avatar is following a predefined dialogue. The predefined dialogue consists of an emotional part, as well as an additional contribution related to the video content. The dialogue is synchronous to the video and can be viewed as another form of video annotation in addition to the hypermedia annotation of the video. The behaviour of this component is constant, meaning every time the video is played, the avatar will get the same dialogue content.

Asynchronous Video Contribution

The conversational interaction unit is realising an indirect access to the video content. This presents the opportunity to get asynchronous access to additional information linked to objects of the video content - independent of objects actually visible in the video. Therefore, the avatar's behaviour is not constant through the video presentation, but adapted to user questions.

Feedback Contribution

Each user interaction on the digital video requires a defined feedback [Bau95]. This feedback can be one of the following types:

- Assistance feedback - if the user has interacted with the video content via the annotations on the content.
- Navigational feedback - based on the video player VCR functionality.

The feedback is given via the avatar in the form of guidance through the whole digital video system. Therefore, the user will always have an emotional companion when he is interacting with the video. The feedback is based on [Pérez96]. It has the following different states:

- Not aware of the utterance
- Aware of the utterance, but did not hear it
- Heard the utterance, but did not understand it

- Fully understood the utterance

The assistance feedback consists of immediate additional information on the video's annotations and the presentation of the available additional content. Figure 6 shows three forms of feedback contribution:

- The upper row shows an accepting reaction
- The middle row shows an incomprehensible reaction
- The lower row shows a non-accepting reaction

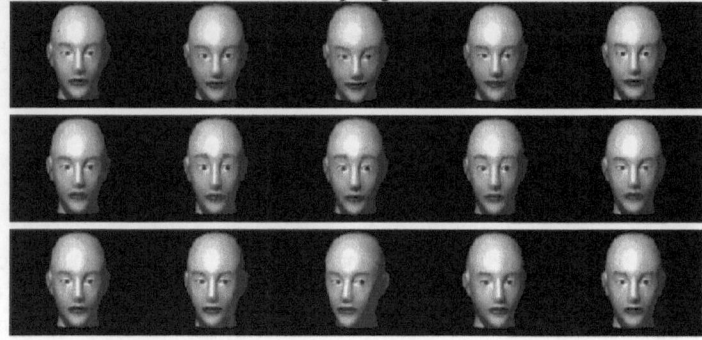

Figure 6: Feedback Contribution of the Video on Demand System

4 System Architecture

The system architecture follows the requirements of an interactive digital video environment that combines a direct manipulation approach based on hypermedia (see paragraph 2) and a conversational interaction approach based on human-like avatars (see paragraph 3). The video itself is annotated in order to provide content-based navigation and additional information to the user. An avatar is designed as an assistance for synchronous, asynchronous and feedback contribution to the video content.

The system consists of three modules: a content server, a presenter client and a client-server connection. In addition, an authoring environment is set up to provide tools to create content based upon our approach. The system uses the Real video server [Hefta97] to stream media content from server to client. In addition, the system uses the SMIL technology [SMIL98] to synchronise media streams. The Real server is used to handle the following media:

- video and its hypermedia annotation
- synchronous and asynchronous avatar behaviour and speech

The media synchronisation is based on SMIL.

The system client is based upon Real Media with a Java-3D rendering plug-in. Figure 7 shows the system architecture with its synchronous and asynchronous data storage on the server side, as well as the Real Media plug-ins for the presentation of the video on the client side.

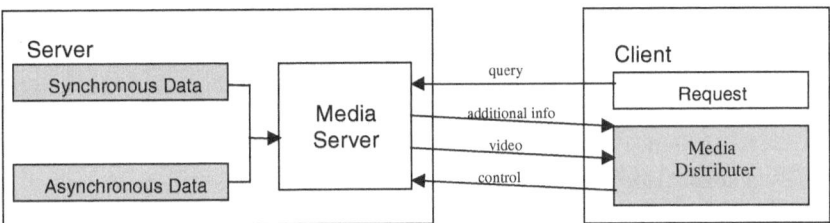

Figure 7: System Architecture

The authoring process is performed with tools that prepare the information to be served by the Real server and to be visualised by the Real client.

Authoring

The authoring of the system is divided into two components: the authoring of the avatar geometry and behavioural animation sequences, as well as the authoring of the video hypermedia annotation, the avatar behaviour triggers, and the additional content presented by the avatar.

The geometry and animation sequences of the avatar are authored in VRML; therefore, every VRML supporting 3D animation tool can be used. The animation of every kind of avatar behaviour is transferred to a feature graph [Alex99]. The Java-3D rendering plug-in is using feature morphing to animate the avatar geometry on the basis of the feature graph of the particular behaviour. If a behaviour is stored as a feature graph in the system, it can easily be triggered by an avatar behaviour trigger. If the behaviour is not stored in the system, a feature graph of the behaviour can be given to the Java-3D rendering plug-in to animate the behaviour.

The authoring of the interaction facilities of the digital video is divided into a synchronous and an asynchronous part. The synchronous annotation of the digital video contains the hypermedia annotations (both graphic and audio annotation), as well as the additional behaviour and information given via the avatar. The asynchronous part contains the avatar's general behaviour towards user questions, as well as a content database with information to answer user questions.

- Video Hyperlinks: The annotation of the graphic content of the video is done in an intramedia way on the basis of a MovieGoer [Gerf98] and Real Video combination. The hypermedia annotation is split into the graphic annotation and the hyperlink target. The graphical annotation is inserted into the original video; the timing and spatial information, as well as the hyperlink target (for example, a URL), is stored separately from the video.

- Sonic Hyperlinks: The annotation of the acoustic content of the video is done in an intramedia way on the basis of the Sonic Hyperlink [Braun98]. The acoustic annotation of the audio information is stored within the video. The temporal and spatial information, as well as the hyperlink target (for example, another part of the video), is stored separately from the video.

- Avatar information and behaviour: The avatar's behaviour is given via an avatar behaviour trigger (or feature graphs) and the text that the avatar has to speak. This content is stored with temporal information (in regard to its synchronous or asynchronous manner related to the video) within a text file.
- The additional content information for answering user questions is stored in a content database.

The avatar's geometry and animation is the basis for interactive videos cooperating with avatars. The video annotation, as well as the additional avatar text and additional content information, are dependent on the given video.

Content Server

The content server consists of a Real video server and a database containing videos and their relating information: hypermedia annotation (temporal, spatial, hypermedia targets), additional information and SMIL – synchronized avatar behavioural triggers and text.

The information located on the server side at the beginning of a session can be divided into three functional parts:

- The first part contains the necessary information to establish an avatar on the client side that is immediately ready for conversation. Therefore, this is the asynchronous avatar information that is responsible for the conversational interaction and the avatar feedback system (see paragraph 3).
- The second part contains the video and the synchronous information, which is the avatar behaviour trigger and the text that the avatar is speaking on the client side. The video and the synchronous information are played simultaneously, controlled by SMIL.
- The evaluation of the user's interaction is done with part three of the server information which is the temporal, spatial and target information of the hypermedia annotation, as well as the general information for user's conversational interaction. This information is kept on the server side until it is requested by a user interaction. Unnecessary datatransfer between server and client is avoided.

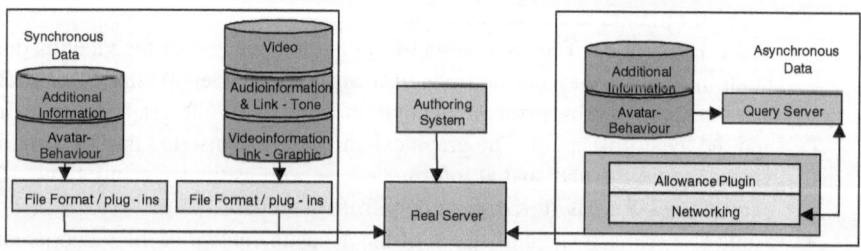

Figure 8: Server Architecture

Figure 8 shows the server architecture from the aspect of data storage. The data storage is hereby divided into the synchronous and asynchronous data. The server contains several RealMedia plug-ins that enable the RealMedia Server to serve the data and to synchronize with SMIL.

Presenter Client

The interaction with the medium video is initiated on the client side. The client side provides the user with the ability to interact with objects occurring in the video in order to retrieve additional information. In addition, users can retrieve information by querying the avatar. The client structure contains six parts, see figure 9:
- an internal media distributor that streams the different media types to the different client plug-ins
- a real video client
- a decision handler, a rule system for computing the avatar behaviour
- the Java 3D avatar renderer
- a speech recognition system that converts user questions to requests on video content
- the request unit that sends a request to the server database

The internal media distributor receives the video and the SMIL-based synchronous and asynchronous avatar behavioural triggers and the synchronous avatar text. The unit transfers the complete asynchronous avatar behavioural triggers to the rule system before it starts to play the video and to transfer the synchronous avatar behavioural triggers and text to the rule system. In the case of a user interaction on the hypermedia annotation of the video, the video client notifies the rule system that an interaction by the user has occurred. The temporal and spatial coordinates of the user interaction is transferred to the content server.

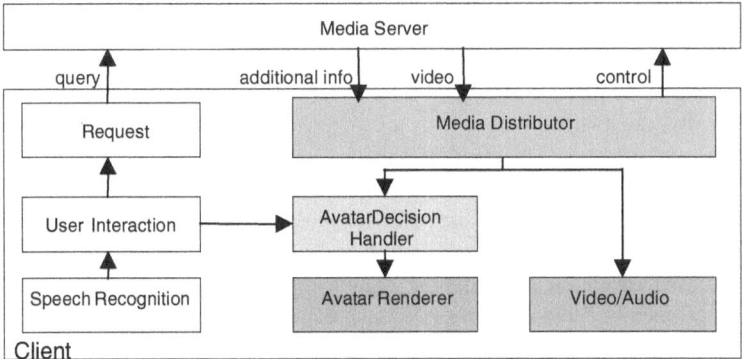

Figure 9: Client Architecture

The rule system calculates the actual behaviour of the avatar. The avatar's behaviour upon user interaction is based on the asynchronous avatar behaviour triggers that are stored before the start of the video presentation. The actual behaviour of the avatar consists of the synchronous avatar behaviour trigger and text, and the actual user interaction behaviour. The calculated behaviour and the avatar's speech in the form of text is transferred to the avatar renderer in the form of avatar behaviour trigger and text.

A speech recognition system calculates the user's speech inputs to send requests on the video content and additional video information which is stored on the server side. These requests are given to the request unit which is performing a query towards the server-side database system.

Client-Server Connection

The client-server connection consists of three RealMedia channels. There are two forward channels from the server to the client, transmitting the video and synchronous avatar information, as well as the additional information channel with the asynchronous avatar information and any other information that is stored in the database. The back channel, the connection from the client to the server, serves the temporal and spatial information of selected hypermedia annotations, as well as requests to the content database.

User Interaction

There are two different ways in which the user can retrieve additional information from the system within the video presentation on the client side.

- One possibility is for the user to interact with the video directly, meaning that he could click onto an annotated object of his interest with a mouse cursor in the video display or activate a Sonic Hyperlink annotated object of the video's acoustic and request additional information stored on the server. Finally, the avatar in his role as a *conferencier* presents the additional information to the user. This user interaction possibility is a direct manipulation onto the objects of the video content.
- The user can also interact with the video content indirectly by asking the avatar questions. This way, additional information can be retrieved through the whole video presentation without any limitation on time-dependent hyperlinks. For this interaction, speech recognition is the interface unit between the user and the video content. This user interaction possibility is conversational onto the objects of the video content.

- Every user interaction, direct as well as indirect, with the video content leads to a feedback reaction of the avatar. In this way, the avatar guides the user through the video from the beginning to the end. As a guide or as a companion within an interactive video presentation, the facial expressions, emotion and speech of the avatar will always be present to encourage the user to work with the video in order to retrieve the desired information.

Figure 5 shows the avatar and the video on the left window side. The result of a user request is displayed on the right side of the window - the avatar is giving some additional information via its voice and facial expressions. Figure 5 shows a prototype application that uses the system presented in this paper to explain some research results of the ZGDV departments.

5   Conclusion

An integrated video on demand system for the authoring, service and presentation of digital interactive video is presented. The digital video's interaction component consists of two parts:
- a direct manipulation component realised with hypermedia continuous media annotation
- a conversational interaction system based on an avatar

The system provides the possibility of time-based (synchronous) and time-independent (asynchronous) interaction on video content to the user. The interaction possibilities are adapted to the video's temporal nature. The user can get direct access to object information independent of the object's visibility. The information may concern objects the user actually sees or objects that were presented previously or will be played back later on within the presentation.

We presented a Server System, a Client Presenter, as well as an authoring environment for an interactive video on demand application.

The next work on the system is the integration of gesture and gesture/speech combination for multimodal user interaction. The digital video interaction will be enhanced to stereo sound and stereo acoustic interaction.

6   References

[Alex99] Alexa, M., and W. Müller: The Morphing Space. *Proceedings of WSCG*, Pilsen, Czech Republic, 1999.

[Bau95] Bauersfeld, P.: *User Interface Challenges of Media Design*. 3. International Multimedia Conference, San Francisco, USA, November 1995.

[Bente97] Bente, G., and P. Vorderer: The Socio-Emotional Dimension of Using Screen Media. Current Perspectives in German Media. PsychologyWinterhoff-Spurk et al. (eds.) *New Horizons in Media Psychology* Westdeutscher Verlag, 1997.

[Böhm93] Böhm, K., Figueiredo, M., and J. Teixeira: Advanced Interaction Techniques in Virtual Environments. *Computers & Graphics*, Journal, Vol. 17. No. 6., Pergamon Press Ltd, pp. 655-661, 1993.

[Braun98] Braun, N., and R. Doerner: Using Sonic Hyperlinks for Web-TV. *International Conference on Auditory Displays ICAD*, Proceedings, Glasgow UK, 1998.

[Braun99] Braun, N., and R. Doerner: Temporal Hypermedia for Multimedia Applications in the World Wide Web. *Proceedings of the International Conference on Computational Intelligence and Multimedia*, ICCIMA99.

[Camp95] Campbell, R.H., Chen, Z., Tan, S.M., and Y. Li: Real Time Video and Audio in the World Wide Web. *Fourth International World Wide Web Conference Proceedings*, Boston, Massachusetts, USA, 1995.

[DeRoure98] DeRoure, D., Blackburn, S., Oades, L., Read, J., and N. Ridgway: Applying Open Hypermedia to Audio. *Ninth ACM conference on Hypertext and hypermedia: links, objects, time and space – structure in hypermedia systems*, Proceedings, ACM Press, Pittsburgh, PA, USA, 1998.

[Finke2000] An Architecture Of a Personalized, Dynamic Interactive Video System, DCC2000, Bradford, Great Britain, April 2000.

[Gerf98] Gerfelder, N., and L. Neumann: Interactive Video on the World Wide Web - Convergence of the WWW and Digital TV. *W3C Workshop "Internet and TV"*, , Sophia Antipolis, France, June 29-30, 1998.

[Hefta97] Hefta-Gaub, B.: The Real Media Platform Architectural Overview. *Real Networks Conference*, San Francisco, USA, 1997.

[Mandel97] Mandel, T.: Social User Interfaces and Intelligent Agents. Chapter 15 in: *The Elements of User Interface Design*, pp.357-382, John Wiley & Sons, New York, 1997.

[Martel96] Martel R.: A Distributed Network Approach and Evolution to HTML for New Media. In: Hoschka, P. (ed) *Real Time Multimedia and the World Wide Web*, No. 25, 1996.

[Pérez96] Pérez-Quiñones, M. A., J. L Sibert: A Collaborative Model of Feedback in Human-Computer Interaction. *International Conference on Computer Human Interaction*, Proceedings, ACM, USA 1996

[SMIL98] World Wide Web Consortium W3C. *Synchronized Multimedia Integration Language (SMIL) 1.0 Specification* http://www.w3.org/TR/REC-smil/ , 1998.

[Saw96] Sawhney, N., Balcom, D. and I Smith: HyperCafe: narrative and aesthetic properties of hypervideo. The seventh ACM conference on HYPERTEXT, Proceedings, Bethesda, MD USA, 1996

[Shne95] Shneiderman, B., and J. Preece: Survival of the Fittest: The Evolution of Multimedia User Interfaces. *Department of Computer Science, Human-Computer Interaction Laboratory, and Institute for Systems Research*, University of Maryland, College Park, MD 20742 USA, 1995.

How to Make a Digital Whiteboard Secure –
Using JAVA-Cards for Multimedia Applications

Rüdiger Weis[1]*, Jürgen Vogel[2], Wolfgang Effelsberg[2], Werner Geyer[2]**, and
Stefan Lucks[3]***

[1] Cryptolabs Amsterdam,
Convergence integrated media GmbH, Berlin, San Francisco, Amsterdam
`ruedi@cryptolabs.org`
[2] Praktische Informatik IV, University of Mannheim,
`{jvogel,effelsberg,geyer}@pi4.informatik.uni-mannheim.de`
[3] Theoretische Informatik, University of Mannheim
`lucks@th.informatik.uni-mannheim.de`

Abstract. In this paper we propose a light-weight, provable secure
smart card integration for the OpenPGP secure message format. The
basic idea is that the secret keys are stored on a smart card and never
leave it. We have integrated this new security approach into an enhanced
whiteboard, the digital lecture board (dlb). Existing whiteboards neglect
security mechanisms almost completely, even though these mechanisms
are extremely important to allow confidential private sessions and billing.
The primary application field of our concept are small and closed groups,
whereas the smart card serves to testify group membership. Our first implementation supports the JAVA i-Button which provides an additional
hardware security.

1 Introduction

Video conferencing via the Internet has become more and more important during
the last few years, its application fields ranging from telemedicine to distributed project meetings in global companies. Typically, a video conferencing system
combines several media types such as audio, video and whiteboard. A whiteboard
offers a shared workspace where slides can be presented to the conference group
and where documents can be edited by all participants. Besides audio, the whiteboard is the most important instrument for sharing knowledge among distributed
participants for many application fields (e.g., distance education [EfGe98]). But
at the same time, the features offered by existing whiteboards often are not sufficient for effective conferencing [GeEf98]. Shortcomings mainly concerned the
user interface, media handling, and the collaborative services needed to support group interaction. To overcome the weaknesses of existing whiteboards, we
developed the digital lecture board (dlb) [Geye98].

---

* A part of this research was done while the author was at the University of Mannheim.
** Since 15. March 2000: IBM, T.J. Watson Research, USA.
*** Supported by the Deutsche Forschungsgemeinschaft (DFG) grant KR 1521

H. Scholten and M. van Sinderen (Eds.): IDMS 2000, LNCS 1905, pp. 187–198, 2000.
© Springer-Verlag Berlin Heidelberg 2000

One important issue we had in mind while designing the dlb was to develop a security concept in order to protect sensitive information and to restrict access to sessions [GeWe98,GeWe00]. In this paper we describe further enhancements to the existing security mechanisms of the dlb in respect of both high security and user-friendliness, especially for small and closed user groups (e.g., personal decisions or business conferences).

The basic idea is to use the new JAVA card technology to provide a secure place for the valuable secret keys that are used to encrypt/decrypt data. One main advantage of JAVA card technology is the ability to load different applications and secret keys very easily. Thus, only one card for is needed different sessions.

In our tests we used the i-Button resp. the JAVA Ring by Dallas Semiconductor [Dall99], which provides a very high hardware security [BWL00].

We also present the integration of the two very promising, fast and free candidates for the DES (Data Encryption Standard) successor Advanced Encryption Standard (AES) Twofish and Rijndael. Additional, we have implemented for the first time the new provable secure DES^2X construction.

The remainder of this paper is structured as follows: after a brief description of the main features of the dlb, we describe the security concept and the new features of the dlb, especially the novel smart card protocols. We conclude this paper with a summary and an outlook.

2   The Digital Lecture Board (Dlb)

Shared Workspace. The user interface of the dlb is depicted in Figure 1. A document created with the dlb consists of an arbitrary number of pages, whereas a page is composed of an arbitrary number of objects such as imported postscript slides and images, graphical objects (e.g., circles, freehand lines, etc.) and text. At a certain point in time exactly one page is displayed in the so–called shared workspace. User actions on the shared workspace (drawing, pointing, etc.) are generally visible to all participants of the session. Furthermore, a private workspace allows the preparation of documents without interfering with the ongoing session. Documents created within the public or private workspace can be stored on disk in an SGML-like file format for later reuse.

Collaboration Tools. Because the number of communication channels normally used in a video conferencing environment is restricted, explicit collaborative services are needed to organize a session and to increase awareness among the participants. For example, technical problems occurring during a session (e.g., bad audio quality due to packet loss) often result in one or more participants writing directly onto the displayed whiteboard page, thereby interrupting the regular session. Furthermore, the designation of a speaker during a lively discussion is awkward in a session with only one audio channel. And the video quality is often not good enough to recognize facial expression or to see a raised hand in a large audience. All these problems increase with the number of

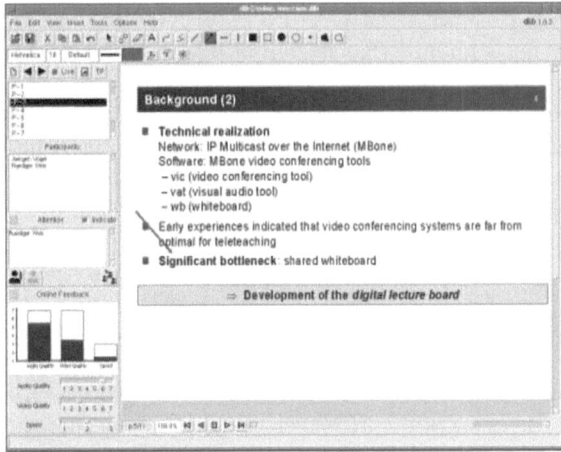

Fig. 1. User Interface of the dlb

participants. Therefore the dlb offers a variety of collaboration tools to support social communication between the participants:

- The integrated chat may be used to discuss technical issues without disturbing the speaker.
- With the help of the attention tool a participant can raise his/her hand to announce his/her intention to speak.
- To create a common point of reference within the shared workspace a telepointer is available.
- Voting allows additional feedback by polling opinions about certain session criteria (e.g., presentation quality).

A more detailed discussion of collaborative services supported by the dlb can be found in [Geye99] and [WeGe99].

Communication Model. The distribution architecture of the dlb is replicated, i.e., each participant's instance of the dlb holds a complete copy of the session data. To maintain a consistent shared state, user events such as drawing and writing are exchanged between all instances. The communication model of the dlb is layered. The *Whiteboard Transfer Protocol* (WTP) is the application protocol of the dlb. WTP defines packet formats and the semantics for creating graphical objects or pages, for telepointer data, etc. WTP packets are the payload of *Realtime Transport Protocol* (RTP) packets [SCFJ96], a protocol that was chosen for several reasons. The timestamps of RTP allow the synchronization with other RTP-compatible data streams (e.g., audio, video). Furthermore RTP makes it possible to use existing MBone recording systems. And RTP provides light-weight session control through RTCP. The security concept described

later on is compatible with the *OpenPGP* (OPGP) standard [Caea97]. OPGP
is an open internet standard that is compatible to the de facto standard Pretty
Good Privacy (PGP). The OPGP layer realizes the encryption/decryption of
the transmitted data, i.e., RTP packets are wrapped into OPGP packets. We
then use either unreliable UDP connections (e.g., for telepointer data) or reliable SMP connections to transmit the OPGP packets. *Scalable Multicast Protocol*
(SMP) is a reliable transport service developed in the context of the dlb project
[Grum97].

Secure Communication. The well–known DES encryption algorithm, which
was originally designed for confidential, not–classified data, is used in many
applications today (e.g., electronic banking). The MBone whiteboard wb also
relies on DES for encryption. However, DES is not secure anymore since e.g.
the non-profit organization Electronic Frontier Foundation (EFF) has built a
hardware DES cracker [EFF98]. In recent years, novel algorithms that perform
better while being similar to the DES scheme have been developed [Weis98].

Due to the US export restrictions, export versions of many software products
have the DES encoding disabled. Thus, outside the U.S., the wb's DES encryption feature has not been avaible for a long time. Moreover the source code of
wb is not publicly available which inhibits the evaluation or modification of the
cryptographic implementation. These security limitations of the MBone whiteboard wb have stimulated the integration of modern encryption algorithms into
the digital lecture board in order to provide secure conferencing with a powerful,
collaborative whiteboard developed outside the US.

User–Orientated Cryptography. The dlb employs a flexible user–oriented
security concept that can be adapted to different user requirements. Users may
choose from predefined security profiles or customize their own security requirements. The choice may be driven, for instance, by legal issues, costs, required
level of security, and performance. We identify the following main profiles or user
groups: *public research*, *financial services*, and *innovative companies*.

Since users who work in *public research* often benefit from license–free employment of patented algorithms, we rely on the *IDEA* cipher. IDEA was designed by Xueja Lai and James Massey. The algorithm has a strong mathematical foundation and possesses good resistance against differential cryptoanalysis. Many cryptographers think that IDEA is the strongest public algorithm
[Schn96]. IDEA was the preferred cipher in PGP (Pretty Good Privacy) until
version 2.63. However, commercial users have to pay high license fees.

In the *financial services* business we find a strong preference for DES–based
systems. Since Single DES has been cracked by brute force attacks, we suggest
to use *Triple–DES* in this application field.

For *innovative companies* which are not afraid of new algorithms, we use
the novel, license–free algorithm *CAST*. Since January 1997, CAST is freely
available. CAST is the preferred cipher in PGP since version 5. The Canadian

government has evaluated and recommended the CAST-128 variation as very secure [Adam98].

In addition to these predefined user profiles, we have implemented options for full compatibility to *PGP 2.63i*, *PGP 5.x* and *GPG* [Koch98]. The "GNU Privacy Guard" (formally known as G10) is a free PGP replacement that does not rely on patented algorithms. GPG prefers *Blowfish* as symmetric algorithm.

3   New Symmetric Approaches

All the algorithms mentioned above are well tested and state–of–the–art in cryptography. In this section we present some innovative approaches we have integrated into the dlb.

3.1   DESX and DES²X

An inexpensive and effective technique to extend the key size of a cipher and to frustrate brute force attacks is whitening. This is the use of a simple key-dependent permutation, such as the bit-wise XOR ($\oplus$).

DESX. A DES variant with whitening is DESX, which was invented by Ron Rivest and is used by RSA Data Security Inc. in their BSAFE toolkit. DESX takes two 64-bit whitening keys M_0 and M_1 and a 56-bit DES-key L as its key:

$$\text{DESX}_{M_0,L,M_1}(P) = \text{DES}_L(P \oplus M_0) \oplus M_1.$$

As proved by Kilian and Rogaway [KiRo96], the DESX construction is sound. Assuming that no new weaknesses of DES itself are discovered, and that the attacker cannot learn an excessively large number of plaintext/ciphertext pairs (more than 2^{32} such pairs would qualify as *excessive*), the result from Kilian and Rogaway implies that the attacker needs an amount of work which is infeasible today (i.e. more than 2^{85} DES encryptions). Note that the actual key size of DESX is 184 bits[1], while the effective key size is more than 85 bits.

DES²X. According to Moore's law, computer hardware gets four times faster every three years. For this reason, one may want to extend the effective key size of DESX for long-term security. We have implemented DES²X [Luck98], which combines whitening and double encryption. DES²X takes three whitening keys M_0, M_1, and M_2 and two DES-keys L_1 and L_2 and is defined by

$$\text{DES}^2\text{X}_{M_0,L_1,M_1,L_2,M_2}(P) = \text{DES}_{L_2}(DES_{L_1}(P \oplus M_0) \oplus M_1) \oplus M_2.$$

---

[1] For some applications, a 184-bit key may be too long. As was pointed out by Kilian and Rogaway [KiRo96], 120 bits of key size are sufficient to achieve the same level of security: just set $M_2 = M_1$ (see also [LuWe00]).

3.2   Twofish

Twofish [Scea98] is the AES candidate of Bruce Schneier, John Kelsey, Doug Whiting, David Wagner, Chris Hall and Niels Ferguson. Twofish is a 128-bit block cipher with a variable-length key up to 256 bits. The cipher is a 16-round Feistel network with a bijective F-function using four key-dependent 8-by-8-bit S-boxes, a fixed 4-by-4 maximum distance separable (MDS) matrix over $GF(2^8)$, a pseudo-Hadamard transform (PHT), fixed length and bitwise rotations ($<<<$). The key schedule uses the same construction elements as the round function. Twofish is carefully designed for high speed encryption on the Intel Pentium family. A fully optimized version encrypts one byte in 18.1 clock cycles on an Intel Pentium. We can use a variety of performance tradeoffs with respect to the key schedule.

Twofish has a conservative design and all elements are well evaluated in many research papers. This construction seems to provide a large margin of security.

3.3   Rijndael

Rijndael is the AES candidate of Joan Daemen and Vincent Rijmen [DaRi98]. Rijndael is not a Feistel cipher but a new structure of a Substitution-Permutation Network. Rijndael is a member of the SQUARE family and offers both variable block length and key size. We use the AES version with 128 bit blocksize, 128 bit key length and 10 rounds.

Rijndael uses a square scheme, combines an 8×8-S-box substitutions with a MDS code oriented linear mapping and cyclic shift operations. The main criteria for using these building blocks are resistance against know attacks (especially Differential and Linear Cryptoanalysis), speed and code compactness on different hardware platforms, and design simplicity.

4   A Flexible Framework for Highly Secure Communication

Any security architecture is only as strong as the protection of its secret keys. Thus one main problem is that there is no "safe place" for secret information for a computer that is connected to the Internet.

4.1   Smart Card Is the Key

Many security-relevant applications store secret keys on a tamper-resistant device, in most cases a *smart card*. Protection of the valuable keys is the card's main purpose. Although in recent years some interesting cryptographic [Weis97], and many very dangerous hardware [WKT97] attacks have been mounted, smart cards provide much higher security than other storage systems [WBL00]. But the physical limitations of smart cards make them typically much to slow for high-bandwidth applications. Fortunately, remotely keyed encryption schemes are designed to allow "High-Bandwidth Encryption with Low-Bandwidth Smart cards" [Blaz96]. (This is accomplished with the help of a fast but untrusted host.)

This considerations leads to the following minimum requirements for a smart card supported encryption protocol [Weis00] :

- The secret master key *must never* leave the card.
- The protocol has to be secure against a master key recovery attack.
- It must be infeasible to decrypt a ciphertext packet without breaking the host cipher or the pseudorandom function on the card.

There are much stronger security models by Lucks [Luck97] (s.a. [Weis99b]), and Blaze, Naor and Feigenbaum [BFN98]. But all of their protocols require a much greater amount of operations on the host side and on the smart card system. In the next section we present a very fast and easy to understand encryption protocol.

4.2   OpenPGP Key Generation

In the Passphrase–Only scenario we use the *StringToKey* (S2K) of the OpenPGP standard. The secret passphrase has to be distributed to the participants in advance. This can be managed by means of secure mail (e.g., GNU Privacy Guard), an alternative channel, or of course by public key cryptography. In this subsection we discuss the two main StringToKey algorithms. These algorithms are used to convert the user's passphrase to a symmetrical encryption key.

Simple S2K. The Simple S2K algorithm directly hashes the string to produce key data: KeyData := HASH(Passphrase).

Salted S2K. The *Salted S2K* includes a 64 bit random number ("Salt") in the S2K specifier that gets hashed along with the passphrase string, to help prevent dictionary attacks: KeyData := HASH(Salt‖Passphrase).

4.3   Integrating Smart Card Support into OPGP

The key idea of our protocol is to send the result of the S2K algorithm to the smart card. The card encrypts the message using the secret card key and sends the result back to the host. The host uses this result as session key. The smart cards with the longterm card key have to be distributed in advance. For cryptographic details please refer to [Weis00].

4.4   Two-Attribute Authentification

It is surprising that the most secure scenario is very easy to integrate into the existing whiteboard application. The main idea is quite simple and provides very high security: *"To have something and to know something."* This strategy is used, e.g., for getting cash with ones credit card: If somebody steals the card the thief cannot get cash if she does not know the PIN. Similarly an intruder who can read the passphrase is not able to encrypt messages if she has no access to the card.

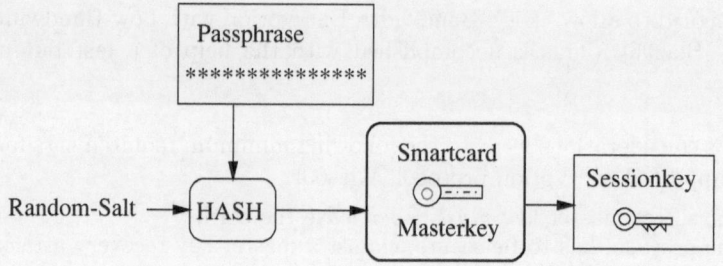

Fig. 2. Simple Two–Attribute Authentification.

5   Smart Card–Only Construction

If we want to have a system that uses only a smart card but no user passphrase, we choose a variation of the *Trivial Host Card Encryption Protocol* [Weis00]. The THCEP was designed by Weis in the context of video encryption schemes [WeBo00] (s. Fig 3).

Fig. 3. Integration of JAVA i-Button with THCEP on CEBIT 2000 [WeBo00].

5.1   Trivial Host Card Encryption Protocol (THCEP)

Encryption Protocol

1. The host generates a random number r-bit R.
2. The host sends R to the card.
3. The card performs the pseudo random mapping $F : \{0,1\}^r \rightarrow \{0,1\}^k$ using master key M

$$S := F_M(R).$$

4. The card sends S back to the host.
5. The host encrypts the stream P using S as session key $C := E_S(P)$.
6. The host sends $\{R, C\}$ to the receiver.

Note that we can use a cheap "non-encryption" (e.g., a signature card) card to implement the pseudo random mapping F [Weis99a].

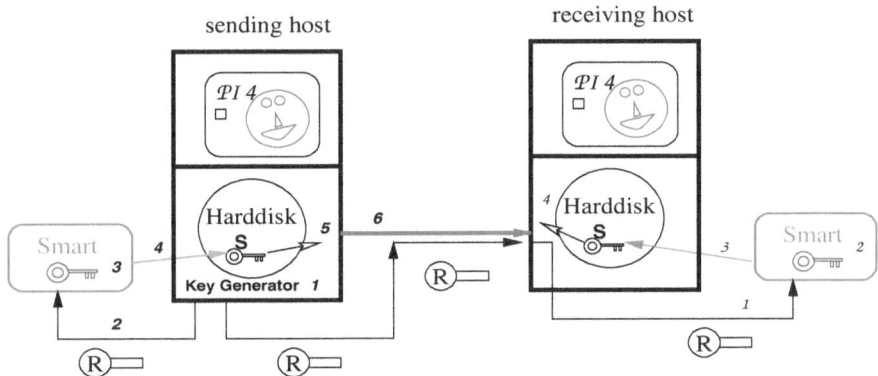

Fig. 4. Trivial Host Card Encryption Protocol [Weis00]

Decryption Protocol

1. The receiver sends R to the card.
2. The card performs the pseudo random mapping $F : \{0,1\}^r \rightarrow \{0,1\}^k$ using master key M

$$S := F_M(R).$$

3. The card sends S back to the host.
4. The receiver decrypts the stream C using the card's response $P = E_S^{-1}(C)$.

5.2   Integrating THCEP into OpenPGP/SmartCardOnly

Now we discuss the integration of THCEP into our OPGP framework. The most trivial idea is to use a fixed, known passphrase together with the salted S2K algorithm and the Two–Attribute scheme discussed in the last section. If we have different random salts, even with a fixed passphrase different keys are generated. But OPGP only uses 64 bit of random salt. Because of this fact we generate random bits as the "passphrase". We transmit these random bits as a user-specific OpenPGP packet.

6 Implementation and Experiences

All of the security profiles, ciphers and hash functions presented in section two and three have been successfully implemented in C++ and Tcl/Tk. They are available with the current version 1.8.3 of the dlb. User handling of these algorithms is quite simple, since only a profile selection has to be done and the correct passphrase has to be entered.

An implementation of THCEP was shown at the CEBIT 2000 in context with the encryption of video streams [WeBo00]. The integration of the smart card protocols is currently under way and should be straightforward, since only minor changes of dlb code are necessary. Next step will be a field trial in a closed group scenario such as a regular teleteaching lecture, preferably using i-Buttons resp. JAVA rings (see Fig. 5).

Fig. 5. JAVA Ring and BlueDot receptor [BWL00].

7 Conclusion

The usage of smart cards as a storage device for the long-term keys offers two major advantages: First, it increases the security since the master key is stored on the smart card and never on the host. The i-Button, for instance, proves to be reasonably resistant against both hardware and software attacks. Second, if the smart card-only protocol is chosen, user-friendliness increases, since a participant does not need to remember a personal key. This improves the security in two ways. First, a key on the card can hardly fond by social engineering and, second, a good random generator will generate a much better (higher) entropy than humans.

Future work will target the successor of the dlb, the so-called multimedia lecture board (mlb). Besides new security mechanisms and payment schemes (s.a. [WEL00b]), the mlb will integrate different media types such as presentation animations, HTML, audio, and video into a combined video conferencing platform.

Acknowledgements. The authors would like to thank Bastiaan Bakker, Andreas Bogk and Peter Honeyman for their help to make the i-Buttons usable and Sascha Kettler for supporting the implementations.

References

[Adam98]   Adams, C.,"CAST-256", AES submission, 1998.
[BWL00]   Bakker, B., Weis, R., Lucks, S., "How to Ring a S/WAN. Adding tamper resistant authentication to Linux IPSec", SANE2000 - 2nd International System Administration and Networking Conference, Maastricht, 2000.
[Blaz96]   Blaze, M., "High-Bandwidth Encryption with Low-Bandwidth Smartcards", Fast Software Encryption , Springer LNCS 1039 1996.
[BFN98]   Blaze, M., Feigenbaum, J., and Naor, M., "A Formal Treatment of Remotely Keyed Encryption" , Eurocrypt '98, Springer LNCS 1403, 1998.
[Caea97]   Callas, J., Donnerhacke, L., Finnley, H., "OP Formats - OpenPGP Message Format", Internet Draft, November 1997.
[DaRi98]   Daemen, J., Rijmen, V., "Rijndael", AES submission, 1998.
            http://www.esat.kuleuven.ac.be/~rijmen/rijndael/
[Dall99]   Dallas Semiconductors, iButton Hompage: http://www.ibutton.com/
[EfGe98]   Effelsberg, W., Geyer, W., "Tools for Digital Lecturing - What We Have and What We Need", Proc. BITE '98, Bringing Information Technology to Education, Integrating Information & Communication Technology in Higher Education, Maastricht, Netherlands, March 1998.
[EFF98]   Electronic Frontier Foundation, "EFF press release (July 17, 1998): EFF Builds DES Cracker that proves that Data Encryption Standard is insecure", http://www.eff.org/descracker/
[Geye98]   Geyer, W., "The digital lecture board (dlb)"
            http://www.informatik.uni-mannheim.de/ geyer/dlb/dlb.eng.html
[Geye99]   Geyer, W., "Das digital lecture board - Konzeption, Design und Entwicklung eines Whiteboards für synchrones Teleteaching" (in German), Reihe DISDBIS, Bd. 58, ISBN 3- 89601-458-7, Infix-Verlag, St. Augustin, 1999.
[GeEf98]   Geyer, W., Effelsberg, W., "The Digital Lecture Board – A Teaching and Learning Tool for Remote Instruction in Higher Education", ED–MEDIA '98, Freiburg, Germany, June 1998.
[GeWe98]   Geyer, W., Weis, R., "A Secure, Accountable, and Collaborative Whiteboard", Workshop on Interactive Distributed Multimedia Systems and Services, IDMS '98, Oslo, September 1998.
[GeWe00]   Geyer, W., Weis, R., "The Design and the Security Concept of a Collaborative Whiteboard", Computer Communications 23, Elsevier, 2000.
[Grum97]   Grumann, M., "Entwurf und Implementierung eines zuverlässigen Multicast–Protokolls zur Unterstützung sicherer Gruppenkommunikation in einer TeleTeaching–Umgebung", Master's Thesis (in German), Lehrstuhl für Praktische Informatik IV, University of Mannheim, 1997.
[KiRo96]   Kilian, J., Rogaway, P., "How to protect DES against exhaustive key search", Proc. Advances in Cryptology–Crypto '96, Springer, 1996.
[Koch98]   Koch, Werner, "The GNU Privacy Guard", 1998. http://www.gnupg.org/
[Luck97]   Lucks, S., "On the Security of Remotely Keyed Encryption", Fast Software Encryption, Springer LNCS, 1997.

[Luck98]   Lucks, S., "On the Power of Whitening", Manuscript, Universtität Mann-
              heim, Fakultät für Mathematik und Informatik, 1998.
[LuWe99]   Lucks, S., Weis, R., "Remotely Keyed Encryption Using Non-Encrypting
              Smart Cards". USENIX Workshop on Smartcard Technology, Chicago,
              May 10-11, 1999
[LuWe00]   Lucks, S., Weis, R., "How to Make DES-based Smartcards fit for the 21-st
              Century", CARDIS2000, Bristol, Kluwer , 2000.
[Schn96]   Schneier, B., "Applied Cryptography Second Edition", John Wiley &
              Sons, New York, NY, 1996.
[Scea98]   Schneier, B., Kelsey, J., Whiting, D., Wagner, D., Hall, C., Ferguson, N.,
              "Twofish", AES submision, 1998.
[SCFJ96]   Schulzrinne, H., Casner, S., Frederick, R., Jacobsen, V., "RTP: A Trans-
              port Protocol for Real–Time Applications", Internet RFC 1889, IETF,
              Audio–Video Transport Working Group, 1996.
[Weis97]   Weis, R., "Combined Cryptoanalytic Attacks against Signature and En-
              cryption schemes", (in German), A la Card aktuell 23/97, S.279, 1997.
[Weis98]   Weis, R., "Moderne Blockchiffrierer" (in German), in: "Kryptographie",
              Weka–Fachzeitschriften–Verlag, Poing, 1998.
[Weis99a]   Weis, R., "Crypto Hacking Export Restrictions", Chaos Communication
              Camp, Berlin, 1999.
[Weis99b]   Weis, R., "A Protocol Improvement for High-Bandwidth Encryption
              Using Non-Encrypting Smart Cards", IFIP TC-11, Working Groups 11.1
              and 11.2, 7 th Annual Working Conference on Information Security Ma-
              nagement & Small Systems Security, Amsterdam, 1999.
[Weis00]   Weis, R., "A Trivial Host Card Encryption Protocol", Technical Report,
              Universität Mannheim, Feb. 2000.
[WBL00]   Weis, R., Bakker, B., Lucks, S., "Security on your hand: secure file systems
              with a "non-cryptographic" Java-Ring", IRUSA/INRIA/JAVA CARD
              FORUM: JAVA CARD Workshop, Cannes, 2000.
[WeBo00]   Weis, R., Bogk, A., "Videoencryption with the JAVA i-button", CE-
              BIT2000, Hannover, 2000.
              http://www.informatik.uni-mannheim.de/~rweis/cebit2000/
[WEL00a]   Weis, R., Effelsberg, W., Lucks, S., "Remotely Keyed Encryption with
              Java Cards: A Secure and Efficient Method to Encrypt Multimedia
              Stream", IEEE International Conference on Multimedia and Expo, New
              York, Kluwer, July 2000.
[WEL00b]   Weis, R., Effelsberg, W., Lucks, S., "Combining Authentication and
              Light–Weight Payment for Active Networks", Smartnet2000, Wien, Klu-
              ver, 2000.
[WKT97]   Weis, R., Kuhn, M., Tron, "Hacking Chipcards", Workshop CCC '97,
              Hamburg 1997.
[WeGe99]   Weis, R., Geyer W., "Cryptographic Concepts for Online-Feedback in
              Teleteaching Applications", Proc. NLT '99, Bern, Switzerland, 1999.
[WGK00]   Weis, R., Geyer, W., Kuhmünch, C., "Architectures for Secure Multicast
              Communication", SANE2000 - 2nd International System Administration
              and Networking Conference, Maastricht, 2000.

How to Keep a Dead Man from Shooting

Martin Mauve

University of Mannheim, Germany
mauve@pi4.informatik.uni-mannheim.de

Abstract. The state-of-the-art approach to realize consistency in distributed virtual environments (e.g., action games, multi-user virtual reality, and battlefield simulations) is dead reckoning. A fundamental problem of dead reckoning is that participants may perceive different states for the same entity. While these inconsistencies will eventually be repaired, the resulting state of an entity may be incorrect. In this position paper we will investigate the reasons for this type of problem and propose an improved consistency approach. Our approach guarantees that all participants of a session will eventually agree on what has really happened and how the correct state of the medium should look like. In order to reach this aim we propose to transmit events as additional information, thereby allowing a simple timewarp algorithm to reconstruct the correct state of the distributed virtual environment.

1 Introduction

Both, in research and commerce, distributed virtual environments (DVEs) attracted much attention over the recent years. Examples for DVEs include networked computer games [1], multi-user virtual reality [3] and military battlefield simulations [2]. One key problem for DVEs is to maintain a consistent shared state. The current state-of-the-art approach to realize consistency in DVEs is to use dead reckoning [6]. In this paper we will demonstrate that traditional dead reckoning mechanisms may fail in ways that cause significant harm to the overall state of the DVE. Dead-reckoning does not allow to identify the situations in which it fails and, even if these situations could be identified, it does not provide the required information to repair the problem. Some DVEs may be able to accept and ignore this problem, e.g., in a large scale battlefield simulation the fate of an individual entity may not be considered decisive. However, many DVEs, such as distributed computer games, can not afford to ignore the problem.

We therefore propose to use a different mechanism to keep the state of a DVE consistent. This mechanism can be used either in combination with, or as a replacement for, dead reckoning. While it does not prevent brief periods of inconsistent state, it does allow applications to recognize and repair the problem based solely on information that is locally available. In particular no state management protocol or centralized authority is required.

2 Dead Reckoning: Why a Dead Man May Be Able to Shoot

Dead reckoning is a combination of state prediction and state transmission that is commonly used to keep the state of a distributed virtual environment consistent. It is a distributed approach that does not require a centralized server. Using a distributed

H. Scholten and M. van Sinderen (Eds.): IDMS 2000, LNCS 1905, pp. 199-204, 2000.

approach is mandatory for many DVEs in order to avoid the well known problems of centralized systems such as increased latency, single-point-of-failure and lack of scalability.

In order to use dead-reckoning the distributed virtual environment is partitioned into entities. Examples for entities are a person, a car, a plane or a bullet. Each of these entities is controlled by exactly one application that participates in the DVE. All applications that are interested in an entity are able to predict how the state of the entity will change over time. For example a plane will be predicted to change its position depending on its velocity and heading. The controller of an entity regularly checks whether the difference between the prediction and the actual state exceeds a certain threshold. If this is the case the controller of the entity transmits the state so that other applications may learn about the correct new state of the entity.

Generally there may be two reasons why the real state of an entity may differ from the predicted state. First, the prediction of remote entities may be inaccurate. This is commonly the case when the behavior of an entity cannot be fully duplicated at remote sites due to its computational complexity. This problem can be solved by providing sufficient computational resources. A much more fundamental reason for divergent states are user actions. For example, a user might have changed the heading or the velocity of an entity. This is a problem that cannot be solved by adding processing power since user actions are not predictable in a precise way. For the remainder of this work we assume that the prediction of state changes is accurate and that the only source of state divergence are user interactions.

Dead reckoning has properties that are desirable for DVEs. The first, and maybe most important one, is that it is self-healing. A lost state transmission will eventually be compensated by another state transmission for the same entity. Furthermore there is no need for a centralized management of states. Each application is able to manage the state of the entities locally by means of prediction and received state updates. Because of these properties dead reckoning is used for many DVEs.

However, there is also an important disadvantage to using dead-reckoning: at certain times applications may hold distinct states for the same entity. Because of network latency and packet loss the difference between the predicted and the real state may significantly exceed the threshold that triggers the transmission of a state for the entity by its controller. While these inconsistencies will eventually be repaired, the resulting state of an entity may be incorrect.

Let us consider two examples for this problem. Both examples are taken from a military background. The same problems can occur for distributed sport games or in multi-user virtual reality, although with a somewhat less dramatic effect.

A Dead Man that Shoots. Consider the following situation that could happen in an action game: player A shoots player B. Using dead reckoning the controlling application of player A will create a bullet entity with a certain heading and velocity. It will then transmit the state of that bullet entity. Upon receiving the state of the bullet, remote applications will start to check whether any entity that they control is hit by the bullet. Unfortunately there is a certain network delay between the time A transmits the initial state of the bullet and the time B receives the state. This network delay may be much larger (in the order of 100ms or more) than the amount of time that the bullet needs to hit its target. During this time player B might take actions, e.g., shoot at another player C, even though he should not be able to do so (because he's dead). In the presence of packet loss, the delay may increase tremendously, giving B an even longer time to take actions. Furthermore, at the time player B receives the state of the bullet he might have moved so that the bullet would not hit him. Players A, B, and C may therefore disagree about whether a hit has been scored on B or not. By simply transmitting states dead reckoning neither allows the detection of this problem nor does it provide any help in solving it. Therefore most DVEs that use dead reckoning simply ignore the problem and let the target decide whether it got hit or not.

A Flying Tank. Another example is depicted in Figure 1. In this example application A is the controller of a tank, which currently moves straight ahead. A mine is controlled by application B. The mine should explode if the tank moves over it. At some time the driver of the tank decides to change the heading to a course that would intersect the mine. As usual this will lead to a divergence between the predicted and the actual state of the tank which will cause A to transmit the new state. However for the time that is required to receive the updated state information, B will predict that the tank continues to move straight ahead. At the time the updated state is received by B, it predicts the position 1. After the new state for the tank has been examined B learns that the actual position of the tank should be 2. However, B does not know how the tank got there, so it may not trigger the explosion of the mine. As with the problem described above DVEs that rely on dead reckoning typically ignore this problem.

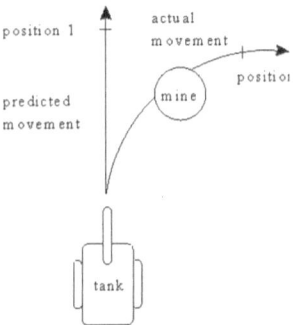

Fig. 1. Flying Tank Problem

The list of examples where dead reckoning leads to unreasonable results can be arbitrarily extended. It is therefore interesting to investigate the fundamental reason for these problems. At a first glance it seems to be the delay introduced by the network that causes the undesirable behavior of entities in DVEs. Unfortunately the delay is very hard to reduce. Steps may be taken to minimize packet loss and optimize buffering delay at routers; however, a certain amount of delay will always be present. In the Internet this amount of delay can be expected to be significant, especially in a geographically wide-spread session.

Therefore we must accept that the state of entities will become inconsistent from time to time, especially when a user controls the entity. This is not the point that we criticize about traditional dead reckoning. What we do view as a shortcoming is that traditional dead reckoning approaches undertake no efforts to detect and repair the problems that arise from these inconsistencies.

3 Timewarp: Preventing That a Dead Man Shoots

Why is a DVE that relies on dead reckoning not able to detect and repair such problems as those described in the two examples above? The answer is simple: because applications do not have adequate information about entities that they do not control.

Consider the flying tank problem. When dead reckoning is used then the application controlling the mine simply gets the information about the tank's new position and velocity. It does not get the information when the tank turned. However without this information it cannot decide whether the tank triggered the mine or not.

In order to detect and repair problems like those described above, we therefore propose to transmit additional information from the controller of an entity to the other participants in the session. Whenever the state of an entity changes because of external influences that cannot be predicted (such as user actions), a piece of information is produced that may be vital to the overall behavior of the DVE. We call this piece of information an event.

Events carry a timestamp of the time they took place. This assumes that a common time base is available for all participants. This assumption is common for almost all DVEs; they typically use NTP [5] or the GPS timer. While such a common time base may not be sufficient to establish a full ordering on events (e.g., several events may have the same timestamp) this does not pose a problem. Events with the same timestamp are either interpreted to be simultaneous or an additional criterion is used (e.g., unique participant identifier) to establish a full ordering relation on events with the same timestamp.

Events are distributed to all participants of a session. There are different ways to do this which we will discuss later. For now let us assume that events will be transmitted in a way so that all participants in a DVE will eventually receive all events that are relevant to them.

With the information about events the problems described above can be avoided by using the timewarp algorithm depicted in Figure 2. It works as follows:

- Each application periodically makes snapshots of the state of all entities. Old snapshots are deleted when it becomes sufficiently unlikely that a delayed event arrives which pre-dates the recorded state (depending on the network this time will be in the order of a few seconds).
- Each application keeps a list of the events that it has received in the past. Events from this list are deleted if a state snapshot that succeeds the event is deleted.
- When a new event arrives it is inserted into the list of events.
- A timewarp is performed to the last recorded state prior to the new event. In the example this would be state 2. The current state is then re-computed based on the last recorded state and the events that have occurred since that state has been recorded, including the new event. Events that are performed after the new event may be modified by the occurrence of that event.

Let us now consider how the two examples would be handled with the timewarp approach. In the „dead man that shoots" example the event of player A firing a gun will still arrive late at player B's machine, possibly after he himself fired an illegal shot at C.

However, since all participants are guaranteed to receive both events and since they use timewarp if any one of the events arrives late, they will be able to agree on the correct state. In this case the correct state require a re-interpretation of the shoot event from participant B. All three applications will ignore this event, once they have been informed about the successful shoot event from A.

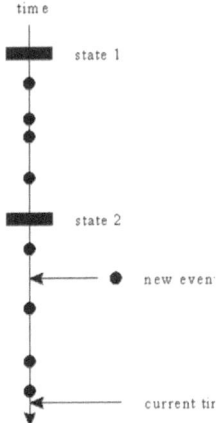

time

state 1

state 2

new even

current tir

Fig. 2. Timewarp

In the „flying tank" example the controlling application of the mine will eventually receive the event that the tank has turned at a certain position. Performing the timewarp algorithm, it learns that it should have exploded and can take the appropriate actions.

The timewarp approach derives its effect from the fact that it ensures that all applications will eventually agree on what has really happened. The state of the DVE will be consistent and correct when all events have been delivered to all participants and all required timewarps have been executed. Even while events are being generated and transmitted the state produced by timewarp converges towards the consistent and correct state. Traditional dead reckoning approaches on the other hand converge towards a consistent state that may be incorrect. Our approach provides this service without requiring any centralized component. As long as each participant will eventually receive all events, only local information is required to perform the timewarp.

Two questions remain open. The first is how to guarantee that each application will really get all events. Basically this problem can be solved in two ways. The first approach is to extend dead reckoning by appending the events to the entity's state. Whenever the entity's state is transmitted, the list of events caused by the entity is also transmitted. Events may be deleted from this list after it becomes unlikely that a participant has not yet learned about the event. This approach can be used as an extension to the regular dead reckoning algorithm. The second approach is to use reliable transmission for events. Compared to the transmission of states this has the additional benefit that multiple participants may control (i.e., generate events for) a single entity in a joint way.

The second question concerns the complexity of the timewarp algorithm. It is clear that the timewarp algorithm has a higher computational complexity than the plain dead reckoning approach. For small numbers of entities (such as in a virtual soccer game) this does not pose an unsolvable problem. For large DVEs, however, it will become infeasible to use the timewarp algorithm for the entire DVE. As the number of entities and events grows it becomes important that the DVE is partitioned so that the timewarp can be performed with acceptable computational resources. Such a partitioning is not only required to make the timewarp less demanding but also to reduce the network traffic that each individual participant needs to receive.

4 The Road Ahead

One challenging problem that remains to be solved is how to prevent that the rendering of inconsistent state information (e.g., the erroneous explosion of a mine) distracts the user. A possible solution for this problem is to delay the rendering of critical state changes (explosion, player death, etc.) until it is likely that the rendered state change is correct. Alternatively one could try to reduce the number of situations where a timewarp is required to repair an inconsistent state.

In [4] we explored one possibility in this direction by introducing a voluntary delay before local user actions are applied to an entity. Thereby the number of short termed inconsistencies caused by the network delay can be significantly reduced. This, in turn, reduces the number of timewarps that need to be performed by the application.

In addition to preventing inconsistent information from distracting the user, a reduction of timewarps also reduces the computational effort for keeping the state consistent. The reduction of the computational complexity is the second challenge that needs to be addressed in future work. We expect that the partitioning of DVEs will play an important partin this area.

References

1) L. Gautier, C. Diot. Design and Evaluation of MiMaze, a Multi-player Game on the Internet. In: Proc. of IEEE ICMCS'98, Austin, Texas, USA, 1998, pp. 233-236.

2) IEEE Computer Society: IEEE standard for information technology –protocols for distributed simulation applications: Entity information and interaction. IEEE Standard 1278-1993. New York: IEEE Computer Society, 1993.

3) M. R. Macedonia. A network software architecture for large scale virtual environments. Ph.D. dissertation, Naval Postgraduate School, Monterey, CA, 1995.

4) M. Mauve. Consistency in Continuous Distributed Interactive Media. Technical Report TR-9-99, Department of Computer Science, University of Mannheim, 1999.

5) D. L. Mills. Network Time Protocol (Version 3) specification, implementation and analysis. DARPA Network Working Group Report RFC-1305, University of Delaware, 1992.

6) S. Singhal, M. Zyda. Networked Virtual Environments Design and implementation, ACM press, New York, 1999.

Building Web Resources for Natural Scientists

Paul van der Vet

Faculty of Computer Science, University of Twente, P.O. Box 217, 7500 AE
Enschede, the Netherlands; voice +31 53 489 3694; email `vet@cs.utwente.nl`

Abstract. Natural scientists increasingly rely on web-based information resources. The speed with which these data will be brought to the
scientist's desktop in the near future, however, makes only too clear that
automated support for efficient and effective use is in its infancy. In fact,
most data are still processed by a combination of manual work and *ad hoc*
programming. We propose to alleviate some of the problems by means of
a research environment: a web site with highly graphical user interface
that allows transparent access to resources and performs a certain degree
of fusion of the information found.

1 Introduction

Natural scientists (biologists, chemists, geologists, ...) increasingly rely on web-
based information resources. In highly competitive, front-line research areas such
as supramolecular chemistry and molecular biology, knowing your way around
in often thousands of web sites can make the difference between a fair and a
famous group. In a compelling scenario, de Jong and Rip [4] sketch the activities
of a research group in molecular biology who just have identified an unexpected
experimental outcome. The group is able to interpret the finding and submit a
paper to an established journal in a few days by making heavy use of Internet.
They have searched data bases and knowledge bases for similar findings and
they have remotely run qualitative simulation programs to predict experimental
outcomes given particular theoretical assumptions. As a matter of important
detail, the group includes a software librarian who knows where to find resources,
what they do, and how they can be operated. Sadly, the scenario still is pure
fiction.

In the present paper, I will take the information chain (or, rather, a simplified
version of it) as a guide for an exploration into the possibilites of web interfaces
that serve researchers' needs and may bring the scenario of De Jong and Rip
closer to reality. We plan to build one for molecular biology, in cooperation with
the Center for Molecular and Biomolecular Information (Nijmegen, the Netherlands) and the European Molecular Biology Research Laboratory (Heidelberg,
Germany).

From the point of view of the end-user, the information chain can be roughly
characterised as a four-step process.

Step (a) Identify needs. Searching information presupposes at least a rough
idea of what is needed. *Step (b) Identify location,* traditionally the province of

H. Scholten and M. van Sinderen (Eds.): IDMS 2000, LNCS 1905, pp. 205–210, 2000.
© Springer-Verlag Berlin Heidelberg 2000

Information Retrieval (hence: IR). On WWW, search engines help sometimes. More often, one knows the location from other sources. *Step (c) Obtain information*, a laborious process in the old days that has changed dramatically through the introduction of WWW. *Step (d) Process information*. Having obtained the information, there remains the frustrating step of putting the information to further use. Data are scattered over many resources and rarely conform to the format demanded by the application at hand. Knowledge is often expressed in a representation language the user's programs cannot handle. Therefore, most processing is still done by hand.

2   Automating the Information Chain

Workers in natural sciences have to cope with a massive information supply. Technical developments will see the merging of steps of the information chain for reasons of both effectiveness and efficiency. Merging steps assumes automated tools that possess awareness of what the information conveys, in other words, of *content*.

ICT research so far has devoted its main efforts on steps (b) and (c) of the information chain, concentrating for step (b) on full-text search. Content providers like Chemical Abstracts Services tend to favour 'classical' approaches to the IR problem. At the same time, they widen their product range to include all kinds of secondary information. For content providers, acquisition of vendors of secondary information is a strategic investment because the company can offer its clients both the means to locate information and the information itself.

Indeed, there is no need to keep steps (b) and (c) separate, as, for instance, the SRS system [1] nicely demonstrates. Once the needed information is located, obtaining it is just a click away. For now, SRS and its likes only provide free information this way. For instance, SRS does not provide links to primary journal articles because accessing the articles themselves takes a subscription. Merging steps (b) and (c) in practice, therefore, presupposes an interface that incorporates an advanced payment system. The end-user has to be shielded from administrative details while at the same time total costs incurred so far have to be known. Also, the end-user may already have a subscription to some of the resources and does not want to pay twice.

Merging steps (b) and (c) leaves step (d). Reuse of information obtained elsewhere requires a dedicated combination of middleware and application software.

3   Systematisation of Domain Knowledge

Automated exchange of scientific information requires a certain degree of systematisation of domain knowledge. Within the field of artificial intelligence, systematisation of knowledge has long been a key issue. One of the important ingredients of any systematisation is a shared commitment to employ particular concepts for well-defined purposes. A concept system or ontology lays down such a commitment (see, for instance, [7], chapter 8). An ontology is a limitative, structured system of concepts. Limitative: the commitment is not to use other concepts. Structured: concepts are related; if possible, concepts are defined in terms of other concepts. Concepts: the contrast intended is with natural-language terms, taking the disambiguation provided by a good thesaurus or keyword system one step further.

A concept has to be identified by a name. Very often, this is a natural-language term, but since ontologies are designed for computer manipulation we can also decide to use another naming system, for instance, a system in which concept names reflect the definitions of the concepts [8]. The translation into something humans understand is then performed by an interface. In fact, developments in imaging and virtual reality allow even further abstraction from natural language. Chemists are already quite used to systems that visualise complex molecules and allow them to grab the molecule and turn it. In such a system, the molecule is depicted as a graph where the nodes stand for atoms and the edges for chemical bonds. The nodes and edges in the picture stand for concepts, and the set of such pictures can be regarded as an ontology of molecules. A further step, using techniques from virtual reality, would be to allow chemists to step into the molecule, inspect parts of it, and (for instance) push or pull to get an idea of the forces governing the molecule's shape. Even more exciting is the possibility to do this while the molecule is involved in a chemical reaction. These possibilities rely in part on the power of pictorial languages that abstract from reality.

4 Research Environments and the Role of Middleware

Processing information obtained from scattered and heterogeneous sources means that the researcher can plug the information she has found into the desktop application she happens to be running. For instance, a paper about the melting point at standard pressure of a particular substance was sought only because that value was needed for a calculation performed by some package. Ideally, the researcher should be able to tell the system so. In response, the system would extract the value from the text in a form suitable for the package and hand it over to the package without further manual intervention. More complicated jobs may require wiring together several databases, knowledge bases, and programs, including the researcher's desktop application. The particular configuration constitutes a research environment. It will only be useful for a short time, say between half an hour and a couple of days, and fine-tuning may occur frequently. We call such systems *coalitions*. A graphical user interface that allows the user to wire the coalition in a manner analogous to making a Lego object will significantly improve the accessibility and manoeuverability of these complex information spaces.

Long-lived research environments are equally attractive. To return to the molecule example, we can make a virtual molecule. (The present proposal is inspired by the Virtual Music Center project [5].) A virtual molecule constitutes an environment in which a chemist can 'live'. Obviously, the environment will have to provide transparent access to a wealth of chemical knowledge. Some knowledge can be taken from existing sources, other knowledge can be derived from existing sources. Still other knowledge is simply not available: these are the places where the researcher can make a contribution, if she wishes.

What such an environment effectively does is to provide an overview of what is known about a particular topic. Steps (b), (c), and (d) of the information chain have been integrated into a single environment in which the boundaries between the steps have disappeared. As an interesting bonus, step (a), Identify needs, is also facilitated. Roaming through the molecule is a natural way of browsing. One cannot get lost because the environment is well-defined: any competent chemist knows her way through a molecule.

Research environments function because they integrate information from heterogeneous and scattered resources, both remote and in-house. Resources and programs tend to employ their own formats so that, behind the scenes, conversion operations are going on all the time. Wiring resources together presupposes that at least directly connected items understand each other's formats. The problem can be solved by standardisation, but standardisation has a bad track record in many domains. For example, in spite of considerable standardisation efforts, there are about twenty different formats in use for files with chemical structural information. Ironically, most of these formats have started their career as proposals for a standard format. There are, I believe, two reasons why standardisation will not work. First, different jobs take different approaches, and no format will ever effectively cater for them all. Second, reaching agreement on a standard is a social process in which the stakes are often high. This also explains why so many standards are unmanageable: they are a compromise between consistency and social acceptability.

A format is a combination of form and content. On closer inspection, disagreement on standards very often involves form rather than content. If parties do not agree on the semantics of information carriers, communication is impossible. Agreement on meta-standards can be formalised by way of a concept system. To make this work, content providers have to provide a syntactic specification of their format accompanied by a semantic part in terms of an explicit ontology. Our own research shows that converters able to convert those data into another format can be generated automatically from unambiguous descriptions of the formats involved. Information is transferred internally in a knowledge representation language that heavily relies on the ontology.

A system of converters programmed by format specifications of the content providers will significantly improve information exchange because it can operate at lower costs, both financial and social, than a system that relies on standardisation. The advent of XML does not change this, if only because current discussions about standardisation of XML tags begin to display some of the symptoms of standardisation trouble discussed earlier: multiplicity of proposals and the tendency to associate particular programs with tag systems, thus destroying the separation of form and content.

5   Research Environments and the Scientific Communication System

The ideal research environment links resources and programs in a way that is shielded from the user. For the user, it appears as a virtual world in which she can move and interact with what is there. Many scientists explore processes that occur at or within three-dimensional structures: molecules, cells, tissues in bodies. For them, a virtual world provides a natural access to information. Realising such environments takes close collaboration between information scientists, computer scientists, and domain experts.

Research environments will function in highly dynamic social contexts which they in turn help shape. We can trace some of these influences by looking at the four functions of the scientific communication system identified by Roosendaal and Geurts [6]: *registration* (registering the research results of an author), *archiving* (making information reliably available), *certification* (assessing the quality of information), and *awareness*. The *awareness* function is the core function of the system. It deals with internalisation of information in an ongoing process of systematisation, comparison, and discovery.

Research environments are built with the awareness function as prime motivation. Researchers feel overwhelmed by the information flood. Indeed, some believe that even without performing experiments (other than, perhaps, confirmatory experiments), many new and important discoveries can be made by exploring what is known already. This, then, is an important issue on the research agenda that will drive both technical developments and changes in the scientific communication system. ICT research should rise to this challenge. Researchers in developed countries find a heavy computer with continuous Internet access a minimal workplace provision. For them, computers and programs serve exclusively as tools to obtain, process, and disseminate content. I am not entirely sure that this is appreciated fully by ICT workers. For one thing, it entails that tools cannot be developed other than in the larger context of the scientific communication system and its organisational aspects.

While developed primarily to serve awareness, research environments affect the other functions as well. There will have to be some form of certification of the resources accessed by the environment. Certification is a matter of trust between author and reader. When access to the sources is effectively shielded from the user, the maintainers of the environment have to take measures to guarantee a certain minimal quality level or at least to tag sources with an indication of their quality. This shifts the trust relationship to one between maintainers and users of the environment. Trust is furthered when users are able to do assessments themselves, so that they can compare their own assessments with those of the maintainers. Technical support of quality assessment is possible to a certain extent [2,3].

There are other consequences when the environment becomes widely known and used. There will be demand for use of the site to 'publish' new results. The enviroment then also serves a registration function. Maintainers will have to time-stamp and authenticate new additions. Incorporation of new results inevitably introduces variations in the quality level of the information offered, which puts extra demands on the certification function. Finally, a stable research environment has to address the archiving function. For instance, when the accessability of a particular remote source is believed to be unreliable, a mirror has to be set up.

A research environment obviously is a useful tool to researchers only if technical and organisational issues are addressed with equal emphasis and in their mutual relations. This makes the construction of research environments an essentially multidisciplinary effort. Building research environments promises the identification of new and challenging problems; using research environments promises new perspectives and discoveries. Let's start building.

Acknowledgement. The author is indebted to Hans Roosendaal and Peter Geurts for comments.

References

1. T. Etzold, A. Ulyanov, and P. Argos, "SRS — an indexing and retrieval tool for flat file data libraries", *Methods in Enzymology* 266 (1996), 114–128.
2. Jérôme Euzenat, "Building consensual knowledge bases: context and architecture", in: *Towards very large knowledge bases. Knowledge Building and Knowledge Sharing 1995*, N.J.I. Mars (ed.), IOS Press, Amsterdam, 1995, 143–155.
3. Hidde de Jong, *Computer-supported analysis of scientific measurements*, Ph.D. thesis, University of Twente, Enschede, the Netherlands, 1998.
4. Hidde de Jong and Arie Rip, "The computer revolution in science: steps towards the realization of computer-supported discovery environments", *Artificial Intelligence* 91 (1997), 225–256.
5. Anton Nijholt, Joris Hulstijn, and Arjan van Hessen, "Speech and language interactions in a web theatre environment", in: *Proceedings of the ESCA Workshop on Interaction Dialogue in Multi-Modal Systems*, P. Dalsgaard, C.-H. Lee, P. Heisterkamp, and R. Cole (eds.), ESCA/Center for PersonKommunikation, Aalborg, Denmark, 1999, 129–132.
6. Hans E.Roosendaal and Peter A.Th.M. Geurts, "Scientific communication and its relevance to research policy", *Scientometrics* 44 (1999), 507–519.
7. Stuart J. Russell and Peter Norvig, *Artificial intelligence. A modern approach*, Prentice Hall, Upper Saddle River NJ, 1995.
8. Paul E. van der Vet and Nicolaas J.I. Mars, "Bottom-up construction of ontologies", *IEEE Transactions on Knowledge and Data Engineering* 10 (1998), 513–526.

On the Failure of Middleware to Support Multimedia Applications

Gordon S. Blair
Distributed Multimedia Research Group,
Department of Computing,
Lancaster University, Bailrigg, Lancaster, LA1 4YR, UK
gordon@comp.lancs.ac.uk

Abstract. In recent years, middleware has emerged as an important architectural element in modern computer systems. For the purposes of this paper, we define middleware to be a layer of software residing on every machine and sitting between the underlying (heterogeneous) operating system platforms and distributed applications/ services, offering a platform-independent programming model to programmers, and masking out the problems relating to distribution. Examples of middleware platforms include CORBA, DCOM, Java RMI and Jini. One notable problem however with such middleware technologies is the complete lack of support for multimedia programming. A number of extensions have been proposed to such platforms, but they are often rather flawed in that they tend to treat multimedia as a service, rather than as a fundamental aspect of the underlying middleware infrastructure.

This short paper addresses this lack of support for multimedia and presents the author's own experiences in overcoming this problem. The discussion is structured as follows. Firstly, in section 2, we examine the problem of supporting multimedia in relatively stable environments, where the end systems technologies and the quality of service offered by the underlying network are fairly predictable. In section 3, we then extend the analysis to consider problems associated with ubiquitous multimedia, where, due to user mobility, both the end systems technologies and networks being employed can vary dramatically over time. Finally, section 4 presents some concluding remarks.

We use the term multimedia to mean the handling of a variety of representation media in an integrated manner. The variety of media types in such systems is often discussed; we therefore emphasise the other part of the above definition, namely that such media types must be integrated into the computing environment. In other words, all media types should be first class entities, and thus be processed, stored and transmitted in the same manner. This conflicts with the view of multimedia as a service, as found in many middleware environments. In such platforms, some media types are supported directly by the middleware, whereas others are only supported by value-added services.

It is now well recognised that multimedia computing places the following additional requirements on distributed system platforms:

- The need to support continuous media, both in terms of the programming model, and also in the underlying system infrastructure;

H. Scholten and M. van Sinderen (Eds.): IDMS 2000, LNCS 1905, pp. 211-213, 2000.

- The need to provide both static and dynamic quality of service (QoS) management functions as an integral part of the middleware platform;
- The need to support both inter- and intra-media synchronisation in a distributed environment.

The Sumo project, a joint initiative between Lancaster University and CNET, France Telecom, addressed these issues in depth and developed an extended middleware platform to support multimedia. Key features of this extended platform included: i) an extended object model featuring three styles of interface, namely stream, signal and operational interfaces, ii) the crucial introduction of explicit binding into the object model, and iii) the use of reactive objects to implement both real-time synchronisation and dynamic quality of service management functions.The results of this project are now reflected in the ISO standard Reference Model for Open DistributedProcessing (RM-ODP). Additional interesting work is now being carried out in the QoS in RM-ODP initiative. The author believes that this standard offers a useful roadmap for how multimedia should be supported in future middleware platforms, but sadly this approach is not being followed by the major vendors in this area.

Further details of the Sumo Project can be found in [1].

In terms of multimedia, the major challenge at present is to be able to preserve multimedia communications in spite of the mobility of end users. The implication of this mobility is that the underlying computer environment is prone to dramatic change in terms of the quality of service attainable from the underlying network, the characteristics of the end system (screen size, processor capacity, battery life, etc), and indeed the physical location of the end system. All these have a profound impact on the (multimedia) service being offered. Given this, it is well recognised that mobility requires support for adaptation. Randy Katz summarises this argument rather succinctly [2]:

> "Mobility requires adaptability. By this we mean that systems must be location- and situation-aware, and must take advantage of this information to dynamically configure themselves in a distributed fashion."

Adaptation is typically required at all levels of a system, from the application potentially right down to the operating system. However, this immediately introduces a number of problems. For example, adaptation at the operating system level can be quite dangerous in terms of affecting integrity and performance. In addition, the programmer would inevitably have to rely on operating system specifics to achieve adaptation, thus compromising the portability of applications. The opposite extreme of leaving all adaptation to the application is also clearly unacceptable, as this would introduce an unacceptable burden for the application writer. Our solution is to introduce a framework for managing adaptation at the middleware level of the system.

We are therefore currently working on a next generation middleware platform capable of supporting adaptation. This is intended to support ubiquitous multimedia, but also has more general applicability in other areas requiring flexibility and support for change. The solution adopted is to exploit a combination of component technology and reflection to provide the levels of configurability and reconfigurability required by modern computer systems. In more detail, the resultant

Open-ORB architecture features a multimedia component model, based on the results of the Sumo Project described above, a per component meta-space enabling open access to the underlying support infrastructure for the component, and the division of meta-space into different meta-models covering both structural and behavioural reflection.Further details of Open-ORB can be found in the literature [3,4,5].

There is growing evidence that the lack of support for multimedia is becoming a significant problem for the middleware industry. The result is that manufacturers are by-passing middleware technology, and developing their own inevitably more proprietary solutions for handling distribution and heterogeneity. It is therefore vital that these weaknesses are resolved quickly. The author claims that RM-ODP provides one path towards more sophisticated support for multimedia applications and services, but believes the longer term solution is to embrace the new opportunities offered by reflective middleware technology. There is growing evidence to support this viewed, e.g. as witnessed by the strong interest in the recent inaugural workshop on Reflective Middleware (RM'2000) [6]. Such technologies offer the potential for extensibility, configurability and re-configurability to meet the increasing demands facing middleware developers.

References

[1]   Blair, G.S., Stefani, J.B., "Open Distributed Processing and Multimedia, Addison-Wesley, 1997.

[2]   Katz, R., "Adaptation and Mobility in Wireless Information Systems," IEEE Personal Communications Magazine, Vol. 1, No. 1, pp. 6-17, 1994.

[3]   Blair, G.S., F. Costa, G. Coulson, H. Duran, N. Parlavantzas, F. Delpiano, B. Dumant, F. Horn, J.B. Stefani, "The Design of a Resource-Aware Reflective Middleware Architecture", Proceedings of the 2nd International Conference on Meta-Level Architectures and Reflection (Reflection'99), St-Malo, France, Springer-Verlag, LNCS, Vol. 1616, pp. 115-134, 1999.

[4]   Blair, G.S., Coulson, G., Robin, P., Papathomas, M., "An Architecture for Next Generation Middleware", Proc. IFIP International Conference on Distributed Systems Platforms and Open Distributed Processing (Middleware'98), Springer, 1998.

[5]   Blair, G.S., Blair, L., Issarny, V., Tuma, P., Zarras, A., "The Role of Software Architecture in Constraining Adaptation in Component-based Middleware Platforms", Proceedings of the IFIP/ACM International Conference on Distributed Systems Platforms and Open Distributed Processing (Middleware'2000), IBM Palisades, New York, April 2000.

[6]   See http://www.comp.lancs.ac.uk/computing/rm2000.

JASMINE: Java Application Sharing in Multiuser INteractive Environments

Abdulmotaleb El Saddik[1], Shervin Shirmohammadi[2],
Nicolas D. Georganas[2], and Ralf Steinmetz[1,3]

1) Industrial Process and System Communications, Dept. of Electrical Eng. & Information
Technology, Darmstadt University of Technology, Darmstadt, Germany
2) Multimedia Communications Research Laboratory, School of Information Technology
and Engineering, University of Ottawa, Ottawa, Canada
3) GMD IPSI, German National Research Center for Information Technology,
Darmstadt, Germany
{Abdulmotaleb.El-Saddik, Ralf.Steinmetz}@kom.tu-darmstadt.de
{Shervin.Shirmohammadi, Nicolas.Georganas}@mcrlab.uottawa.ca

Abstract. In this paper, we describe an approach for *transparent* collaboration
with java applets. The main idea behind our system is that user events occurring
through the interactions with the application can be caught, distributed, and reconstructed, hence allowing Java applications to be shared transparently. Our
approach differs from other collaborative systems in the fact that we make use
of already existing applets and applications in a collaborative way, with no
modifications to their source-code. We also prove the feasibility of our architecture presented in this paper with the implementation of the JASMINE prototype.

1 Introduction

The simplicity of access to a variety of information stored on remote locations led to
the fact that the World Wide Web has gained popularity over the last decade. In this
context, Computer Supported Collaborative Learning (CSCL) is becoming more and
more important. Collaborative systems allow users to view and interact with a distributed application during a session. The use of collaborative systems increases in research and business as well as in education. A problem of many cooperative applications is their platform dependence, leading to the fact that users communicating in
heterogeneous environments are restricted in their choice of a cooperative application.
For example, a user might choose a UNIX-workstation, while another might prefer
Windows 95/98/NT or a Macintosh. The introduction of the platform-independent
programming language, Java, made it possible to overcome these problems. Diverse
approaches were used to develop Java-based collaborative systems. Almost every
system described in the literature requires the use of an Application Programming

H. Scholten and M. van Sinderen (Eds.): IDMS 2000, LNCS 1905, pp. 214-226, 2000.

The approach presented in this paper differs from other approaches in the way that we neither propose a new API for developing collaborative systems nor try to replace core components at run time. In fact, a great variety of well-designed applets already exist on the World Wide Web, which were developed to be run as stand-alone and it would not be acceptable or possible for many developers to re-implement or change these programs to make them work in a collaborative way. In our architecture, we make use of the Java Events Delegation Model [13] to extend the capabilities of Java applications in a way that stand-alone applets can be used collaboratively. The delegation event model of JDK1.1 provides a standard mechanism for a source component to generate an event and send it to a set of listeners. Furthermore, the event model also allows to send the event to an adapter, which then works as an event listener for the source and as a source for the listener. Because the handling of events is a crucial task in developing an application, this enhancement makes the development of applets much more flexible and the control of the events much more easy.

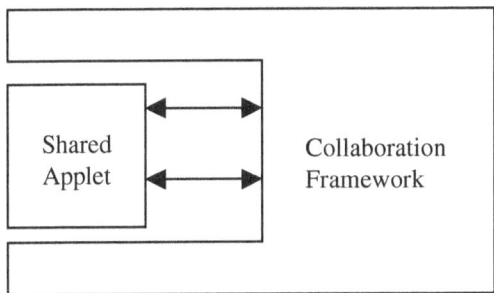

Fig. 1. Illustration of the main idea

The approach behinds our concept, which is illustrated in Figure 1, underlies the following requirements:
- ☐ No restrictions in the source code are required to share an applet. Both AWT- and Swing-based applets should be supported. A solution restricted to only one kind of components is not acceptable.
- ☐ Applications using the standard Java-Core API should be supported.
- ☐ No new API should be developed.
- ☐ As less as possible of the network's bandwidth should be consumed.

The practicality of our architecture is proven by an implementation. We have developed a collaboration system, called JASMINE (Java Application Sharing in Multiuser INteractive Environments), which facilitates the creation of multimedia collaboration sessions and enables users to share Java applets and applications, which are either preloaded or brought into the session "live". The system also provides basic utilities for session moderation and floor control. Our approach applies to both applets and applications and hence these terms are sometimes used interchangeably in this document.

The rest of the paper is organized as follows. Section 2 discusses the system architecture, while section 3 describes the implementation of JASMINE, followed by discussion of related work in section 4. Finally, section 5 concludes the paper and gives an out-look for future work.

2 JASMINE Architecture

The principal idea of JASMINE is that user events occurring through the interaction with the GUI of an applet can be caught, distributed, and reconstructed, hence allowing for Java applets to be shared transparently. This form of collaboration which is supported as long as a learning-session takes part, enables users to interact in real-time, working remotely as a team without caring about low-level issues, such as networking details.

Figure 2 illustrates the overall concept of JASMINE, where our collaboration framework wraps around an applet that is to be shared. The framework listens to all events occurring in the graphical user interface of the applet and transmits these events to all other participants in order to be reconstructed there. The framework captures both AWT-based and Swing-based events. After capturing the event, it is sent to the communication module (JASMINE-Server) where the event is sent to all other participants in the session.

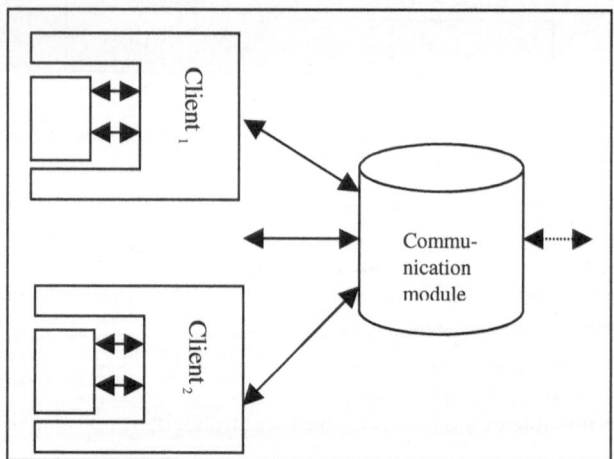

Fig. 2. Overall system architecture of JASMINE

In the next sections we are going to discuss the architecture in more details, first the client side, and then the communication module.

2.1 JASMINE-Client

The JASMINE client can be seen as a component adapter. Every event occurring at the graphical user interface of the application is sent to this adapter, which then sends the events to the collaboration server (JASMINE-Server). The client is a Java application, which consists of the following components:
☐ Collaboration Manager
☐ Component Adapter
☐ Listener Adapter
☐ Event Adapter
These components are discussed next.

2.1.1 Collaboration Manager

The Collaboration Manager is the main component on the client side and provides the user with a graphical interface offering options such as joining the session, starting and sharing applications/applets and chatting with other participants. The collaboration manager is also responsible for dispatching external events coming from the communication module and forwarding them to the component adapter, as well as receiving internal events from the component adapter and sending them to the communication module.

2.1.2 Component Adapter

The Component Adapter maintains a list of the GUI-components of all applications and applets. This list is created with the help of the *java.awt.Container* class, which allows us to get references of all applet components [13]. With the help of the main window of an application, a list of the GUI components in the application can directly be created. Therefore, the main window of an application loaded by the Collaboration Manager is registered by the Component Adapter. However, Java applets do not use stand-alone windows. They are an extension of the class *java.applet.Applet* and thus of *java.awt.Panel*. Hence, applets can be easily placed into a window, which can then be registered as the main window for the applet. All these registrations are done at the Component Adapter. An example syntax of the registration by the Component Adapter is shown in Figure 3.

```
.....
Class cl = Class.forName(className);
// If it is an applet, instantiate and locate
// it in a Frame
myApplet = (Applet)cl.newInstance();
myApplet.init();
myWindow = new Frame("Titel");
myWindow.add("Center", myApplet);
// Otherwise (if it is an instance of Window) just
// instantiate it
myWindow = (Window)cl.newInstance();
// Register this Frame as main Frame
// by Components Adapter
ComponentsAdapter.addContainer(myWindow);
....
```

Fig. 3. Excerpt of the instantiation method

After the registration is completed, a list of all Swing and/or AWT-components within the loaded application/applet is created. This task is done in the same order on each client, so that a component has the same reference identification at all clients. These references are used to point to specific components, which are the source of the events generated internally and the recipient of the events generated externally. With the help of the references, the recipient of an incoming event is located and the event is reconstructed on each client, as if it occurred locally.

2.1.3 Listener Adapter

The Listener Adapter implements several AWT listeners, which listen to *MouseEvent* and *KeyEvent* for all AWT-components except of *java.awt.Scrollbar*, *java.awt.Choice* and *java.awt.List*. For these components the Listener Adapter listens to *AdjustmentEvent*, *ItemEvent* and *ActionEvent*. When an event occurs on the GUI of the application, the Listener Adapter catches it, converts it to an external event, and forwards it to the Collaboration Manager. The Collaboration Manager in turn sends this event to the communication module, which propagates the event to all other participants.

2.1.4 Event Adapter

The Event Adapter works opposite to the Listener Adapter: it converts incoming external events to AWT events, which then can be processed locally.

2.1.5 Data Flow

Let us summarize the client side's architecture through the following data flow diagram. Figure 4 shows the overall event circulation of the system.

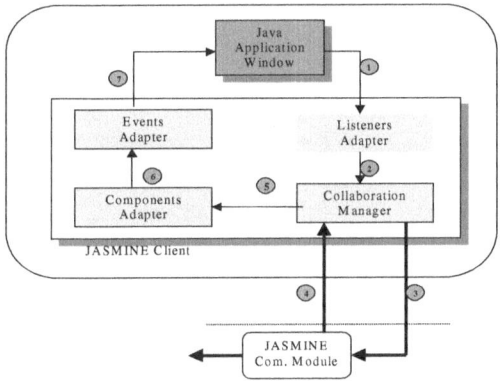

Fig.4. Event Circulation

There are two main data paths in the system: the first path is labeled with numbers 1,2 and 3. This path is used to send the internal AWT events to the communication module, and it works as follows: any Event occurred in a Java-application is caught by the Listener Adapter. The Listener Adapter first tests whether the event is an external or an internal event. It then sends only the internal events, which were not received from other clients, to the Collaboration Manager, which in turn sends the events to the communication module.

Via the second data path shown in figure 4 with numbers 4, 5, 6 and 7, the external AWT events received from the communication module are captured by the Collaboration Manager and the Component Adapter in order to reconstruct the event locally. After receiving the remote event, the Component Adapter extracts the information about its target component and sends this information together with the events to the Event Adapter. The Event Adapter converts the event to normal AWT events and sends them to the application, which then reacts to the event in the same manner as it would to a local user's interaction with the application's GUI.

2.2 JASMINE Server

JASMINE uses a multithreaded server, where the main server launches a sub-server for each user joining the session. The sub-server is responsible for processing only the update messages or requests coming in from its own client. Once the sub-server receives the update message, it will send it to all other clients in the session (figure 5). This will create a fast system response, at the expense of more resources utilized due to sub-server threads. However, usually only one client at a time can control and interact with an application (due to floor control as we will see), and most threads will simply be waiting and won't consume too many resources.

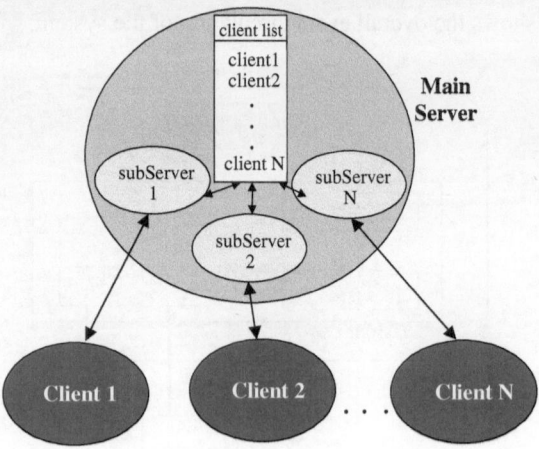

Fig. 5. JASMINE Server

The server's main job is to propagate the incoming events from a user to all other users. But it also provides other services, which are necessary for maintaining a collaboration session. It provides services for session moderation and management, floor control, and data exchange. Data exchange is of particular importance for multimedia sessions as we will see next.

2.3   Advanced Multimedia Applications

As discussed in the literature, a pure transparent collaboration system is not sufficient for multimedia applications [12]. This fact is due to specific services that are required by multimedia applications such as synchronization, quality of service, etc. For example, think of a collaboration session where a video applet is being played. When one user presses the pause button, simply capturing the "pause event' and sending it to all other clients is not sufficient because when other clients receive the pause event and apply it to their video player, at each client the video player will pause on a different frame and clients will not be synchronized. Hence there is a need to send control messages between clients, such as "pause on frame number 57" to maintain consistency among all users. The JASMINE server provides a high-level API that can be used for this type of advanced requirements. However, an application must specifically use the API to take advantage of these functionalities, hence the transparent feature of the system is somewhat diminished.

3 Implementation

Figure 6 illustrates a sample screenshot of a typical JASMINE session. It shows the client's Collaboration Manager and some shared applets and applications running in the session.

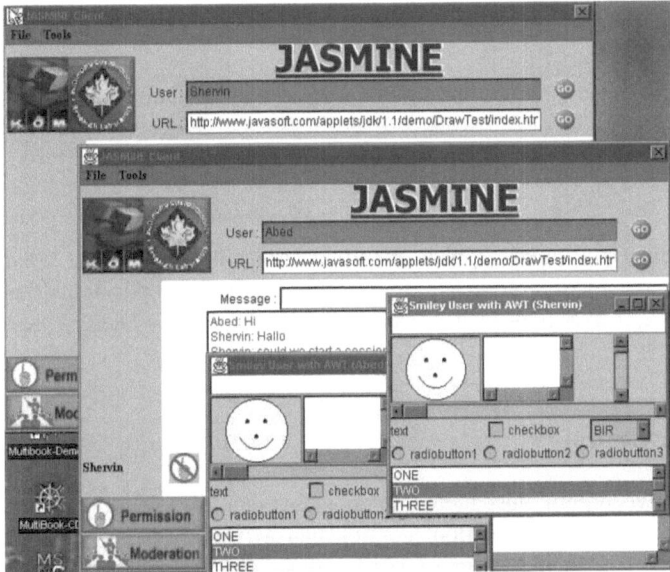

Fig. 6. A screenshot of a sample JASMINE session

3.1 Configuration File

Information about locally available applications and applets, which can be used in a collaborative way, are read from a configuration file. The configuration file, which is organized as a properties file, contains the names of the applications/applets, which will be presented in the menu and the full names of their main class or URL. The entries have the following syntax:

application.[n].name = [name]
application.[n].class = [class]

where:

n: number of the application in the list.

name: a suitable name for the application to be shown in the menu.

class: full name of the main class.

An example configuration file is illustrated in Figure 7.

```
#Application entries
application.1.name=myTestApplication
application.1.class=kom.develop.apps.MyApp
# Applet entries
applet.1.name=myTestApplet
applet.1.class=kom.develop.applets.TestApplet
# URL entries
url.1.name= TestUrl
url.1.address=http://desiered.server/test.html
```

Fig. 7. Excerpt from a configuration file

Before starting the session, applets and applications that are thought to be useful can be placed in this configuration file. Additional applets and applications can be brought into the session live as needed by typing the corresponding URL in the appropraite field.

3.2   Floor Control

A collaborative system must address many issues such as synchronization, latecomers, management or moderation, floor control, and awareness [12]. Among these, floor control is perhaps the most primary issue without which a collaborative session won't function properly. In short, floor control ensures that only one person at a time controls the shared application. Without floor control, there will be collisions of events, which leads to unwanted results in the shared application.

In JASMINE, floor control is achieved by means of locking. Each application has a corresponding *semaphore* on the server. When a user wants to interact with the shared application, the system first locks the application by locking a semaphore. At this point, any other users trying to interact with the application will be denied access. When the first user is finished, the system releases the semaphore and others can take control of the application.

```
public void mouseDragged(MouseEvent e) {
    //user is dragging the mouse, so ask for control
    if (getControl()==true) {
        // do whatever must be done for a mouse drag
        releaseControl();
    }
    else displayMessage("Access Denied!");
}
```

Fig. 8. Intuitive floor control

For a specific shared application, most developers prefer an "intuitive" implementation of the floor control capability; i.e., as soon as the user tries to interact with the application, the client automatically asks for floor control and allows or disallows its user to interact. After the user is finished, the client releases the lock automatically. Figure 8 shows sample Java code that demonstrates how the floor control is used in an intuitive way. This approach is in contrast to the "direct" approach, where a client must specifically ask for control, for example by pressing a "control-request" button.

3.3   Moderation

Although floor control addresses the issue of event collisions, it works on a first-come-first-serve basis. This in turn leads to the possibility of a participant to abuse or disrupt the session by feeding unwanted events into the session. There is therefore a need to have a moderator in order for a session to be more productive, for example, a teacher moderating a distance learning session. The moderator is usually the person who calls for a collaborative session and starts the server. In JASMINE we have two types of sessions: moderated, and non-moderated. The server can be started by specifying a login name and password for the moderator.

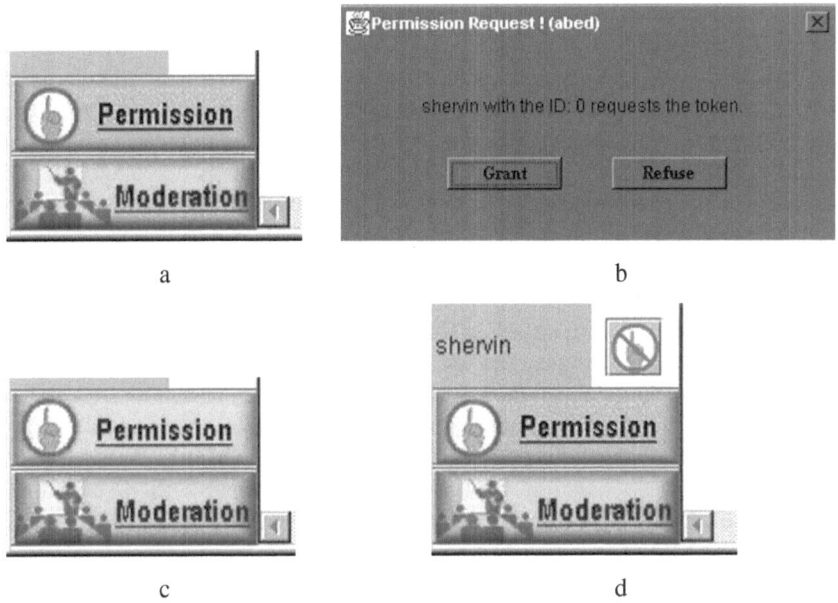

a

b

c

d

Fig. 9. Moderation capabilities in JASMINE.

Once the session starts, the moderator can login at any time and take control of the session. When the session is moderated, no one can send any events to the server. A participant wishing to do so must ask for permission from the moderator as shown in Figure 9a. The moderator will subsequently receive a message indicating the participant's request to interact (Figure 9b) which the moderator can allow or refuse. Upon moderator's acceptance of the user's request, the user will receive a green light, which indicates that he or she can now send events to the session (Figure 9c). The moderator can also dynamically "cut off" a user's permission to interact if needed (Figure 9d). In JASMINE, we allow only one user at a time to have permission to send events, although this number can be increased based on the application.

4 Related Work

There are many Java-based collaboration systems, none of which offer a management or moderation feature similar to ours. Kuhmünch [10] at the University of Mannheim developed a Java Remote Control Tool, which allows the control and synchronization of distributed Java applications and applets. The drawback of this approach is that it is necessary to have access to the original source code of the application or applets in order to make it collaborative. That means every applet must initiate a Remote-Control-Client object, which is usually done in the constructor of the applet. Also, the event handling within the applet must be modified in order to receive and/ or send events from / to remote applets. The Java Shared Data Toolkit (JSDT) from JavaSoft is also an API-based framework [15]. Habanero [1] is an approach that supports the development of collaborative environments. Habanero is in its terms a framework that helps developers to create shared applications, either by developing a new one from scratch or by altering an existing single-user application which has to be modified to integrate the new collaborative functionality. Instead of using applets, which can be embedded in almost every browser, the Habanero system uses so-called "Happlets" which need a proprietary browser to be downloaded and installed on the client site. Java Collaborative Environment (JCE) has been developed at the National Institute of Standards and Technology (NIST) coming up with an extended version of the Java-AWT [2] called Collaborative AWT (C-AWT). In this approach AWT-components must be replaced by the corresponding C-AWT components [3].

All these approaches propose the use of an API, which has the cost of modifying the source-code of an application, re-implementing it or to design and implement a new application from scratch in order to make it collaborative.

Java Applets Made Multiuser (JAMM) [8] is a system, which is similar to our approach in terms of its objective: the transparent collaboration of single-user applications. The difference between JAMM and JASMINE is the way collaboration is achieved. In JAMM [6], the set of applications that can be shared is constrained to those that are developed using Swing user interface components as part of Java Foundation Classes, which are part of the standard JDK since version 1.2. JAMM's set of applications is furthermore restricted to those which implement the Java serializable interface.

5 Conclusion

We presented the architecture and implementation of our transparent collaboration framework for Java applets and applications. We developed this architecture in order for users to be able to collaborate via collaborative-unaware applications and applets without modifying the source code. Our architecture enables us to use almost all single-user applets and applications in a collaborative way. We have successfully tested our system on a number of applets. We also observed that using the TCP-client-server approach of our communication module can support relatively large number of users.

There are two outstanding issues remaining. These issues are not directly related to JASMINE but are research areas of the transparent collaboration paradigm. The first issue is that of latecomer-support. When a user starts a session later than other participants, there is a need to bring this user up-to-date as opposed to start from scratch. This can be achieved either by sending the entire object state of the shared application to the newcomer using object serialization, or by sending all the events occurred up to now to the new user so that it follows the same sequence of events that other participants have gone through [12]. We're currently using JASMINE to experiment with these methods.

Another issue was brought up in Section 2.3: multimedia inter-client synchronization and control. Transparent collaboration cannot address this issue alone and we believe that using an API is necessary to achieve such functionality for multimedia applications.

Today, computing environments where Java applications and applets are running over IP have become very popular and widespread. Our architecture helps people to collaborate in such environments easier.

Acknowledgments

The authors acknowledge the financial assistance of the Volkswagen Stifftung, Germany, as well as the Telelearning Network of Centers of Excellence Canada (TL-NCE) and the Natural Sciences and Engineering Research Council Canada (NSERC).

References

1. Chabert et al, , "Java Object Sharing in Habanero", Communications of the ACM, Volume 41, No. 6, June 1998, pp. 69-76.
2. H. Abdel-Wahab et al "An Internet Collaborative environment for Sharing Java Applications" IEEE Computer Society Workshop on Future Trends of Distributed Computing Systems (FTDCS'97), October 29 - 31, 1997, pp. 112-117.
3. H. Abdel-Wahab et al, *"Using Java for Multimedia Collaborative Applications"* Proc. PROMS'96, Madrid, Spain, 1996.

4. *Handheld IP Connectivity for 1998*, IEEE Internet Computing, Vol. 2, No. 1, January/February 1998, pp. 12-14.
5. International Data Corporation, "IDC's Forecast of the Worldwide Information Appliance Marketplace 1996-2001", IDC Bulletin #w15080, December 1997, (screen phone revisions 5/7/98).
6. Abdulmotaleb El Saddik, Oguzhan Karaduman, Stephan Fischer, and Ralf Steinmetz. "Collaborative Working with Stand-Alone Applets". In Proc. of the 12th International Symposium on Intelligent Multimedia and Distance Education (ISIMADE'99), August 1999.
7. J. Begole et al, "Leveraging Java Applets: Toward Collaboration Transparency in Java", IEEE Internet Computing, March-April 1997, pp. 57-64.
8. J. Begole et al, "Transparent Sharing of Java Applets: A Replicated Approach". Proc. Symposium on User Interface Software and Technology, ACM Press, NY, 1997, pp. 55-64.
9. J. Grudin, "Computer-Supported Cooperative Work: History and Focus", IEEE Computer, Vol. 27, No. 5, May 1994, pp. 19-26.
10. Kuhmünch et al, "Java Teachware - The Java Remote Control Tool and its Applications", Proc. ED-MEDIA'98, 1998.
11. Multimedia Communication Forum Inc., "Multimedia Communication Quality of Service", MMCF document MMCF/95-010, Approved Rev 1.0, September 24, 1995.
12. S. Shirmohammadi et al, "Applet-Based Telecollaboration: A Network-Centric Approach", IEEE Multimedia, Vol. 5, No. 2, April-June 1998, pp. 64-73.
13. Stephan Fischer and Abdulmotaleb El Saddik, <u>Open Java:</u> Von den Grundlagen zu den Anwendungen. Springer-Verlag, ISBN: 3540654461 (1999).
14. K. Obraczka, "Multicast Transport Protcols: A Survey and Taxonomy", IEEE Communications, Vol. 36, No. 1, 1998, pp. 94-102.
15. Javasoft (for Java, JINI, RMI, and JSDT technologies) http://www.javasoft.com

Design and Implementation of a Framework for Monitoring Distributed Component Interactions

Nikolay K. Diakov[1], Harold J. Batteram[2], Hans Zandbelt[3], and
Marten J. van Sinderen[1]

[1] CTIT, P.O. Box 217, 7500 AE, Enschede, The Netherlands
{diakov, sinderen}@ctit.utwente.nl
[2] Lucent Technologies, P.O. Box 18, 1270 AA, Huizen, The Netherlands
batteram@lucent.com
[3] Telematica Instituut, P.O. Box 589, 7500 AE, Enschede, The Netherlands
zandbelt@telin.nl

Abstract. This paper presents a framework for monitoring component interactions. It is part of a larger component framework built on top of the CORBA distributed processing environment that supports development and testing of distributed software applications. The proposed framework considers an OMG IDL specification as a contract for distributed interactions and allows precise monitoring of interaction activities between application components. The developer is not burdened with monitoring issues because all the necessary code instrumentation is done automatically. The tester is given the opportunity to use monitoring facilities for observing interactions between distributed component applications. This paper explains the monitoring framework and reasons about its expressive power, accuracy and applicability. The approach is validated in a platform for design, development and deployment of on-line services.

1. Introduction

Distributed multimedia applications are becoming one of the major types of software used nowadays. At the same time, the fast-paced software market demands for rapid development of multimedia applications. This leads to reshaping of the software development methodology towards the usage of off-the-shelf components for quick assembly of applications of arbitrary complexity. Unfortunately, the ability to quickly manufacture software through assembly and configuration of available components does not guarantee correctness of the solution nor quality of the product.

One way to enhance quality is through thorough testing. However, testing of applications that run in a distributed environment is not an easy task. Distributed environments usually consist of several physical machines with different hardware configurations, having installed different operating systems and middleware software, and different characteristics of the network connections between them. Thus, testing in distributed environments introduces additional aspects to the single-computer case.

This paper presents a framework for monitoring component interactions. It is part of a larger component framework that is built on top of the CORBA distributed processing environment, and that supports development and testing of distributed software appli-

H. Scholten and M. van Sinderen (Eds.): IDMS 2000, LNCS 1905, pp. 227-240, 2000.

cations. The proposed framework considers an OMG IDL specification as a contract for distributed interactions and allows precise monitoring of interaction activities between application components. The developer is not burdened with monitoring issues because all the necessary code instrumentation is done automatically. The tester is given the opportunity to use monitoring facilities for observing interactions between distributed component applications. Monitoring a distributed system requires both observing system behavior at specific observation (or monitoring) points and effectively representing the observed information in a graphical or textual manner. In this paper, we only address the first aspect.

The remainder of this paper is organized as follows. Section 2 gives some background and motivates this paper. Section 3 identifies and analyses basic issues regarding the design of a monitoring framework. Section 4 outlines the architecture of the framework. Section 5 discusses related work. Section 6 presents our conclusions and exposes our plans for future work in this area.

2. Background and Motivation

New technologies, in particular broadband and multimedia technologies, open many possibilities for the telecommunications market. In this dynamic environment, new and innovative services are being invented and introduced at a high speed. These services exist in a large variety of application domains, such as on-line entertainment, tele-conferencing, tele-education, tele-medicine, CSCW, and electronic commerce. In general, service providers, service developers, and service users struggle to keep up with the rate of change and constantly have to adapt their way of work.

Producing high quality software becomes a serious challenge, since development and testing of distributed applications requires new methodologies and more powerful tools. Quality testing involves diverse static and dynamic analysis techniques, which can handle a large number of monitoring states, visualize program behavior, support off-line log analysis, e.g., for the purpose of conformance testing, and others [5].

An important aspect of system design using (de-) composition is that the system must be verifiable and testable at the abstraction level of its design. For a component based design this means that not only the behavior of each separate component must be testable, but also the interactions between the components. For complex systems, language level debuggers are not very suitable for testing because they have no knowledge of components and manage distribution in proprietary ways. This means that tools and techniques are needed that allow a tester to verify and test distributed systems at higher levels of abstraction, e.g., at the component level. Once a component has been identified that does not behave as designed, implementation language level debuggers can be used to pinpoint the problem within the component. The majority of software development processes consider software quality during a testing phase and often alternate testing with the implementation phase. Exercising the program code in order to observe its real behavior is the ultimate test for every software product. Better tools and techniques are needed to help the tester carry out the observation activities through the cyclic development process of distributed applications.

Our approach allows a tester to test individual operations on component interfaces and to generate a sequence diagram as an invocation propagates through the system. The monitoring system observes invocations at component interface levels and generates all the necessary information needed to identify the component, the involved interface, the operation, the parameter values, etc. The collected monitoring information can be used for generating graphical representations of complete sequence diagrams of the component interactions. During system testing, the output of the monitoring process can be used for conformance tests with the system design specification.

In order to contribute to solutions in this area, we set ourselves the following goals:

1. Provide a basic *monitoring* framework for dynamic analysis of distributed component applications, enabling tracing of distributed component interactions through the system at runtime.
2. Automate any code instrumentation needed for monitoring in the development process, so that the developer will not be burdened with monitoring issues.
3. Prove that the approach is realistic by implementing a prototype of the monitoring system.

The monitoring framework has been designed and implemented within the co-operating research projects FRIENDS [10] and AMIDST [11]. A monitoring prototype has been developed within the FRIENDS project. This prototype has been integrated with the Distributed Software Component (DSC) [1, 2] model used in the FRIENDS integrated services platform.

3. Problem Analysis

This section identifies and analyses the following basic requirements, which are essential to our goals:

- Accuracy - relates directly to the granularity of the observed information, the frequency of the measurements and the information model of monitoring events;
- Expressive power - enables precise verification techniques like conformance testing using formal methods. The expressive power of the monitoring system rests on, for example, the ordering and causality information that can be retrieved from the generated monitoring events;
- Applicability - has to do with the software development process, how easy it is to incorporate, manage and use the monitoring framework. Another important issue is the performance and flexibility of the system at runtime.

We use the following terminology to further explain and discuss these requirements.

The monitoring system operates on applications - the *targets* that have been modified to facilitate a monitoring framework. A target consists of software units of distribution called *entities*. The framework monitors application *behavior* and as a result produces *monitoring events*. With application *behavior* we refer to the interaction patterns between the entities of an application. A *process* in this paper is equivalent to a *thread of execution*. An *execution environment* corresponds to the operating system running on a computer that is wired to a communication network.

3.1 Interaction Contracts

We assume that the monitored distributed application consists of *entities* that interact with each other via a communication network. Further on in this paper, we specialize entities into DSC components for the purpose of the prototype of our framework.

The contemporary middleware platforms employ interaction contracts for standardizing the interactions between software entities. An example of such a contract is the OMG IDL specification. An example of a middleware platform is CORBA. The monitoring framework we offer is built around the notion of an interaction contract. For the purpose of building our prototype we chose CORBA to provide the contract for interactions between the entities of the distributed application.

In Fig. 1, entity A interacts with entity B using an interaction contract. The contract contains a set of services (IDL operations) that one of the entities explicitly provides (implements) and the other entity uses (invokes). In CORBA, the definition of interactions through a contract is unidirectional, unlike, for example, a bi-directional communication channel. Referring again to Fig. 1, Entity A has the role of initiator that calls an operation on the implementation of the contract in entity B.

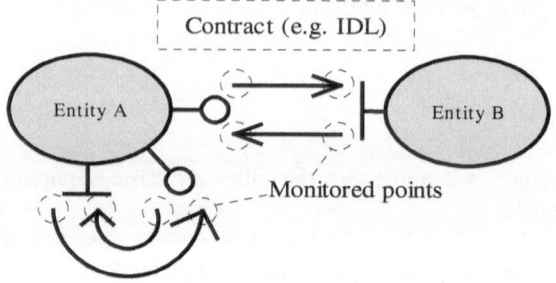

Fig. 1. Points of observation. Entity A interacts with B and itself.

We define the monitoring framework to observe entity interactions at the level of the contract. Every service in the contract is monitored and every value related to this service is recorded. The frequency of the measurements performed by the framework is determined by the monitored entities, which invoke services using the contract. Fig. 1 depicts how the framework observes entity interactions at monitored points.

Since the design and the implementation of the application entities follow exactly the contract definition, the monitoring framework will be able to monitor any application designed and implemented with this middleware. The prerequisites are such that the implementation of the monitoring framework conforms to the computational model of the middleware and the developers conform to the interaction contracts used with the particular middleware.

3.2 Execution Model, Order, and Causality Relations of Events

At any moment, distributed applications consist of a number of asynchronous processes that execute different tasks. Each process generates a sequential stream of events. The processes communicate by passing messages using the available communication environment [8]. The exchange of a message also generates events in each one of the participating processes. Without restrictions we assume that process execution and message transfer are asynchronous, and the communication environment is finite and unpredictable. The processes, messages and events are main actors in the execution model of the distributed application.

The computational model of the middleware may influence the execution model of the distributed application. For example, a particular middleware product may allow usage of different policies for assigning of processes to interactions between entities.

In general, an interaction between two entities can be mapped to a message exchange between two processes within these entities. Nevertheless, there are cases, in which entity interactions cannot be mapped to process message exchange, as for example when two entities share the same execution environment.

Processes are called collocated if they act on the same entity. Entities are called collocated if they share the same execution environment. Normally, an execution environment corresponds to the operating system running on a physical node, however, in some cases, e.g. as for the Java Virtual Machine, it is possible to run several execution environments on one physical node.

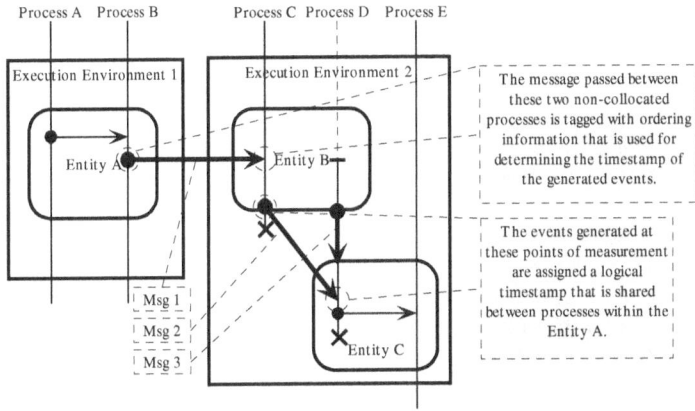

Fig. 2. Entities and processes deployed over execution environments and processes.

We take collocation into account as it relates directly to sharing common resources, e.g., memory space. Fig. 2 shows how several entities interact. Processes A and B are collocated processes, as are C, D and E, whereas entity B and C are collocated entities. Process C finishes after sending "Msg 2" to process D, which is created at the time the message is sent to it and dies after sending a message to process E. Process D is shared between entities and performs a direct method call ("Msg 3") across the boundaries of entities. This example comes from our experience using ORB implementations that optimize interactions between collocated entities, so that the usage of the communication environment is circumvented. Note that only the thick arrows depict entity interactions that are observed by our monitoring framework.

The monitoring framework captures its measurements into standalone events that can be stored to persistent media or can be sent to a consumer application (monitor) that further analyses them. In the monitored system, these events are generated in a particular order. Distributed systems often operate in an environment of network delays and other unpredictable factors that may lead to receiving events in a different order than the order in which they were generated. There are different approaches that deal with this ordering problem. We employ solutions using a logical timestamp that is recorded in the event (See [8])

The monitoring of causality is useful to assess the correct working of applications. For example, it helps the tester to reason about the actual source of runtime errors during the testing phase of the software development process.

In an execution environment of asynchronous processes, the order relation provided by logical clocks is also a causal relation. However, the shared memory solution to retrieve ordering information within an entity does not allow one to keep track of causal relations anymore, since delays are introduced as a result of applying a scheme for sequential access to the shared logical clock.

After investigating several particular technologies, like the Java platform and the CORBA Interceptor standard, we have developed a technique for propagating the ordering information between collocated processes within the same entity. This technique allows us to restore causality information in a generic way, without entering a conflict with a black-box approach.

4. Technology Solutions

In this section we investigate available technologies that can be used to build the monitoring framework, and we outline the software architecture of our prototype.

4.1 Black-Box Approach

An entity can be instantiated and configured as a part of an application. Deciding on what is an entity in the application influences the granularity of the monitoring system. Granularity does not really put restrictions on the monitoring system, as the *entity*-based observation model does not have a notion of what is inside the entity. Entity candidates in our system are CORBA objects (each one implementing a single interface) and software components (based on a particular component model [4]).

When we started this work, we had the DSC framework [2] available. DSC offers a component model similar to the CORBA component model [7], allowing rapid development of distributed components. Although components can be large entities, we have chosen granularity for three reasons: (1) component technology enters the software industry at a fast pace; (2) we had an advanced component framework available for experiments; and (3) components can be approached as black boxes [4].

DSC components use the CORBA middleware as distributed processing environment. In CORBA, interaction is done through invoking operations on an interface implementation. Furthermore, invocations can be synchronous and asynchronous.

The process semantics behind the synchronous and asynchronous invocations is not explicitly defined in the CORBA standard. Fig. 2 shows that sometimes messages passed between entities are not messages between processes but rather method calls in the same process and this depends only on the deployment of the entities, e.g. whether they are collocated within the same execution environment or not.

We consider a component as a black-box designed and implemented by a third-party following a particular component model. Observable events, which occur in a component, are *lifecycle* events and *interaction* events. Lifecycle events relate to creation and destruction of a component instance, announcement of explicit dependencies and implemented interfaces, suspend or resume of component instances, etc. Lifecycle events occur within one process and are not related to the communication between processes. These events receive their timestamps by using the shared memory scheme Interaction events are a result from an invocation of an operation from one component on an IDL interface implemented by another component.

To order the monitoring events, the monitoring framework relies on propagating monitoring information along with the invocations between components.

4.2 Context Propagation

In order to trace the propagation of an invocation through the system, each invocation is transparently tagged with context information, which is updated at each monitoring point. The context is a data structure that encapsulates value of a logical clock, causality information, etc. The context is propagated between interacting components.

Two particular problems have to be solved with respect to the context propagation:
• Sending context from one component to another, in a generic way;
• Propagating context through the black box of the custom component implementation, in a generic way.

Reflective technology allows us to isolate the monitoring specific code into separate facilities and libraries, that will leave component developers free of concerns about the monitoring during design and implementation phases. In our approach we have investigated CORBA Interceptors, reflection on the thread model through Java and CORBA Portable Object Adapter (POA).

CORBA provides the *interceptor* mechanism to reflect on the invocation model by offering low level access to the CORBA request/reply object. Our monitoring scheme uses interceptors (message and process level) to pass monitoring context between components.

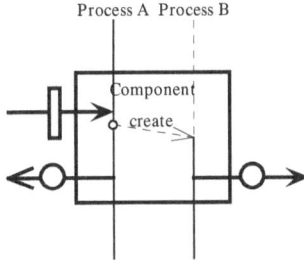

Fig. 3. Process assigned to an incoming invocation creates a second process.

The black box approach interferes with the mechanism for propagating context inside the components of the distributed application. For example, when an invocation enters a component, it may be assigned to a process executing custom code that creates a number of parallel processes, each one following arbitrary interaction patterns with other components (Fig. 3). This custom code may be built by a third-party or in general is encapsulated in an off-the-shelf component.

We use the ThreadLocal Java class to assign context to processes (threads of execution) in a generic way [9]. The advance of the logical clock follows the original rule. We also use the InheritableThreadLocal Java class to propagate the monitoring information when a process creates another process (typical 'fork' in the execution) inside the black box.

Nevertheless, a generic scheme for tagging messages between threads of execution within the Java Virtual Machine is not available. For example, reflection on the message exchange (i.e. synchronization) between threads of execution allows us to propagate context within the component implementation. The current prototype of our monitoring framework can only partially restore causality relations between monitoring events.

The CORBA POA specification defines the execution semantics of the CORBA invocation. ORB implementations that conform to POA always call the stubs in skeletons. This allows us to correctly generate interaction events when the ORB performs invocation optimization for collocated components, i.e. one process handles the component interaction.

4.3 Configurability

Our implementation of the monitoring framework employs techniques like intercepting and tagging interactions between application components, generating and sending monitoring events to consumer applications (Fig. 4), and others. This behavior translates into high CPU utilization and frequent network transfers of monitoring events. Thus depending on the number of application components being monitored per execution environment and in the whole system, the overall performance of the target application is lower and the system may degrade to an unacceptable level.

We believe the solution to the performance problem is in a flexible monitoring framework, that does not employ bottleneck solutions. Configurability of the monitoring architecture in the component can be used to lower the CPU utilization in the physical node where components execute.

Inside the component, we define two types of configurability: static and dynamic. Static configurability comprises reconfiguration activities that require stopping, reconfiguring (e.g. recompiling) and restarting of the component instances. The process is supported by monitoring tools. Depending on the IDL and a static description of what should be monitored inside, such tools determine whether component code will be modified to produce monitoring events or not.

Dynamic configurability does not require stopping and restarting of the component instances. Instead of this, the component implements a special interface that allows dynamic reconfiguration of the internal monitoring architecture, e.g., what for message types will be produced, turning on/off interception of particular interfaces, etc.

Dynamic configuration is preferable, because it does not put restrictions on the lifecycle of the components. Nevertheless, we choose to split configuration into static and dynamic, because some of the technologies (particular ORB implementations, Java, C++ compilers) allow only a static approach to some of the reconfiguration schemes, as for example the activation of interceptors. Moreover, when we choose not to monitor a system, a performance increase may be achieved with the help of static reconfiguration by removing all monitoring hooks. Another way of reducing workload at the components is to decouple components that act as event producers from event consumers.

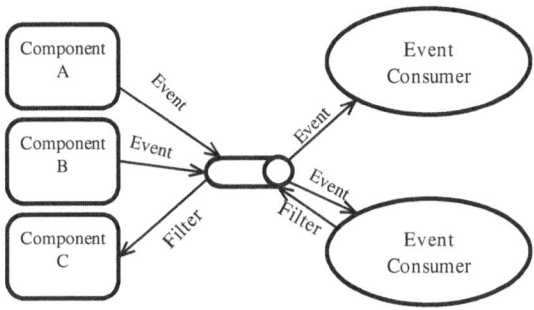

Fig. 4. Events and Event filters are delivered through the notification service.

Decoupling can be achieved by using an event service like the CORBA notification service, such that components do not have to hold references to the event consumers anymore. Instead, a reference to the event channel is held and events are sent only once to the channel, compared to the solution where several consumers must explicitly be notified of the same event. Decoupling also increases the scalability of the monitoring framework and thus of the applications facilitated by the framework.

The CORBA notification service standard defines event filters. These filters encapsulate the request of the consumer application for particular events. Furthermore, event filters are propagated from the consumer through the notification service, to the sources of events - the components (Fig. 4). The monitoring framework can use the event filters for dynamic reconfiguration at the components.

4.4 Modular Architecture

The architecture of the monitoring framework is modular and scalable. Inside the components, the monitoring architecture is based on the Portable Stub and Skeleton architecture as defined in [12].

The IDL compiler generates a pair of *Stub* and *Skeleton* classes for each interface type encountered in the IDL specification file. The Skeleton class is the base class from which the user-created implementation class must inherit. The Skeleton class contains an _invoke() operation that has the following signature:

```
public org.omg.CORBA.portable.OutputStream
_invoke(String method,
org.omg.CORBA.portable.InputStream input,
org.omg.CORBA.portable.ResponseHandler handler) throws
org.omg.CORBA.SystemException;
```

The body of the _invoke operation tests the value of parameter method, extracts the method parameters from the input parameter and invokes the method in the implementation class.

The Stub class is used as a local proxy for a (possibly) remote object. It implements all the operations of the target CORBA object. Within user-created implementation code a reference to a Stub object is obtained by narrowing a CORBA object reference to the appropriate type. When an operation is invoked on the Stub object, a CORBA request is assembled which contains the name of the operation, its parameter values and the destination. The request is further translated into an IIOP request and transmitted over the network by the ORB. The Skeleton and Stub classes mark the entry and exit points of an invocation on a component interface. Custom monitoring interceptor code executes just before and after each invocation, generates context information and prepares a monitoring event. Once the system is sufficiently tested, the monitoring code can be removed by recompiling the IDL files without using the monitoring IDL compiler tool.

Local monitor objects encapsulate functionality for processing the monitoring events like, assigning the proper context to an event, queuing and scheduling events for delivery to the consumer applications (CentralizedMonitor), switching between different event delivery policies (direct send, using notification service, logging to local persistent media) and others.

The CentralizedMonitor is an event consumer application that is capable of analyzing, storing, and presenting monitoring events to the component tester. The Centralized-Monitor component maintains several analysis tools (ObserverGUI objects) that analyze and represent event information properly to the component tester.

4.5 Information Model

A monitoring event is a persistent object that contains a number of data fields.

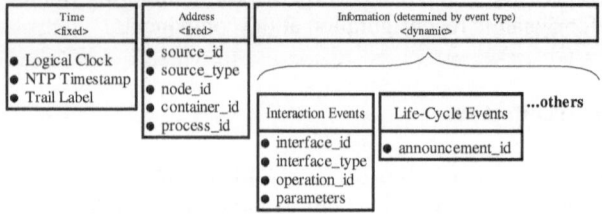

Fig. 5. Monitoring event structure.

The information model consists of three groups of fields: time, address, and information (Fig. 5). The time group contains a fixed set of fields that captures ordering and causality information. The address group contains a fixed set of fields with information about the source of the event. The information group contains variable fields, depending on the event type. For example, an interaction event captures information about interface (interaction contract) id and type, operation name, and parameter values.

4.6 Development Process and Test Cases

A distributed application can be a large composition of components, interacting with each other through their CORBA interfaces in a distributed environment.

Fig. 6 shows the DSC component development process.

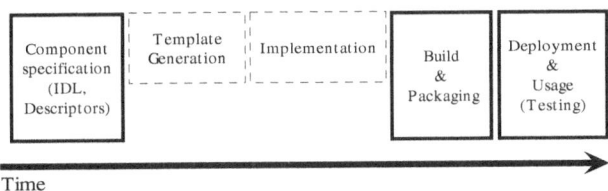

Fig. 6. Phases of the component development process essential to the monitoring framework.

DSC components are described in component specifications. A component specification is the input for the automated monitoring tools. Explicit dependencies and IDL types are the most important entries from the component specification, used during the component build and packaging phases for generation of the augmented with monitoring code stubs and skeletons (Fig. 6).

Once the components are packaged they can be deployed in an execution environment where monitoring tools are executed to observe the behavior of the distributed component application. Application testers follow a common scenario: starting of all centralized monitor components, running the application and usage of the analysis tools provided by the centralized monitoring application.

The central monitoring application is an event consumer specialized in collecting all events from the monitored system. From a GUI at the central monitor, the tester is able to select entry or exit (interaction) points of the component test subjects and to assign them textual label values.

The label is then propagated with every interaction at each interaction point.

The monitor records the events it receives to a persistent storage. One of the graphical representations of the collected events is a Message Sequence Diagrams (MSD) viewer. Note that our notation is an extension of a subset of the ITU notation for message sequence charts [6]. We have also developed a visual tool allowing the application tester browse the complex parameters of the CORBA invocations.

4.7 Prototype

This section reports on different technology specific issues encountered during the implementation of the monitoring framework prototype.

The prototype currently does not use a CORBA Notification Service because no suitable commercial or freeware implementations were available to us.

The monitoring framework supports the following ORBs: Orbacus for Java 3.3, JacORB 1.0 beta 15, JavaORB 2.2.4, Visibroker for Java 3.4.

The monitoring framework makes use of the POA and the Interceptors interfaces. We encountered many problems with incorrect implementations of the POA and Interceptor specifications resulting in a submission of a number of reports to the vendors (Visibroker, JacORB, JavaORB, Orbacus).

The mechanism for propagating context between the components has been implemented using CORBA Interceptors. The specification allows generic access to the assembled request object of the CORBA invocation. The CORBA request format contains a specific field for this purpose: the service context field. This field can hold an array of service contexts that is transparently passed from client to server and vice versa. Below is the Java declaration of the service context field in a CORBA request:

```
Public org.omg.IOP.ServiceContext[]
service_context;
```

Each service context consists of a numeric identifier and a sequence of bytes. In case of a request from client to server it is used to propagate information about the execution context of the client object to the server; in case of a CORBA reply it contains information about the execution context of the server object that may be examined by the client.

Below is a part of the Java source code of the ServiceContext class:

```
package org.omg.IOP;
public final class ServiceContext implements
org.omg.CORBA.portable.IDLEntity
{
   public int context_id;
   public byte[] context_data;
```

A unique number was chosen to identify the DSC monitoring context in a sequence of service contexts.

At some point before the actual execution of a CORBA request (or reply) we need to insert the service context into the message. In a similar way we need to analyze the service context that was propagated back from server to client in the reply to the request. The custom CORBA interceptors we provide, insert the context in the request at the outgoing points and retrieve context at the incoming points. For a detailed explanation of interceptors the reader is referred to the ORB manuals or the OMG documents about interceptors [14].

DSC monitoring uses message-level interceptors, which provide low-level access to the CORBA request/reply at the following four points:
- at the client side - before sending the request and after receiving the reply;
- at the server side - after receiving the request and before sending the reply.

Our framework performs instrumentation of the Stubs and Skeletons that can be replaced by the recently accepted Portable Interceptors submission [13], once implementations become available.

5. Related Work

In [8], Raynal gives an overview of the methods for capturing causality in distributed systems. He defines an execution model of processes that pass messages between each other. In this environment Raynal describes the logical clock schemes for capturing order and causality of messages. Our monitoring framework uses Raynal's technique, however, our execution environment has constraints such that not all messages can be tagged with the necessary information. After applying a shared memory solution for propagating context between the processes within an entity, our framework is capable to employ the logical clock technique and restore order. Nevertheless, because of the shared memory approach, the causality relation in the execution model is broken and a separate solution is sought in order to track causality in the system.

In [15], Logean describes a method for run-time monitoring of distributed applications that supports a software development process. His approach employs ordering technique deriving from logical clocks. The level of entity granularity is the CORBA object. Our framework enhances this approach with introduction of the entity.

The configuration of the monitoring code in the Logean's method can be dynamic and supported by tools. By applying notification service, our approach introduces additional scheme for decoupling event sources from event consumers. Additionally, the monitoring framework supports several event types, ordering within multithreaded components, a tester tool for labeling invocations, support for different ORB implementations, and support for optimized invocations between collocated entities sharing one execution environment.

6. Conclusions

We succeeded to define a framework for the monitoring of component interactions, which has sufficient expressive power, allowing formal analyses to be performed. The information generated by the framework accurately follows the interaction contract, which allows monitoring of most applications implemented using contracts.

The monitoring framework seamlessly integrates in the component development process, by providing automated tool support for all activities related to monitoring. The technologies used to implement the prototype are standard, the framework is flexible and configurable. The emphasis of the proposed monitoring architecture is on dynamic reconfiguration of the monitoring output (through event filters) as opposed to static reconfiguration broadly suggested by previous work done in this area.

The prototype of the monitoring framework is used for testing of services developed for a large application framework [10]. The component tester can make use of several facilities to track invocations, like the centralized monitor and the Message Sequence Diagram viewer.

The current framework supports only partial capturing of causality. In order to improve the opportunities for performing formal analysis, we need to extend the causality support in the framework.

Provided a formal model of the behavior of a distributed system, conformance testing can be done assisted by a tool. For this purpose, a mapping has to be created from the execution model of the monitoring system to the formal model describing the system behavior [16].

The monitoring framework can be specialized to support extension of distributed applications with generic monitoring functionality. The ultimate goal is to define a flexible framework that allows introduction of new application components into legacy distributed systems at reduced cost. All observation-related functionality necessary for the operation of the new application components will be defined as reconfiguration of the monitoring code in the legacy system. For example, using our framework, introduction of an accounting facility can be achieved by reconfiguring the legacy components to produce only events related to the accounting domain.

We intend to develop a generic architecture for development of monitoring application components. Such components use the basic event information model, combine it with a static data mapping and translate the monitoring information to a specific application domain. For example, monitoring events delivered to an accounting application components can be translated through static mapping to the terminology of the accounting domain.

References

1. Batteram, H., H.P. Idzenga. A generic software component framework for distributed communication architectures. *ECSCW'97 Workshop on OO Groupware Platforms*, 1997, Lancaster, UK, 68-73.
2. Bakker, J.L., H. Batteram. Design and evaluation of the Distributed Software Component framework for distributed communication architectures. *2ⁿᵈ Intl. Workshop on Enterprise Distributed Object Computing (EDOC'98),* San Diego, USA, Nov. 3–5, 1998, 282-288.
3. The MESH project, see http://www.mesh.nl/
4. Szypersrki, C. *Component software: Beyond OO programming.* Addison-Wesley, 1998.
5. Krawczyk, H., B. Wiszniewski. *Analysis and testing of distributed software applications.* Research Studies Press Ltd., Baldock, Hertfordshire, England, 1998.
6. *Message Sequence Chart (MSC).* ITU-T Z.120, ITU-TSS, Oct. 1996.
7. *CORBA component model.* RFP orbos/97-06-12, Revised Submission, Feb. 2000.
8. Raynal, M., M. Singhal. Logical time: Capturing causality in distributed systems. *IEEE Computer* 29, Feb. 1996, 49-57.
9. *Java 2 platform.* Standard Edition, API spec., http://java.sun.com/products/ jdk/1.2/ docs/api/
10. The FRIENDS project, see http://friends.gigaport.nl/
11. The AMIDST project, see http://amidst.ctit.utwente.nl/
12. *IDL to Java mapping. Java ORB portability interfaces.* OMG TC Document formal/99-07-53, June 1999.
13. *Portable interceptors.* OMG TC Document orbosl/99-12-02, December 1999.
14. *CORBA 2.3 chapter 21, Interceptors.* OMG TC Document formal/99-07-25, July 1999.
15. Logean, X., F. Dietrich, H. Karamyan, S. Koppenhoefer. Run-time monitoring of distributed applications. *Middleware'98, IFIP Intl. Conf. on Distributed Systems Platforms and Open Distributed Processing*, Springer-Verlag, London, 2000, 459-473.
16. Quartel, D., M. van Sinderen, L. Ferreira Pires. A model-based approach to serv ice creation. *7ᵗʰ IEEE Comp. Society Workshop on Future Trends of Distributed Computing Systems (FTDCS'99),* IEEE Computer Society, 1999, 102-110.

Specification and Implementation of an Extensible Multimedia System

Jürgen Hauser and Kurt Rothermel

University of Stuttgart

Institute of Parallel and Distributed High-Performance Systems (IPVR)

Breitwiesenstr. 20-22

70565 Stuttgart, Germany

e-mail: {hauser, rothermel}@informatik.uni-stuttgart.de

Abstract. Multimedia authoring does not consist solely of the specification of the temporal and spatial layout of a document. Application specific concepts (e.g. for computer based training) or animation effects (e.g. moving media items) are often required by authors. Existing multimedia authoring tools support only one specific concept for the temporal specification, spatial specification and interaction. This paper presents research results for an extensible multimedia system called MAVA (*M*ultimedi*A* document *V*ersatile *A*rchitecture). It allows to integrate new and alternative concepts for multimedia documents. MAVA uses an operator based approach as meta document model, what means that all extensions will define concrete operators that will be used by authors.

1. Introduction

The support for multimedia authoring provided by multimedia systems can be divided into the following two areas. Area one is the specification of multimedia documents using „standard" concepts for the specification of the temporal and spatial layout of documents. These concepts are well understood and base of several different multimedia authoring tools. For example, temporal interval operators are used Madeus [8] and TIEMPO [16]. Also these approaches use different concepts to define the spatial layout, for example absolute positioning. Multimedia systems provide one concept for the temporal layout, one concept for the spatial layout and sometimes a further concept for interaction. The second area of support is for the adjustment on specific application areas for multimedia documents. If such an adjustment is possible it is done by scripting languages, like in MHEG [11] or Macromedia Director [3].

This paper presents an approach to simplify multimedia document authoring for different application areas (e.g. computer based training or particular animation effects during presentation). The presented system uses an operator based approach to specify multimedia documents. That means an author will use operators to define different types of relations between media items (e.g. spatial, temporal, etc.). The presented system is extensible by different sets of operators. Main aspect of this paper is the formal specification of the framework for an extensible multimedia system. This

H. Scholten and M. van Sinderen (Eds.): IDMS 2000, LNCS 1905, pp. 241-253, 2000.

special interest for the formal specification is caused by the extensibility. For example a new concept has to be used with existing media items or a new media item has to be used with existing concepts. Because it is not likely that all media items can be used with all concepts, the correct usage has to be verified by the presentation system. This verification is part of the consistency checking of documents.

The developed multimedia system is called MAVA (*Multimedi*A document *Ver-satile A*rchitecture). The research project is funded by the German Research Foundation (DFG) in the research initiative „Distributed Processing and Exchange of Digital Documents (V3D2)" [14].

The second section of this paper describes two scenarios in which an extensible multimedia system helps to simplify authoring of multimedia documents. The third section introduces an appropriate document model to describe multimedia documents. Section four contains the formal specification of the realization of the semantics. Section five explains the architecture of the implementation of the presented approach. Section six describes related work and finally section seven contains a conclusion.

2. Multimedia Authoring Scenarios

A menu of a multimedia travel guide has to appear animated on the screen (Figure 1). The menu consists of a background image, for example a map of a particular region, and four buttons. The buttons can be used to jump to further scenes in the travel guide. These scenes present for example interesting places for visitors to travel to. At the beginning of the presentation, only the map is visible. As an animation effect the four buttons (initially invisible), will move from the left into the presentation area. The buttons move until they have reached their final position on the map. After reaching the final position on the map the menu is fully built up.

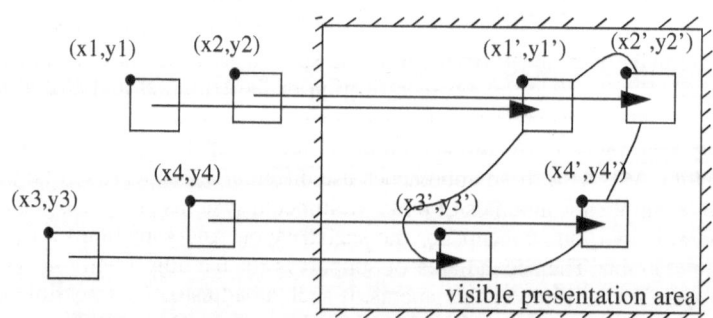

Figure 1 Scenario of moving media items.

For the realization of the described scenario the author wants to use a language construct to describe the movement of the buttons (media items). An extensible multimedia system allows to add such a language construct. The system provides concepts for the semantically realization of the new construct. Once a new construct is realized, it can be reused by other authors. Thus, freeing them from realizing the semantics again.

Another example for the need of extensibility is the usage of alternative concepts for example to specify the spatial layout of a multimedia document. The simplest and most often chosen approach is to place media items into a two dimensional coordinate system. But there are other alternative approaches possible, like the relative positioning. Figure 2 shows a limitation if only one approach to specify the spatial layout is provided by a multimedia system. In Figure 2a, a text is horizontally centered below a picture. If there is a language construct to express this relation between the text and media item directly, the author has not to recompute the positions of media items if for example the size of the image is changed. In case that such a construct is not available the author has to compute new positions of media items after each change. On the other hand, it is no solution to provide only relative positioning for simplicity during authoring. The scenario where buttons have to be placed on particular positions on the map (Figure 2b), is not realizable with only relative positioning. In this case absolute positioning is required. Hence, a multimedia system that simplifies authoring will provide both concepts and allows to integrate further concepts (e.g. relative positioning to one of the four corners of the presentation area).

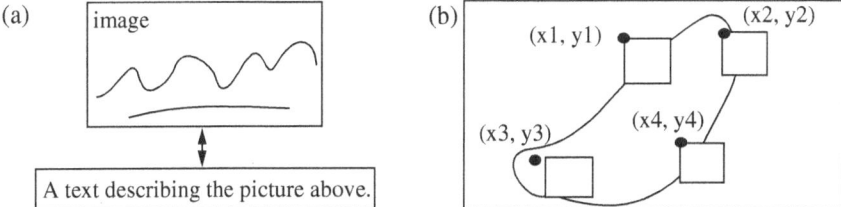

Figure2 Relative positioning (a) and absolute positioning (b) of media items.

These examples show that multimedia authoring can be simplified if a variety of different concepts is available to the author. Like alternatives for the spatial layout there are also several alternatives for the temporal layout, for example different interval operators or a simple time line. And it is even thinkable that application specific concepts, like a concept for computer based training or alternative interaction forms, have to be integrated. Hence, a multimedia system is required that is extensible with new language constructs to simplify multimedia authoring.

3. The MAVA Document Model

The structure of documents is described with a document model. Normally, the author uses a suitable visualization of the document model. Ideally, the visualization provides the „What You See Is What You Get" principle, what means that the visualization for editing is exactly the same as the output on a printer. Because of the temporal and interactive components of a multimedia document the WYSIWYG principle is in general not possible. For single aspects (e.g. the spatial layout) a (sometimes simplified) WYSIWYG visualization for editing is possible. For example a snapshot of media positions at a certain time point or interval can be visualized and edited. The temporal course of a presentation can be visualized with bars over a time line. Existing multimedia systems, like TIEMPO or Madeus, provide such views.

Regarding the standard for multimedia documents propagated by the World Wide Web Consortium, called SMIL [7], it shows that the provided two temporal elements (seq, par) match the tree like structure of XML documents [2] very well. But in general, temporal interval operators do not match the XML structure exactly. They have to provide references to media items as attributes of XML elements to model relations between media items. Hence, a tree like structure is not a suitable document model for all kind of relations that can be used in multimedia documents.

During authoring with interval operators an author creates a temporal relation graph for media items. Thus, the structure of multimedia documents cannot be mapped directly onto a tree.

Further approaches like timed Petri nets (e.g. TSPN [13]) or channels from CMIF [6] cannot be generalized and used to describe other concepts than temporal ones. The duration of a token remaining in a place of a timed Petri net implicitly defines a temporal description. Interaction or spatial relations cannot be described with timed Petri nets. The channel approach of CMIF implies a sequential chronology in a channel.

General approaches like the specification of reactions on certain events (e.g. in [11] (MHEG link-action mechanism) and in [15] (an event based scenery description)) are hard to visualize and they are on a low abstraction level. This complicates authoring of multimedia documents.

The requirements for a meta document model can be summarized as follows. It

• has to be independent of any (application) concept that has to be described and

• has to allow a simple visualization on a high abstraction level to enable easy document authoring.

Especially the simplicity of document authoring is of practical importance, because multimedia documents have to be created by designers and not by programming specialists. In practice this is so far not the case. Authoring of multimedia documents is often done by programmers, which can be recognized by the variety of books on Lingo, the scripting language of Macromedia Director, and the demand of lingo programmers. Thus an approach that simplifies document authoring is needed.

3.1   Specification of a Document Model

MAVA uses an operator approach as document model. Operators can be used to describe relations between media items. Operators are easy to use and permit easy graphical visualization and manipulation. This approach is used by all application concepts used with MAVA. Once an author has learned the editing principles (i.e. direct manipulation of an operator graph) he can use them for all application areas. Besides authoring, the development of extensions will be simplified too. Programming of a manager, a component responsible for the semantic realization of a set of operators, can be based on an application programming interface (API) for the document model.

Therefore all concepts that have to be integrated into MAVA have to be expressible on base of the operator approach. This has the disadvantage, that a few concepts that cannot be expressed with operators cannot be integrated into MAVA. For example temporal Petri nets can hardly be expressed with operators.

For the MAVA document model the operands of an operator are media items. Media items represent information units of a particular type and format, e.g. audio, video or simple text. The graphical visualization of operators becomes in particular effective if an operator has more than two operands. A textual representation of a graph becomes very unreadable because of the references to media items (used to model edges) and documents based on textual representations can hardly be created by hand.

Figure 3 shows a simple MAVA document. The document consists of one container (i.e. the document), that includes three media items and three operators. The first media item is an image (representing a map for example), the second one is an audio sequence (for example background music) and the third is a button (for example to quit the presentation). The first operator is the *setPosition* operator. It is used to position the map in the presentation area. The parameters of the operator determine the position of the upper left corner of the image relative to the upper left corner of the presentation area. The second operator is a temporal interval operator called *while* (see [1] for the description of interval operators). The intention of the author was that 3 seconds after the beginning of the presentation of the image also the audio can be heard. The second parameter of the *while* operator says that five seconds after the audio finished also the presentation of the image ends. The *move-to* operator is used to move the button to a particular position (specified as parameters).

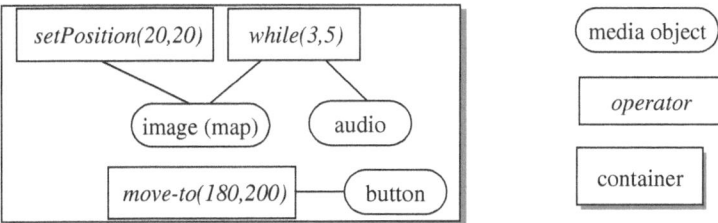

Figure 3 A MAVA document.

For the description of the document model, media data and appropriate viewers are not required. It is sufficient to consider attributes (e.g. media item type, location of the media data). Hence, the document model stores only the attributes in media objects.

Containers permit a partitioning of a document into logical units. Thereby a container includes media items and operators. To allow recursive structures a container is considered as a media item. The root container of the container hierarchy is the document itself.

The presented document model uses the operator approach as a meta document model. A meta document model describes the schematic structure of documents and not particular media items or operators. Document models for concrete application areas (i.e. concepts) base on the meta document model and provide particular operators and media items. So far, operator based approaches provide a fixed set of operators of a particular concept. In contrast to that MAVA uses the approach of a meta document model and allows to integrate new concepts with their operators and media items.

Recently, there have been several publications on research concerning the description of concepts. Main topic of interest is the temporal specification of multimedia documents. In [12] different concepts for the temporal specification were described.

Some of them can be represented by operators for the usage in MAVA. In [16] a system basing on ten temporal interval relations and relations for interaction is presented. Compared to the research work done in this area MAVA does not define new temporal or interaction concepts. MAVA wants to integrate existing concepts.

3.2   Media Items in MAVA

Media items are mono media items (i.e. they are not a container). Their media data (information that has to be presented) may be computed during presentation or may be available before presentation. They are active components during the presentation what means that they can influence the course of the presentation. Therefore, media items may contain state information.

3.3   Operators in MAVA

As previously described, an operator describes a relation between media items. It is necessary to distinguish between source and destination media items. This distinction is required to describe the semantics completely, what can be shown with an *inFrontof* operator. This operator describes the overlapping of media items during presentation. It can be used to specify that a text media item is displayed in front of an image. With the reverse order the image would cover the text and thus make it invisible. Figure 4 shows the usage of the *inFrontOf* operator.

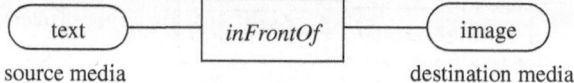

Figure 4 Example for source and destination media items.

The meta document model does not make any constraints regarding cardinality of operators. Basically an operator can have an arbitrary number of source and destination media items. Operators of a particular concept however determine their number.

4. Formal Specification of the Framework

The formal specification of the framework has two purposes: (1) specification of the communication between managers and media items and (2) basis to verify if documents are consistent. Further, this formal specification is the basis for a later implementation of the introduced concepts and the MAVA framework.

The framework and elements of the meta document model are presented in Figure 5. The architecture comprises three layers. The MAVA engine is the first layer, the second layer contains manager components and the third layer includes media items. The second and third layer are extensible with new components. A component (manager or media item) can only be integrated into the framework if it supports the required protocols as described below. The architecture does not consider any operators because they are passive components regarding communication. They describe only the structure of the document but the implementation of the semantics is task of the managers during the presentation (active part).

For the extensibility the most important layer is the second one. A manager component π representing a particular concept is a 3-tuple

$$\pi = (\Sigma, \Delta, \Phi)$$

where Σ is a set of supported operators, Δ is a set of notification events and Φ is a set of control events. Δ and Φ define the communication protocol between a manager component and a media item like it was introduced in Figure 4.

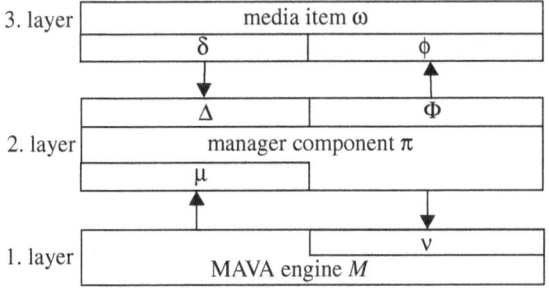

Figure 5 Three layers of the framework.

Let $\Sigma(\pi)$ be the set of operators of a particular manager component π. For the specification of the semantics of an operator σ with

$$\sigma = \Sigma(\pi)$$

the following two event sets

$$\delta = (\sigma)$$

$$\phi = (\sigma)$$

are required. Let $\delta(\sigma)$ be the set of notification events - i.e. notifications from a media item to a manager - and $\phi(\sigma)$ is the set of control events - i.e. control events from a manager to a media item. These two sets define which notifications from the media item and which control events from the manager are required for the realization of the semantics of the operator σ. An example is given in Figure 6.

$$\sigma = while$$
$$\delta(\sigma) = \{started, stopped\}$$
$$\phi(\sigma) = \{start, stop\}$$

Figure 6 Operator and event set example.

With the definition of $\delta(\sigma)$ and $\phi(\sigma)$ for all $\sigma \in \Sigma(\pi)$ it is possible to define $\Delta(\pi)$ and $\Phi(\pi)$ for a manager π. Δ is the union of all notification events that are required for the realization of the operators. Φ is accordingly the union of control events.

$$\Delta(\pi) = \bigcup_{\sigma \in \Sigma(\pi)} \delta(\sigma)$$

$$\Phi(\pi) = \bigcup_{\sigma \in \Sigma(\pi)} \phi(\sigma)$$

Now it is possible to describe the requirements for a media item to be integrated into the MAVA presentation system. Let Ω be the set of all media items. A media item $\omega \in \Omega$ is represented as a tuple

$$\omega = (\delta, \phi)$$

where $\phi(\omega)$ is the set of all control events and $\delta(\omega)$ is a set of all notification events a media item ω supports.

A media item has to support both event sets - $\delta(\sigma)$ and $\phi(\sigma)$ - of an operator σ of a particular concept may be use in conjunction with the media item. Hence, the following two equations must hold for each operator that may be applied to a media item ω.

$$\delta(\sigma) \subseteq \delta(\omega)$$

$$\phi(\sigma) \subseteq \phi(\omega)$$

The event sets of an operator are subsets of the event sets a media item supports because a media item may be used in conjunction with several different operators or managers.

Usually, there are several managers that will use the same event sets. For example there are different managers for temporal synchronization. The event sets for media items that support temporal scheduling may be the following: the notification events are {*started*, *stopped*} and the control events are {*start*, *stop*}. Hence, a media item that supports these two event sets may be used in conjunction with different temporal managers (e.g. a time line manager or any manager of an interval based approach).

After the specification of the elements of the document model, the document itself will be considered next. Let $\Sigma(\Pi)$ be the set of all possible operators in the document. This set depends on the set Π of all managers that are required to implement the semantics of the used operators. Hence, Π is defined by the used operators because each operator determines its manager.

The function *opcard*(σ) returns the number of operands of the operator σ. Because each operator has its own cardinality the cartesian product cannot be used to define the relation between the operators and the media items. Instead, the $\otimes$-operator will be defined as follows.

Let Ψ be the set of all possible combinations of operators and media items. It can be defined as follows.

$$\Psi = \Sigma(\Pi) \otimes \Omega = \bigcup_{\pi \in \Pi} \bigcup_{\sigma \in \Sigma(\pi)} \{\sigma\} \times \Omega^{opcard(\sigma)}$$

A subset ψ of Ψ is the set of all actual used operators with their media items in a document. For example, an element $y \in \psi$ has the following form: y=(while, audio, video). Consider, that y is an ordered tupel and not a set.

After the definition of all components, a document D can be defined as a 3-tuple

$$D = (\Sigma, \Omega, \Psi)$$

where Σ is the set of all operators used in the document, Ω is the set of all media items of the document and ψ are the relations between the operators and the media items.

The manager components also need a basis wherein they have to be integrated. The MAVA engine is the basis of that framework. Formally, the MAVA-Engine M is a tuple

$$M = (\nu, \mu)$$

where ν provides the base functionality and μ defines the control events sent to managers. Therefore, a component that has to be integrated as a manager has to support the protocol μ and may use the base functionality ν. The base functionality includes support for the scheduling of the presentation - in form of a scheduling graph - which is regarded in the next section.

4.1 Consistency of MAVA Documents

Because not all operators can be used with all media items - depending on the event sets the operators require - the following condition must hold true that a document is valid. The equation describes that all media items have to support (i.e. implement) the event sets of the operators. In other words, an operator must not be used with a media item that does not support the event sets of that operator.

To verify the validity of a document the following two auxiliary functions are needed:

operands(y): This function returns the set of the operands of an element $y \in \psi \subseteq \Psi$.

operator(y): This function returns the operator of an element $y \in \psi \subseteq \Psi$.

For example, let $y=(while, audio, video)$ then *operands(y)* would be $\{audio, video\}$ and *operator(y)* would be *while*.

Def: A document D is valid if and only if the following equation holds true

$$\forall (\gamma \in \Psi) \; \forall (\omega \in operands(y))$$
$$((\delta(operator(y))) \subseteq \delta(\omega)) \land \phi(operator(y))) \subseteq \phi(\omega))$$

For the presentation of a document it is not only necessary to get the document but also the implementation of all managers and media items used in the document have to be available to the MAVA engine. This leads to the definition of a loadable document.

Def: A document $D=(\Pi, \Omega, \psi)$ is loadable if the document is valid and the MAVA-Engine M can locate and load all necessary classes and media data for the presentation of the document.

The consistency checking of a MAVA document is two phased. First of all, a MAVA document has to be loadable and valid (i.e. the first phase). A document is only presentable, if the operators of all managers are used properly. Therefore it is up to a manager to verify that its operators are properly used (i.e. second phase). The realization of the first phase can be done in the MAVA engine because it is the same for all managers.

Def: A document is presentable if it is (1) loadable (2) valid and (3) if all managers used in the document have verified the proper use of their operators.

5. Implementation

Figure 7 shows the architecture of the MAVA presentation system. It consists of the following components:

- MAVA engine: The MAVA engine is the core of the presentation system. It handles all basic tasks of a presentation system like document loading, class loading and controlling the presentation.
- Class loader: Loads Java classes that are required for the presentation of a particular document (managers, media items and media viewers). Because the classes can be distributed over several servers in the network the class loader of Java cannot be used. The class loader is not described in this paper.
- Document loader: The document loader is responsible for loading documents from various sources like local disks or network servers. It generates the document model from the storage format of MAVA documents.
- Manager: Managers are responsible for the realization of semantics of particular concepts (i.e. a set of operators).
- Operator: An operator is a component of the document model. It is used to describe relations between media items.
- Internal representation: Internal representations are models for particular concepts that are realized with managers. They are intermediate representations of the relations described by operators. They are used to simplify the generation of the required control events. The usage of internal representations is explained in detail in [5].
- Media objects: A media object is a component of the document model. It is used to realize different types of media items.
- Media viewer: A media viewer is responsible for the visualization of particular media formats (e.g. a mp3 audio).

Besides the Java runtime, MAVA uses the Java Media Framework [9] to realize media viewers for continuous media items like video or audio media items.

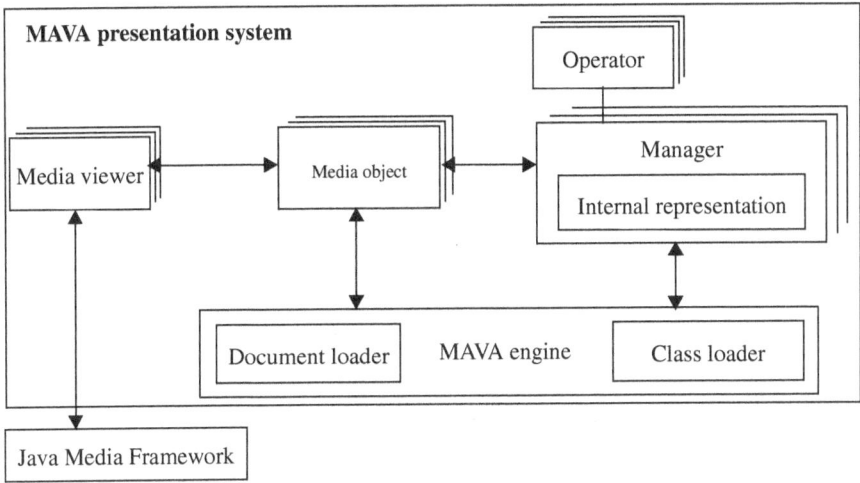

Figure 7 Architecture of the MAVA presentation system.

6. Related Work

Present multimedia systems allow two different ways to extend their functionality. The first way is only possible if the source code of the system is available. Then a programmer could add new functionality. This approach is very insufficient because documents belong to particular versions of the system (which has to be available) and merging different extensions is difficult. Another way, very similar to programming is the usage of a multimedia middleware. A multimedia middleware provides programming abstractions for the development of multimedia applications. This simplifies the development of presentation systems for particular concepts. In [10] it was shown how such a middleware can be used to realize a SMIL player or a presentation system for Madeus.

A second alternative is to provide a scripting language which enables the author to program particular effects into the presentation (e.g. Macromedia Director). From the point of view of simplified authoring this approach has several severe disadvantages: (1) the author has to have programming experience, (2) expressiveness and performance of the scripting language are limited, (3) reuse of scripts is difficult or impossible because they are integrated into a document and solutions for a particular problem and (4) integration of the scripting language into the document model or authoring paradigm. Especially the last point shows that if scripting languages are used, authoring tends to be more a kind of programming and thus lets the original document model loose importance. In the end, the creative part of the design task is vanishing.

Compared to existing approaches, MAVA provides a clear concept for the separation between documents and code for extensions in contrast to scripting language approaches. Compared to approaches using a multimedia middleware, MAVA provides a higher abstraction level for the extension (e.g. a manager and operator concept). Reuse of existing concepts and a common document model require less effort for the realization of a new concept than programming a new presentation system or using a middleware.

7. Conclusion

The MAVA approach is used to simplify document authoring. The simple intuitive approach by using operators to specify multimedia documents requires only little programming knowledge by authors. And due to the fact that all extensions are based on operators, authors do not have to learn different authoring paradigms for different application areas, i.e. a simple presentation is created exactly the same way as a computer based training.

Before authors can use operators for a particular application area the concept for the application area has to be specified (e.g. definition of operators) and an appropriate manager has to be implemented in Java. This requires a MAVA developer who is familiar with Java and MAVA concepts. Compared to other authoring tools this is not a disadvantage, because basic managers for temporal and spatial concepts are already available for MAVA. If more concepts are available the MAVA approach has an advantage. Multimedia Systems that do not support extensibility will not support these concepts. Systems that provide a scripting language require the programming of the extension. Once an extension of MAVA is programmed by a specialist other authors can simply reuse it. They do not have to program the same again.

The platform independent realization of MAVA cares for a large audience and permits the integration into the Internet. Especially in the commercial sector multimedia players are in most cases platform dependent, i.e. browser plug-ins only exist for popular operating systems (e.g. Macromedia Flash [4]).

8. References

1. Allen, J.F. „Maintaining Knowledge about Temporal Intervals". Communication of the ACM, Vol 26, Num 11. 11/1983.
2. Bray, T., Paoli J. Sperberg-McQueen, C.M. „Extensible Mark-up Language (XML) 1.0". Working Recommendation. W3C. 1998.
3. „Director version 8.0". URL: http://www.macromedia.com/products/director. Macromedia. 1999.
4. „Flash 4.0". URL: http://www.macromedia.com/software/flash. Macromedia. 1999.
5. Hauser, J. „Realization of an Extensible Multimedia Document Model„, In Proceedings of Eurographics Multimedia '99". 9/1999.
6. Hardman, L., Van Rossum, G. and Bulterman, D.C.A. „Structured Multimedia Authoring" In Proceedings of the ACM Multimedia '93. 6/1993.
7. Hoschka, P. (Ed.) „Synchronized Multimedia Integration Language (SMIL) 1.0 Specification". W3C Proposed Recommendation. W3C. 4/1998.
8. Jourdan, M., Layaida N., Roisin, C., Sabry-ismail, L., Tardif, L. „Madeus, An Authoring Environment for Interactive Multimedia Documents". In Proceedings of the ACM Multimedia 98. 1998.
9. „Java Media Framework". URL: http://java.sun.com/products/java-media/jmf/index.html.

10. Jourdan, M., Roisin, C., Tardif, L. „A scalable Toolkit for Designing Multimedia Authoring Environments". In special Issue „Multimedia Authoring and Presentation: Strategies, Tools and Experiences" Multimedia Tools and Applications Journal. Kluwer Academic Publishers.1999.
11. ISO/IEC DIS 13522-5. "Information Technology - Coding of Multimedia and Hypermedia Information - Part 5". 1995.
12. Pérez-Luque, M.J.; Little, T.D.C. „A temporal Reference Framework for Multimedia Synchronization". IEEE Journal on Selected Areas in Communication, 14 (1996).
13. Senac, P., Diaz, M. Leger, A. and de Saqui-Sannes, P. „Modeling Logical and Temporal Synchronization in Hypermedia Systems". IEEE Journal on Selected Areas in Communication, Vol 14, No. 1. 1996.
14. German Research Foundation (DFG). „Distributed Processing and Exchange of Digital Documents (V3D2),,. URL: http://www.cg.cs.tu-bs.de/dfgspp.VVVDD.
15. Vazirgiannis, M., Sellis, T. „Event and Action Representation and Composition for Multimedia Application Scenario Modeling". ERCIM Workshop on Interactive Distributed Multimedia Systems and Services. 3/1996.
16. Wirag, S. „Specification and Scheduling of Adaptive Multimedia Documents,,, Faculty Report 1999/04, University of Stuttgart, 1/1999.

Communication Protocol Implementation in Java

Gyula Csopaki, Gábor András Horváth, and Gábor Kovács

Department of Telecommunications and Telematics
Budapest University of Technology and Economy (BUTE)
H-1117 Budapest Pázmány Péter sétány 1/D, Hungary
Phone: +3614633119
{csopaki, horvathg, kovacsg}@ttt-atm.ttt.bme.hu

Abstract. In this paper we examine the possibility of applying Java in the field of telecommunications. We present a method based on formal descriptions we worked out for implementation of communication protocols in Java. We demonstrate how SDL descriptions can be mapped to Java code. We bring out the code of some SDL/PR constructions common for every implementation and parts of SDL descriptions that are implementation specific and apply predefined code patterns. We touch on the problem of mapping abstract data types. This conception is presented by the example of sample telecom protocol INRES and possible connections to the environment are introduced. Then we raise and discuss the matter of automatic compilation and present an SDL/PR to Java compiler. We appraise efficiency of this compilation method and the performance of the realized system. Finally we talk over the advantages and drawbacks of applying Java and give some possible fields of applications.

Keywords: communication protocols, formal description, SDL, Java, compilation

1 Introduction

By the end of the protocol implementation process it is a critical question whether the generated software corresponds to the specification. To achieve the best accordance it is worth considering to create the high level language code by a compiler based on formal description instead of creating it directly from textual description (from unstructured prose). But in this case also the formal description should be constructed if it does not exist. In the last few years several methods related to parts of this process have been worked out.

Nowadays, by the growing significance of the Internet and especially the web, problems like platform independence have got high priority. The use of the Java technology can be an answer to this problem. It is worth investigating whether the advantages of Java can be applied in the field of protocol implementation. In contrast to other prevalent programming languages, through the platform independence of Java, software (in this case protocols) are portable, i.e. transporting and optimising the executable code from one architecture to another is far simpler, adaptation to implementation contexts is better. Applying Java has also the advantages of one-time

H. Scholten and M. van Sinderen (Eds.): IDMS 2000, LNCS 1905, pp. 254-265, 2000.
© Springer-Verlag Berlin Heidelberg 2000

compiling and testing. This fact may cause the substantial reduce of time and cost, because tested Java based software shall not be tested again by other companies or if transported to other architecture.

This paper describes a method for protocol implementation using Java technology that was developed at the Budapest University of Technology and Economy. It presents a method for mapping SDL entities to Java and a compiler. Finally an example demonstrates the application of them.

2 Introducing SDL and Java

2.1 SDL

SDL [1], recommendation Z.100 of ITU-T [2], is primarily used for formal description of telecom systems. Using SDL – as its name indicates – the structure and behaviour of any real-time, distributed, layered communication system can be described. The elaborate syntax and semantic of SDL allows unambiguous system description. Basically SDL is description language; it can not be adapted for creating applications. It has two representations: graphic and textual ones. Graphic representation (SDL/GR) makes specification easier for the users, on the grounds of that textual representation (SDL/PR) is by generated developer tools. These two representations correspond to each other conversely and unambiguously. Mathematically SDL is based on the model of communicating extended finite state machine (CEFSM) that describes objects by responses given to stimulation and state transitions.

The operating of an SDL system is shown in Figure 1. The modelled object is an SDL system, what does not belong to the object is the environment of the system. The system contains sub-parts, so-called blocks that behave like a system on their abstraction level. These blocks may also contain other blocks and so on. This is called hierarchical specification. Blocks on the lowest level of this hierarchy contain processes that realise the working of system. On every abstraction level the environment represents an external process that communicates with the processes of the system on and below this level. These processes are the active elements of the system and operate by the model of CEFSM. On the whole the system is set of processes communicating with each other and with the environment. Channels and signal routes carrying parameterised signals provide the connections between processes. The arrangement, grouping and connection of these automata is the static description, the behaviour of them is the dynamic description. Static description does not play role in the working of the system.

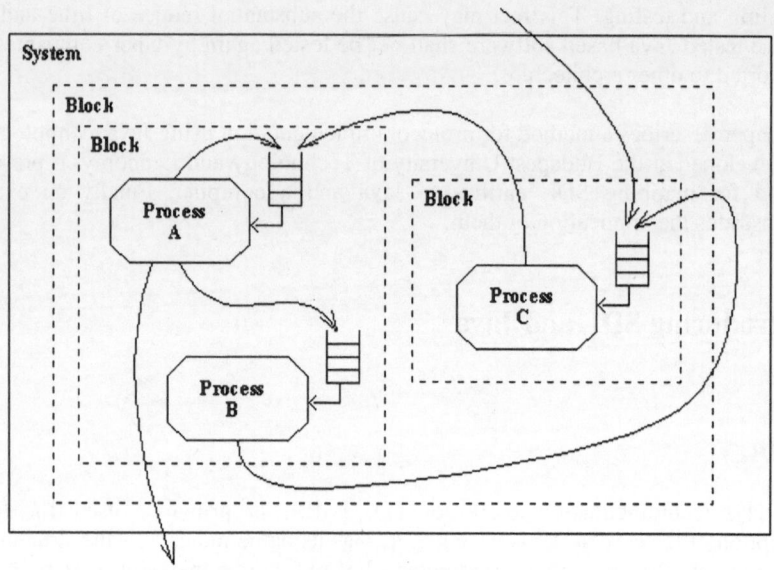

Fig. 1. Operating of SDL systems

2.2 Java

Java is a modern, object-oriented, parallel, platform independent programming technology. Java's built-in threads provide higher efficiency than any scheduling algorithm necessary if using any other languages. However, applying Java has some drawbacks. Attaining the same performance requires faster hardware (the price of portability is paid with speed), and also the non-deterministic garbage collection algorithm also may cause slower working of the software. But the significance of these drawbacks will taper off by the present advancement of hardware devices.

2.3 Analogies and Differences of SDL and Java

Beside portability of Java programs the property of object-orientation is also important, because from the version of 1992 SDL on is also object-oriented. The analogy of SDL types and - Java - classes can be easily observed [3]. In Java type definition (classes) and instantiation can not be sidestepped, so the type-instance paradigm of SDL is automatically fulfilled. Also the inheritance of SDL types can be traced back to the inheritance of Java classes. Due to the possibility of parallel programming the parallel and asynchronous run of processes is provided. The differences can be attributed mostly to the semantics of the two languages.

3 Protocol Implementation Based on Formal Description Techniques

The protocol standards used in the field of telecommunications are given mostly only in textual form, so formal (SDL) description is not accessible and its elaboration without proper methods is a lengthy and uneasy process [4]. The applying of formal descriptions causes overhead in the generated code, so the realisation is not optimal and slower [5]. Until developing efficient methods this problem will balk its permeation. But by this indirect method the numbers of possible errors remaining in the software can be reduced, hereby the reliability of the product may get larger and testing is easier. So in the course of this implementation process time is spent on creating the formal description instead of direct realisation of textual description.

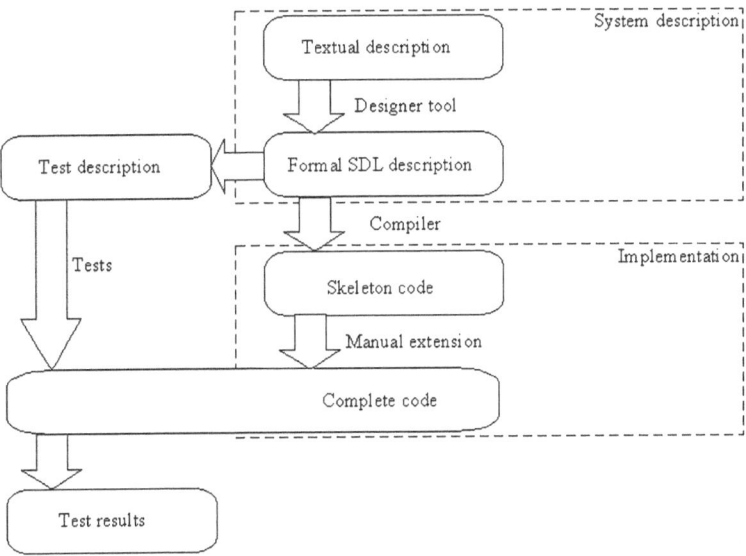

Fig. 2. Protocol implementation based on formal description techniques

Formal descriptions can be created on the grounds of textual ones using a designer tool. Nowadays several commercial products, designer tools (like SDT by Telelogic or ObjectGEODE by Verilog) are available. By the formal description using a compiler the target code can be easily and quick generated. After manual completing the code is ready for testing and application.

Up to the present numerous researches have been done targeting C, C++, CHILL etc. languages. These have had the aim to get a solution to portability, decrease the overhead of the formal description to achieve decent speed and bridge the semantic differences (most of all abstract data types) of SDL and the target language [5]. Using Java we have to deal only with the latter ones, but the investigation of performance parameters shall be emphasised.

4 SDL to Java Mapping

To create a compiler from SDL to Java – first – mapping methods based on SDL/PR are needed. Each SDL syntactic element corresponds to a Java code pattern [6]. In this section this mapping is presented.

4.1 Base Class Hierarchy, A Library

Since trying to implement the SDL concepts in Java on comparison of the source and destination file functional components shall be on same structural levels. For this purpose the use of predefined classes (Figure 3) is needed, which provide all functions assessed in the Z.100 recommendation [2]. Some SDL components have special functions besides existing that can be recognised in each instance. These entities are set to classes. SDL properties are realised by methods and mediately by instance- or class variables. Above all we have to mind perspicuity of the code structure, the execution, and standard realisations of component-component connections.

Java's property of object-oriented programming can be utilised efficiently when arranging all conceived classes in a hierarchy. The principle of the arranging is based on common points of the specified features of SDL issued in the standard.

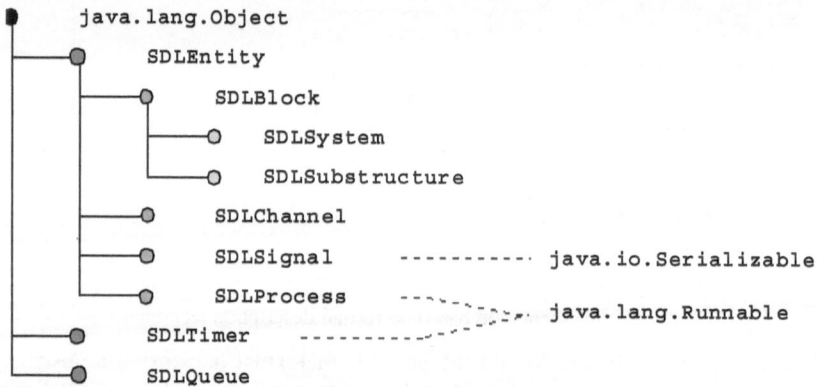

Fig. 3. Hierarchy of base classes

Basic SDL components, like container components, processes or channels, have same attributes. Each instance of them has a name, an identifier, that is the Java reference to the instance, and one direct environment. Direct environment means that – except for the system entity – there is one and only one container entity that contains the component. These attributes can be realised in a common base class (SDLEntity):

```
class SDLEntity {
    String name;       // name of entity
    SDLEntity env;     // reference to container entity
}
```

The hierarchical specification is realised by container components that are block-like components (system, block and substructures). The base class (SDLBlock) of this components has the properties of an SDL entity and in addition array-type variables to store the references to the contained entities, like blocks, processes, channels and signal routes. Channel and signal route connections, and the handling of the contained entities are realised by methods in this class.

```
class SDLBlock extends SDLEntity{
  SDLChannel channels [];   // contained channels
  SDLBlock blocks [];       // contained blocks
  SDLProcess processes [];  // contained processes
  void connect (channel_name, singalroute_name) { ... }
// connecting channels and signal routes
}
```

A common class (SDLChannel) corresponds to channels and signal routes that realise the connections of entities on the level of blocks. Since the delaying of the SDL channels should not be implemented these entities do not needed to be distinguished. Each channel (and signal route) has the properties of an SDL entity and in extension variables of the entities at the endpoints and the carried signals.

```
class SDLChannel extends SDLEntity {
  SDLSignal signals_fwd, signalnames_bwd;
// signals carried by the channel
  SDLEntity connections_to_src[], connections_to_dst[];
// entities (process, channel) connected by the channel
}
```

All signals transported by channels or signal routes are plain entities (SDLSignal) having three more properties: priority and references to the sender and destination processes. The identification of signals can be managed only by their names.

The dynamic description (operating of the automata) is implemented in a class that provides parallel running of SDL processes using the multithreaded environment of Java. SDL processes have several additional attributes: the actual state, input queue and functions (input, output, save, state transition) that realise the working of CEFSM automaton, and references to dynamic created processes and parent process:

```
class SDLProcess extends SDLEntity implements
java.lang.Runnable {
  int state;            // actual state of the process
  SDLQueue queue;       // the input queue of the process
  SDLProcess instances []; // instances
  SDLProcess self, sender, parent, offspring;
// process references
  void create (processtype) {...}
// dynamic creation of processes
  void output (signal, to_process, via_channel) {...}
// output of signals
  SDLSignal input () {...}
// reading a signal from the input queue
  void save (signal) {...}  // saving a specified signal
```

```
    void nextState (state) {...}
 // transiting the process to the next state
 }
```

Although the SDL recommendation does not specify input queue of processes explicitly, but its functionality is exactly described. Queues, which also appear as a separate class (SDLQueue), have arrays to store received and saved signals, and methods to receive or save signals and read the first element of them. The owner process is referred through a variable.

```
 class SDLQueue {
    SDLSignal queue[], saved[]; // all and saved signals
    SDLProcess owner;        // owner of the queue instance
    void save (signal) {...}      // saving the signals
    void put (signal) {...} // put a signal into the queue
    SDLSignal get() {...}          // get the first signal
 }
```

Due to the multithreaded environment of Java realisation of timers is easier than in other programming languages. Properties of the class java.lang.Thread [7] are utilisable. All we have to do is call the sleep (long millis) method in case of setting the timer, and interrupt () method and catch the exception we caused in case of resetting the timer. Timer has to send a signal (named after timer) to the owner process after time is up.

```
 class SDLTimer extends Thread {
    long duration;          // duration of timing
    SDLProcess owner;       // reference to the owner process
    String name;            // name of timer signal
    void set (long) {...}    // set the timer
    void reset () {...}      // reset the timer
 }
```

4.2 Other Code Patterns

There are some SDL constructions whose realization can not be determined exactly. For these cases there are usable Java patterns. The most obvious patterns of these are tasks, decisions and the type-determination of input signals. SDL value assignments are mapped simply into Java value assignments. SDL decisions and the selection of received signals can be implemented using if (...) else if (...) constructions in Java:

```
 if (signal.name.equals ("CR")) { /* transition */ }
 else if (signal.name.equals ("DR")) { /* transition */
 }
```

4.3 Mapping Abstract Data Types

Since the abstract data types of SDL allows of more universal data description there is no general method to translate them to Java, only prescription can be given for these cases. Maybe the most conspicuous problem is the mapping the domains of basic data types. While *integer* in SDL is an integer value between minus and plus infinity, in Java it is limited to values represented on 32 bits. This problem can not be by-passed. There are other traverses of the automatic mapping that can be attributed to Java (reference-based, operator definitions). This part of the implementation has to be done by hand.

Naturally data types shall be mapped into classes, because Java classes are abstract data types as well. In simple cases, if there are only variable and method definitions, the mapping is determined unambiguously: elements of an SDL structure shall be Java variables, SDL operator definitions and axioms together define the Java methods.

5 Compiler

Our compiler is based on 1992 version of SDL/PR description. The semantic differences of Java and SDL cause limitations that have to be set. The limited compiler does not translate data types, so does not perform semantic analysis related to data types.

As matters stand the automatic implementation of formal descriptions in high level languages is not possible, because efficient mapping of abstract data types is not cleared up and can be implemented only with resulting overhead. So some parts of the formal description shall be implemented using a compiler, others manually by predetermined principles.

For lexical analysis and parsing we used the lexical analyser and parser generating tools JLEX [8] and CUP [9] developed at the Princeton University. Tokens used during creating the scanner are mapped by JLEX. These tokens, that are mostly SDL keywords and special symbols, are the terminal symbols of CUP. CUP is a system for generating LALR parsers from specifications. Production rules of SDL defined in the recommendation Z.100 in BNF are transformed to a grammar accepted by CUP. To generate the target code code-strings containing the Java code are filled with the code patterns presented in the previous section. The generated code consists of classes for signals, blocks and processes that are based on predefined classes. The environment and the classes of data types have to be added manually.

6 A Simple Example

This section shows an example of the practical use of transformation elements presented before. For the sake of the cause we describe a system with its structure and then build it beginning with the largest unit finishing by the smallest ones. Finally the connection of a simple test environment to the constructed system will be presented. Let the system be a well-known sample telecommunication protocol called INRES (Figure 4) [1].

6.1 Compiler Generated Code

The construction of code is determined by the arranging of the system, the contained blocks and processes. Static structure is realised by inner classes that are class definitions inside classes (it is possible from JDK 1.1 on). Visibility conditions of the declared variables and the perspicuity of the code are made sure by applying them. For example, if some static components with the same name are placed on the same structural level (not by inner classes) we could be faced by naming problems. This would be a difficulty factor on the occasion of creating a compiler as well. After first step, channels and signal routes will be translated then connected.

The operation of the dynamic description is implemented in the run method (inherited from java.lang.Thread [6]) of the processes of the system using the methods defined in the class SDLProcess. The declarations of timers and variables are Java variable declarations. States of processes are distinguished by integer variables. The selection of actual state is implemented by switch ... case constructions and transition parts are branches. These all result good perspicuity in the structured code.

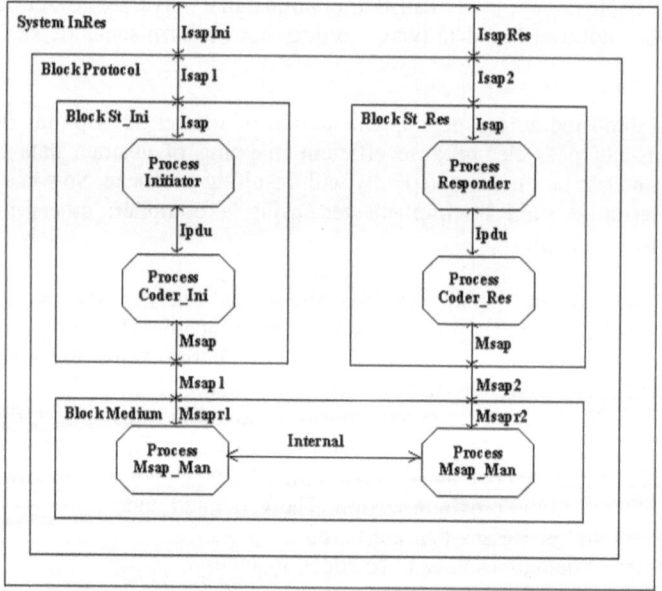

Fig. 4. SDL diagram of system INRES (static description) [1]

There are two ways of sending signals. If the signal does not contain parameters, there is only one data (name of the signal) to be transferred. For this purpose instantiation of class SDLSignal of the predefined hierarchy is enough:
```
output (new SDLSignal ("CR")); // Connection Request
```
If the signal has parameters, then the extension of the class SDLSignal is needed. Parameters are transferred as instance variables:
```
output (new DT (sn, isdu)); // Data for Transmission
```

6.2 Setting Up the Environment and Executing the Code

The main thread represents the environment of the translated system. That is also derived from the predefined classes. The functioning of the environment process has to be comprised by a process (`SDLProcess`) placed in a block and connected to the system by a super-channel. Process, which represents the environment, can be set up using the same method presented for processes in the previous section.

While executing the main thread in the first step initialisation part is done. This time static parts are registered: executor environment detects all entities of the system and registers their arranging and relations. During the functioning of the system these entities have roles only in dynamic creation of processes. After the first step channels and signal routes are detected and then connected, so the connections of processes are established. If this procedure is finished without errors the environment starts all processes of the system.

7 Evaluation

Because we have to compromise between the ease of compilation and the performance of the generated software, the protocols are not optimal, but implementation time, especially the time of Java code generation, is reduced. The duration of the implementation process considerably depends on the size of the protocol.

Using the presented method we implemented protocols from the protocol stack of Signalling System No. 7 [10]. The foregoing observations show, that if this compiler is applied, manual written parts would total up to about 10…20% of the whole code. This ratio is better for larger protocols than smaller ones. The length of the resultant code is about twice as much as the code written manually. The time required for implementation is the time needed to create the SDL description and add data types and their references manually, which are all the work that has to be done. Mostly all these values are dependent on the difficulty of the used abstract data types and the size of the protocol. After executing tests we can state that if manual extension is done correctly, the generated software is operable and contains no errors.

The most critical question of using Java is whether the performance parameters of a specification can be fulfilled. Our findings indicate that if the right combination of non-default priorities is assigned to the processes, i.e. the scheduling is optimized, – counting in the advance of hardware devices – in one or two years protocols implemented this way can provide enough performance to service slow links (64 kb/s).

8 Conclusions

In this paper an alternative method of protocol implementation in Java has been presented. One advantage of the presented protocol implementation method is that owing to validation and testing the software suits the requirements of specification. This way the number of remaining errors in protocol realisations may be much fewer as if it were realised directly from - textual - specification. On different platforms protocol realisations may have different errors increasing the difficulties of testing. Due to having the advantage of portability – as already mentioned – Java is a solution to this problem.

Possible applications of Java implementations may be the precise speed evaluation, simulation and testing. In the far future it is possible that also real-life network devices will operate with Java based protocols, but it would need faster hardware and more exact transformation methods.

Acknowledgements

This work was in part supported by the ETIK organisation. ETIK partners are: ERICSSON Hungary Ltd., Westel 900 Inc., Hungarian Telecommunications Inc., Sun Microsystems Hungary Ltd., KFKI Computer Systems Corp. We extend our gratitude to Géza Gordos, Zsolt Werner and Tamás Kerecsen for all their help.

Abbreviations

BNF	Backus-Naur Form
CEFSM	Communicating Extended Finite State Machine
CUP	Constructor of Useful Parsers
INRES	INitiator RESpronder
JDK	Java Development Kit
SDL	Specification and Description Language
SDL/GR	SDL Graphical Representation
SDL/PR	SDL Phrase Representation
SDT	SDL Designing Tool (product of Telelogic)

References

1. J. Ellsberger, D. Hogrefe, A. Sarma: SDL Formal object-oriented language for communicating systems, Prentice Hall, (1997)
2. ITU-T recommendation Z.100 Specification and Description Language, (1993)
3. B. Wydaeghe: „Translating OMT* to SDL", Proceedings of Methods Engineering '96 -- IFIP WG 8.1/8.2, (1996)

4. ITU-T recommendation Z.100s1 Specification and Description Language, SDL+ methodology: Use of MSC and SDL (with ASN.1) (1993)
5. P. Langendörfer, H. König: „A Configurable Code Generation Tool for SDL", 6th International Conference on Software in Telecommunications and Computer Networks, Split, Bari, October (1998)
6. Java Platform 1.2 API Specification, JDK 1.2 Documentation, Sun Microsystems Inc., (1999)
7. JLex: A Lexical Analyzer Generator for Java, Princeton University, http://www.cs.princeton.edu/~appel/modern/java/JLex/, (2000)
8. CUP Parser Generator for Java, Princeton University, http://www.cs.princeton.edu/~appel/modern/java/CUP/, (2000)
9. Dr. G. Gordos, Dr. Gy. Csopaki, Zs. Werner, G. Horváth, T. Kerecsen, G. Kovács: „Automatic conversion of SDL into modern languages with a Java-based compiler", ETIK Report, Budapest, (1999)
10. ITU-T recommendation Q.0700 Specifications of Signalling System No. 7, Introduction to CCITT Signalling System No. 7

Active Component Driven Network Handoff for Mobile Multimedia Systems

Stefan Schmid, Joe Finney, Andrew Scott, and Doug Shepherd

DMRG, Computing Department, Lancaster University,
Lancaster LA1 4YR, U.K.
{sschmid, joe, acs, doug}@comp.lancs.ac.uk
http://www.LandMARC.net/

Abstract. This paper presents novel solutions to some of the QoS problems associated with delay sensitive multimedia applications in next generation mobile environments. Based upon a set of requirements drawn from mobile multimedia applications, we speculate that the emerging field of active networking can help address these QoS issues. We introduce a set of active software components which, when executed on *active* network routers, are capable of significantly reducing network handoff times for mobile devices. This paper also documents the innovative component-based active network architecture used to support and develop the QoS enhancing components described above. Finally, we demonstrate the flexible and extensible properties of the active router design through a detailed description of the workings of the components described above.

1 Introduction

Quality of Service (QoS) properties such as throughput, latency and error rate are critical to any network providing multimedia services, and it is becoming increasingly apparent that the Internet protocols will play a major role in future multimedia communications. QoS issues are already widely discussed within the Internet research community, and solutions to the problems are evolving from modifications to network and transport protocols (for example, DiffServ, IntServ, and RSVP).

However, when the provision of QoS support for mobile devices is also considered, additional QoS issues become apparent – most notably the latency introduced by network handoffs of mobile devices. This is of particular relevance to Mobile IPv6 [1] (the IETF standardized mobile routing protocol for the next generation Internet protocol). These issues are only now being addressed within the research community.

We believe that current networks are not flexible and extensible enough to cope with this problem in an efficient and elegant manner. This paper introduces a novel component-based approach to active networks providing a highly flexible and dynamically extensible architecture for programmable nodes. In particular, we will show how this architecture can be deployed to resolve some of the QoS problems associated with Mobile IPv6 enabled networks.

H. Scholten and M. van Sinderen (Eds.): IDMS 2000, LNCS 1905, pp. 266–278, 2000.
© Springer-Verlag Berlin Heidelberg 2000

The remainder of this paper describes in more detail the QoS problems associated with Mobile IPv6 handoffs and derives a set of requirements for the solutions to those problems. We then continue by introducing our component-based architecture, LARA++, and the components we have designed to optimise the handoff performance for mobile devices. Finally, we conclude by reviewing how active networking benefits these handoff enhancing components.

2 Motivation

We have recently analysed the performance of multimedia streaming systems in Mobile IPv6 environments. Several problems become apparent when frequent network handoffs are considered. Since existing solutions to these problems rely on enhanced support from network routers, we believe that they can be addressed more elegantly with the use of active network technology than without.

In particular, we consider the two dominant factors that govern the performance of network handoffs in QoS sensitive systems – the speed of address acquisition and the mobile routing convergence time [2]. The following sections outline the issues affecting these factors and show how they can be improved by means of active networks.

Address Acquisition
The stateless address auto configuration mechanism [3] of IPv6 is an excellent means for mobile devices utilising the Mobile IPv6 protocol to dynamically allocate IP addresses, as this greatly simplifies the process of network detection and care-of address acquisition. However, to prevent multiple devices from accidentally using the same IP address in a network environment, a *Duplicate Address Detection (DAD)* mechanism has been mandated [4] for IPv6. This demands devices first verify that a newly acquired address is not already in use before actively utilising that address. Experiments have shown that while addresses can be acquired using stateless auto-configuration within a few milliseconds, the DAD verification process can take up to five seconds to complete [2]. Network outages of this order are, however, not tolerable for multimedia applications. Fig. 1 illustrates the address acquisition problem of mobile devices.

We aim to show that, by means of active networks, routers can be dynamically programmed to monitor the availability of a mobile node's potential care-of addresses within a network domain. When a mobile device roams to a new network within the domain, the device could simply ask a router on the link if the care-of address is still available. If so, the mobile node can immediately utilize the address without incurring the latency introduced by the DAD verification mechanism, and still remain confident that no IPv6 address collision exists on the network.

Mobile IPv6 Convergence Time

The second inefficiency we identified, while analysing mobility support protocols, is that Mobile IPv6 is purely a routing protocol for end systems; network routers (apart from the home agent) are not involved. As a result, handoff times can be unnecessarily high due to the fact that handoffs require a correspondent node (any IPv6 device conversing with a mobile device) to receive a *Binding Update[1]* message notifying the correspondent node about the network change before the route change can take place. This adds at least the delay of one round trip time (between the mobile node and the correspondent node) to the handoff latency. As typical latencies of wide area links can be as high as 400ms, this can result in a handoff performance that again, is unaccept-able for multimedia applications. Fig. 2 illustrates this inefficiency and proposes a solution involving network routers.

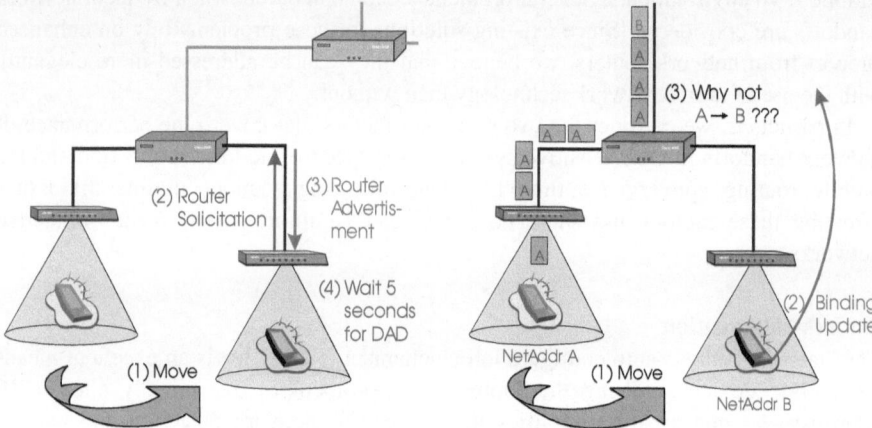

Fig. 1. *Address Acquisition:* After move of mobile device (1), it discovers new network address and prefix (2 & 3). Based on this, it generates its new IPv6 address. To guarantee address uniqueness, it must wait 5 seconds to carry out DAD (3).

Fig. 2. *Mobile IPv6 Convergence Time:* After move of mobile device (1), it sends binding update to inform correspondent node (2). It takes • RTT until data flow reaches the new destination. Why should not the router reroute immediately (3)?

Although several research groups are already addressing this inefficiency [5,6,7] (and there is limited support within the Mobile IPv6 standard), all these solutions require extensions to the Mobile IPv6 standard and/or additional control messaging to operate. By utilising active networks, this problem can be *transparently* resolved, minimising the impact on network utilisation and protocol complexity.

Active routers can be programmed to reroute any data traffic incorrectly routed to the mobile device's old location to its new point of attachment without delay. This provides a means to repair the incorrect routing locally, until the Mobile IPv6 protocol responds to the route change.

---

[1] A Mobile IPv6 control message sent to the correspondent node and the home agent to indicate a network handoff [1].

3 Requirements

Considering the QoS problems of mobile handoffs and the proposed solutions as outlined in the last section, it becomes clear that active support in the network becomes desirable. Routers that actively process on behalf of the mobile end systems can elegantly monitor the utilisation of IP addresses and efficiently perform flow routing optimisation.

This call for active processing in the network is a classic example for the need of active support in today's networks. In recent years, the distinction between programming paradigms of network devices and end systems has led to an ever growing gap between capabilities of end nodes and intermediate devices. While, for example, programmable end systems have included sophisticated support for multimedia streaming for many years, multimedia support for network devices is still in the process of standardization, and deployment is held back by the problems of upgrading significant amounts of network hardware and software.

Building on lessons learned from the development of LARA, the *Lancaster Active Router Architecture* [8] developed as a prototype implementation for active network research, we are currently in the process of building its successor, LARA++, which has a major focus on supporting the needs of current and future network technologies, and in particular mobile multimedia systems. Our new component-based active router architecture, described in the next section, has the advantage of being more flexible and extensible, facilitating the implementation of the active services introduced above.

From the outlined solutions to the mobile handoff problems, we can already derive a set of characteristics and features, which must be provided by the active network in order to solve these problems in an efficient and elegant manner. The following items define the requirements for the active network nodes:

- *Dynamic extensibility of router functionality:* Active routers that can be programmed by network users enable mobile devices to upgrade the network services as they enter a new domain.
- *Flexible packet filtering and insertion mechanisms:* Active routers require sophisticated packet filtering and (re)injection mechanisms in order to act upon certain events in the network (for example, binding update messages).
- *Provision to carry out routing tasks:* Active software components need the ability to route individual flows or even protocol families (for example, to temporary reroute a data stream to a mobile device's new address).
- *Support for active programs to "program" active routers:* Active programs should be able to control (i.e. load, instantiate and remove) other software components as long as the safety and security policies are not violated (for example, a component identifying a handoff should be able to instantiate a routing component).
- *Efficient processing of active programs:* The management overhead for the execution of active programs must be very small. Since the latency of the active processing increases the overall end-to-end delay of data streams, only very small delays are tolerable.

- *Lightweight instantiation and removal mechanisms for active programs:* Since active programs are often very short lived (for example, the flow routing optimisation for mobile handoffs compensates only until the binding update notifies the corresponded node), the instantiation and removal of active programs should be lightweight and fast.

4   Component-Based Active Node Architecture

LARA++ provides a programmable platform for active services development based on the composition of many small components. Components are dynamically loadable onto LARA++ active routers where they provide additional or extended services for dedicated data streams or even for whole protocol families. Fig. 3 provides a conceptual view of the component-based approach used for LARA++.

LARA++ components can be either active or passive. Active components perform the active computations on a node based on one or more execution threads, whereas passive components provide merely static functionality, similar to support libraries.

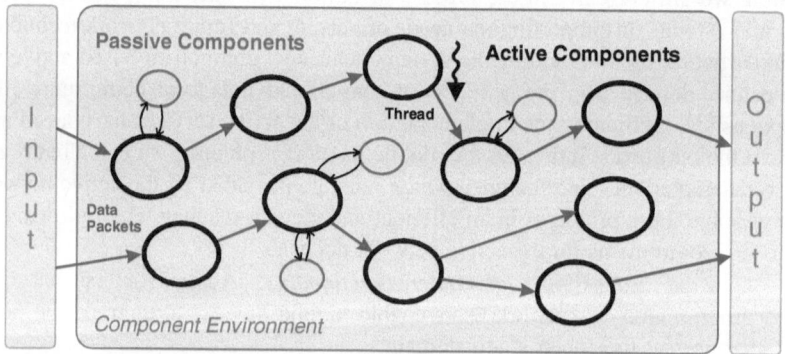

Fig. 3. A Conceptual View of the Active Component Space

Components can be further differentiated by their function into *user components* and *system components*. User components are those active programs or libraries that are injected by users of the active network. System components, in contrast, are the components that constitute LARA++ and expose the API to low-level resources and system calls.

The remainder of this section provides a short introduction to LARA++, its design goals and component-based architecture. Due to the very restricted space in this report, we limit the description to the very basics. Further details on the component architecture and an in-depth discussion of LARA++ internals such as the installation and registration of active components, component interfaces, inter-component communication, service composition, policing, as well as the security and safety framework are freely available in the LARA++ design report [9].

4.1 Design Goals

The motivation for our component-based approach arises from the fact that conventional active node architectures can be classified into one of two broad categories – either fixed execution environments, which can "only" be programmed through in-band active programs (also referred to as active capsules), or a programmable switch platform, which enables users to download and install active programs. While the former approach is limited by the programming capabilities offered by the execution environments, the latter is usually restricted by the support the programmable switch provides for interaction between single active software components and service composition.

LARA++ tries to resolve these limitations by providing a sophisticated composition framework for active components. It enables us to split the processing tasks of active services into many simple and easy to develop functional components and yet enable the active router to compose the services needed. This "divide and conquer" approach also simplifies extensibility of the router functionality. Individual software components can be upgraded more easily and new components can be gradually added.

As a result, our component-based active node architecture is clearly advantageous over traditional implementations in terms of reusability of active code and customisability of user tailored services. The critical part is to provide a sufficiently flexible composition method, which is further discussed in the following sections.

4.2 Node Architecture

A key difficulty in designing active networks is to allow nodes to execute user-defined programs while providing reliable network services for everyone. An active node must therefore protect co-existing network protocols and services from each other and securely control shared resources.

As a result, LARA++ executes active code only within restricted processing environments that limit access to low-level service routines and shared resources. The framework for our safety model is based on a layered architecture, where higher levels are limited and controlled by lower level policies.

The active node is assembled out of the following four layers[2]. Fig. 4 illustrates the layered node architecture.

(i.) The *NodeOS* provides a set of low-level service routines and system policies to access and manage local resources and device configurations.

(ii.) *Execution Environments (EE)* form the management unit for resource access and security policies, which are enforced on every active program executed within that environment.

(iii.) *Processing Environments (PE)* provide the code space where trusted components are processed. Trust relations among components are defined

---

[2] This layered architecture extents the basic active network node architecture specified by the DARPA Active Networking Group [13].

by the code producers (i.e. code signatures) and the network users (i.e. user authentication) instantiating the components.

(iv.)   *Active Components (AC)* comprise the active programs. ACs are processed within PEs alongside with other trusted components. The active code is executed by means of one or several user-level threads.

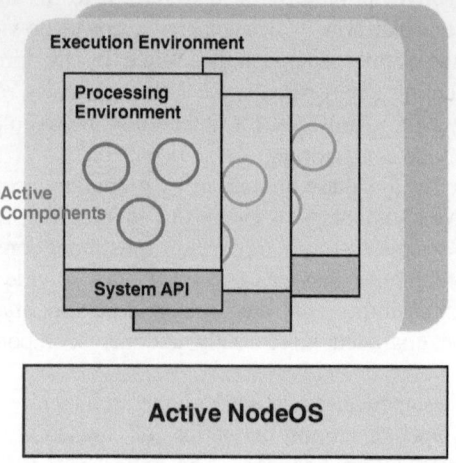

Fig. 4. The layered architecture of LARA++

In contrast to other active network architectures, LARA++ introduces the concept of processing environments as an extra protection layer between the actual active programs and the execution environment. Although this might seem to make the execution of active code more complex and heavyweight, the PEs were introduced for exactly the opposite reason. PEs allow components with trust relationships to be processed very efficiently in a single process-like environment. This unique characteristic of the LARA++ architecture makes use of the performance benefits arising from trust relations.

Trusted active components, which are either pre-compiled executables or "on-the-fly" compiled programs, are directly loaded into the PE's address space and then executed by creating one or more active threads. Trusted components can be executed within the same protection environment (i.e. address space) without creating security threats. The performance boost results from the fact that user-level thread scheduling is remarkably faster than kernel threading, since only a lightweight context switch is involved. Early experiments with our LARA++ implementation have shown that user-level thread scheduling is approximately one order of magnitude faster than the scheduling of conventional active processes.

The intermediate processing environment on top of the EE has besides the obvious performance boost the advantage that the concept of *module thinning*[3] can be applied on a fine-grain basis. Each PE can provide a specially customized system API depending on the trust relation between its components and the NodeOS.

---

[3] Module thinning secures programmable system by individually tailoring the programming interface exposed to a software process based on the privileges of the user or program.

4.3 Service Composition

The structure and make-up of active services depends on the composition method. The method defines the set of components used for an active service and the bindings linking the components.

Within LARA++, active services are composed by means of packet filters. During component instantiation time, active components register their filters with the packet classifier. If a filter matches a data packet passing through the active router, the packet classifier forwards the packet to the corresponding active component. The composition method deployed within LARA++ is thus based on the conditional binding of active components. Only if the filter for a binding matches a data packet, is that binding in force. This composition model leaves the responsibility for the service composition to the packet classifier, or in other words, the classifier is held accountable for the proper "routing" of data packets while they are passing the active component space.

The basic structure for component bindings is defined by the TCP/IP layer model, which ensures that active components providing low-level support are executed before components dealing with higher level computations. For example, network protocol options are processed prior to transport protocol operations. In addition to the ordering imposed by the layer model, the individual protocol stacks (for example, IPv6, TCP, UDP) also need to process their headers, options and so forth in a protocol specific order. For example, the extension headers of IPv6 packets must be processed in the defined order.

In order to enable active components to flexibly extend the functionality of active services, LARA++ requires an "elastic" means to describe at which point in the processing chain a component should be inserted. For this purpose, we chose a *Management Information Base (MIB)* style representation, referred to as classification hierarchy, which is sufficiently flexible to extend current protocol stacks and has provision to incorporate new protocols.

For performance reasons, the packet classifier maintains the packet filters in a tree style representation, called classification graph. Its nodes represent the packet filters of the active components that are instantiated at a given point in time. The graph is built according to the structure imposed by the classification hierarchy. Fig. 5 illustrates a simplistic representation of the classification graph with special emphasize on the network-layer part of the input path.

The packet classifier filters incoming data packets based on the packet filters of the classification graph. The *n-nary* tree structure is parsed according to a left-to-right depth-search approach. If a filter matches a packet, it is "forwarded" to the active component associated with the filter. When the component has completed the active processing, it either discards the packet or reinserts it at the same point or any other point of the graph (depending on the system policies and privileges of the component). The classifier continues the classification process until it runs out of filter branches in the root node.

Finally, we can conclude that the service composition method deployed within LARA++ is *dynamic* and *public*. Since active services are composed through insertion of packet filters in the classification graph, which can be done at any time during normal operation of the active node, the composite changes dynamically over time. Furthermore, the fact that several network users might be allowed (depending on the local security policies) to install packet filters, which might match the same data streams, makes this method of service composition a public process.

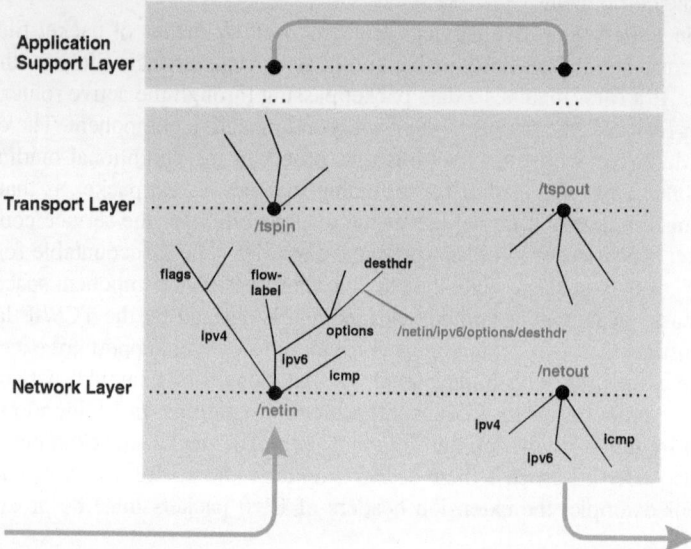

Fig. 5. The Classification Graph. The grey arrow represents the general route data packets take when passing through an active node. The classification graph on the network layer part of the input and output paths show how the classifier manages the packet filters of the corresponding active components.

5   Components for Mobile Multimedia Systems

This section describes the active components we have designed to resolve the two inefficiencies in the Mobile IPv6 handoff mechanism as discussed in earlier sections – latency introduced from duplicate address detection and poor response to route changes due to high latency network links.

Optimised Stateless Address Auto-Configuration
The handoff latency introduced by duplicate address detection (DAD) is solved by means of an *Address Placeholder Component (APC)*, which actively monitors for the use of mobile devices' potential care-of addresses within a network domain, making it possible to optimise the DAD procedure. Fig. 6 illustrates this procedure.

Mobile nodes, upon roaming to a new network segment, inject APCs into each active network node in the local domain, by sending the APC component to the IPv6 site-local all-routers multicast address. One piece of information is carried as a parameter to the APC component – the interface identifier of the mobile's network device (i.e. a EUI-64 number [3] or pseudo random identifier [10]).

Upon reception at an active router, one APC is instantiated for each interface on that router. Each APC then calculates what the care-of address of the mobile node would be, based on the interface identifier supplied and the network prefix of the corresponding interface. This process is essentially the same operation as an IPv6 node uses upon receipt of a router advertisement. The component then registers two LARA++ packet filters, in order to monitor neighbour solicitation and neighbour advertisement messages relevant to that address. This allows the component to perform DAD on the potential care-of address, on behalf of the mobile node. If at any time the DAD procedure fails, the component quietly removes itself from the active router.

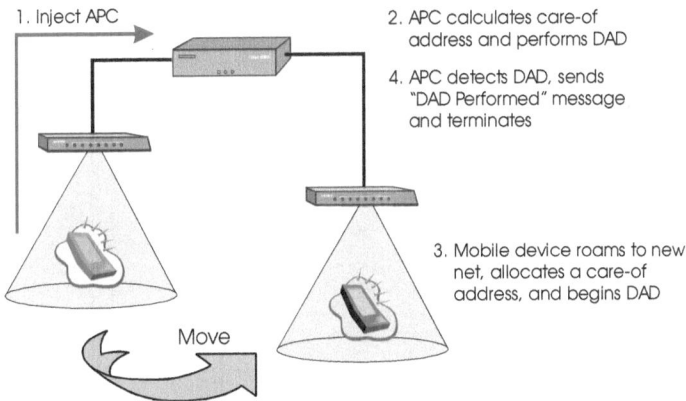

Fig. 6. Optimised Address Auto-Configuration

Once the potential care-of address is verified as unused, the component enters a passive state, waiting to receive a neighbour solicitation for that address. Receipt of such a message implies an IPv6 node is performing DAD on the potential care-of address. At this point, the APC transmits an IPv6 datagram to the source of the neighbour solicitation, containing new "DAD Performed" destination option, and immediately removes itself. A mobile device that understands the DAD Performed option then knows the address is safe to use without the need to perform DAD, and can therefore immediately begin to use its newly acquired address without further delay.

In order to minimize the number of stale components in the network, the system uses soft state – APCs timeout a fixed period of time after instantiation, and remove themselves. Mobile nodes wishing to continue to use the address auto-configuration optimisation continually refresh the active routers by re-injecting APCs.

This simple soft state approach also has the advantage of being resilient to router failures and packet loss – if a mobile node does not receive the positive acknowledgment for address availability, it falls back into default DAD mode.

Optimising Handoff over High Latency Links

The network handoff latency of mobile devices can be minimized by means of a generic LARA++ *Flow Routing Component (FRC)* which introduces Mobile IPv6 functionality into active routers, such that they actively take part in mobile routing as illustrated in Fig. 7.

Upon network handoff, mobile devices are mandated to inform the home agent and correspondent nodes about the network change by sending binding update messages. In order to reduce the mobile routing convergence time, the mobile device injects (or simply invokes) an FRC into active LARA++ routers along the reverse transmission paths[4]. After instantiation on a LARA++ router, the FRC immediately re-routes any packets misrouted to the mobile node's old location to its real location by means of network address translation [11].

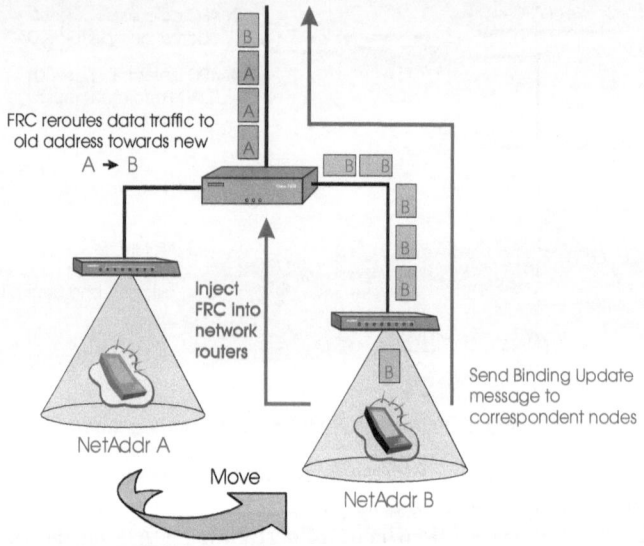

Fig. 7. Optimised handoff for high latency links

The FRC re-routes the data streams by registering a packet filter specifying the relevant filter information (i.e. old care-of address, home address) with the classifier. Since the flow routing optimisation is only required temporarily (until all relevant correspondent nodes have received a binding update), the FRC is programmed to timeout and remove itself after a period of inactivity (typically within a few seconds of instantiation). This complementary routing optimisation to Mobile IPv6 has again the advantage that it is based on a simple soft-state mechanism, which falls back into conventional operation mode if the active code accidentally stops.

---

[4] Note that although active network support might be restricted to the local network domain, the handoff optimisation is nevertheless effective, since mobile route changes occur primarily close to the mobile's network location.

6 Conclusion

The intrinsic lack of active support in current IP networks has hindered the development of efficient solutions for the QoS problems arising from network handoff of mobile devices – resulting in more complex protocols and greater bandwidth overhead than is otherwise necessary. To address this failing, we have designed and are currently developing LARA++, a component-based active network node architecture. We have demonstrated the flexibility and extensibility properties of LARA++ by solving the handoff problem based on simple to develop, highly dynamic and efficient active components, which provide a more simple and transparent service than any other solution.

Further work in the LARA++ area will involve completing the implementation and evaluation of the Mobile IPv6 handoff enhancing components. The work will be based around the LandMARC implementation of Mobile IPv6 for Microsoft Windows 2000 and WinCE [12]. In the longer term, research is also planned into the benefits of LARA++ component technology for both tethered and mobile end-stations, in addition to network routers.

Acknowledgements

The LARA++ and mobility support work documented in this paper was undertaken as part of the Microsoft funded LandMARC collaborative project between Lancaster University, U.K. and Microsoft Research, Cambridge U.K. The LARA platform was developed under U.K. EPSRC (grant no. GR/L59603).

References

1. Johnson, D., Perkins, C.: Mobility Support within IPv6. IETF Internet draft draft-ietf-mobileip-ipv6-11.txt, (2000). Work in progress.
2. Finney, J.: Supporting Continuous Multimedia Services in Next Generation Mobile Systems. Ph.D. Thesis, Lancaster University. (1999)
3. Thomson, S., Narten, T.: IPv6 Stateless Address Autocofiguration. Internet RFC 2462. (1998)
4. Narten, T., Nordmark, E., Simpson, W.: Neighbor Discovery for IP version 6 (IPv6). Internet RFC 2461. (1998)
5. Caceres, R., et al.: Fast and Scalable Handoffs for Wireless Internetworks. ACM Mobicom '96, (1996) 56-66
6. Campbell, A., et al.: Cellular IP. IETF draft draft-ietf-mobileip-cellularip-00.txt, (2000). Work in progress.
7. El Malki, K., Soliman, H.: Hierarchical Mobile IPv4/v6 and Fast Handoffs. IETF draft draft-elmalki-soliman-hmipv4v6-00.txt, (2000). Work in progress.

8.   Cardoe, R., Finney, J., Scott, A.C., Shepherd, W.D.: LARA: A Prototype System for Supporting High Performance Active Networking. In Proceedings of IWAN '99. (1999)

9.   Schmid, S.: LARA++ Design Specification. Lancaster University DMRG Internal Report, MPG-00-03. (2000)

10. Narten, T., Draves, R.: "Privacy Extensions for Stateless Address Auto-configuration in IPv6", IETF Internet draft draft-ietf-ipngwg-addrconf-privacy-01.txt. (1999). Work in progress.

11. Egevang, K., et al.: The IP Network Address Translator (NAT). Internet RFC 1631. (1994)

12. The LandMARC Project, available via the Internet at "http://www.LandMARC.net". (2000)

13. Calvert, K. (ed.): Architectural Framework for Active Networks. Active Network Working Group Draft. (1998). Work in progress.

"Mix and Match" Media Servers

Mauricio Cortes and J. Robert Ensor

Bell Laboratories, Lucent Technologies
600 Mountain Avenue, Murray Hill, New Jersey 07974, USA
{mcortes, jre}@lucent.com

Abstract. Network appliances are specialized computing units attached to one or more communication networks. They encompass a wide range of devices, including pagers, cellular telephones, and personal digital assistants, as well as cameras, refrigerators, and other devices with network interfaces. These appliances have limited display and control capabilities, and they exchange information using fixed transport protocols and data encodings/formats. Therefore, to be accessible from various network appliances, multimedia applications must be able to send and receive data through a variety of transport protocols and be able to handle several data encodings/formats. This paper describes an architecture that allows multimedia applications to be built from collections of media servers. It also shows how these applications can exchange data with various network appliances by "mixing and matching" appropriate media servers.

Introduction

Network appliances are computing devices that exchange data over one or more networks. Common examples include pagers, cell phones, and personal digital assistants (PDAs), while less common examples include cameras, refrigerators, and other appliances with network attachments. These devices encompass a wide range of processing and input/output capabilities, and they are becoming more diverse as their popularity grows. This diversity imposes new demands on multimedia applications. To be accessible from a variety of network appliances, applications must produce and consume data in many forms. They must handle information represented in several media, and, for each medium, they must handle several formats and encodings. Application servers must also be able to exchange these data with the appliances using appropriate protocols.

Applications meet these demands by using media servers to handle data transformations and transport. Media servers are hardware or software processing units that create, store, transform, or transmit data. A media server can be an integral part of a specific application. However, media servers that are distinct modules, capable of functioning in multiple applications, provide important advantages. When media servers are used in multiple applications, the cost of their development can be spread over all of these applications. Also, by invoking specialized sets of media

H. Scholten and M. van Sinderen (Eds.): IDMS 2000, LNCS 1905, pp. 279-284, 2000.
© Springer-Verlag Berlin Heidelberg 2000

servers, multimedia applications can become accessible from a variety of network appliances. For example, a messaging application could allow its users to retrieve facsimiles from personal digital assistants and telephones. When a user accesses the application from a PDA, facsimile images could be sent directly to a facsimile printer, or they could be sent to a character recognizer for conversion into text. The text could then be transmitted to the PDA for display. When a user accesses the application from a telephone, images could be sent to a facsimile printer. Alternatively, they could be sent to a character recognizer; the text from the recognizer could be sent to a text-to-speech server, and the resulting speech could be presented on the phone.

In the remainder of this paper, we describe a collection of software modules and its architecture. This middleware allows distinct media servers (also called media processing units) to be used in multiple applications. It also allows applications to be built from collections of these processing units. Applications can invoke different sets of these units—*i.e.*, "mix and match" these media servers—to accommodate access from different network appliances. The middleware contains a specialized resource manager—called the *media flow manager*—which creates and manages media flows. We define a *media flow* to be the movement of a set of data—through a collection of media processing units—from a set of sources to a set of sinks. Applications communicate with the media flow manager through a well-defined interface, the media flow control protocol. The media flow manager and the media processing units interact through another well-defined interface, forming a two-level support structure for application developers.

The media flow management middleware is designed to encourage specific characteristics in multimedia applications. Re-use of media processing units is encouraged by standardizing their descriptions as resources. This goal is also encouraged by not permitting sources and sinks to know about each other. Division of labor in creating applications is encouraged by the media flow manager's management of media processing resources. An application uses the media flow manager to establish a flow by describing what types of media servers should be included and how they should be connected. However, the media flow manager insulates the application server from resource management details. Finally, application servers manage application semantics and execution efficiency by issuing specialized control commands to media processing units.

Related Work

The middleware described in this paper is related to several other systems and proposals. The common goal of this related work is to produce and consume data with formats, encodings, and transport protocols that meet the requirements imposed by endpoint devices and their network connections.

Fox et.al. [3] discuss a way to add proxies and "distillers" into networks. A proxy is a complex software application that retrieves information from a network server, manages the transforms of the data by local distillers, and sends data to its clients.

Distillers are similar to our media processing units. They can transform multimedia data, according to the format and encoding needs of endpoint devices. In our architecture, we have also included a resource manager (media flow manager) that locates, allocates, and frees processing units. Furthermore, we use processing units not only to transform data, but also to retrieve data from network servers and send those data to endpoint devices. Our architecture allows application programmers to build simple application servers, which let the MFM manage the processing units that retrieve, transform, and send multimedia data.

John Smith, et.al. [8] describe the InfoPyramid system that manages multimedia documents using different modalities (e.g. audio, video, image, text) and fidelities (e.g. summarized, encodings, bitrates). Given the client's device and network bandwidth, the system selects one or more files with the appropriate modality and fidelity for the device to render. Their system could be implemented as an application on our framework.

Amini et.al.[1] describe an infrastructure that is similar to our framework. Their system contains a control server and graph managers, which work together to perform the tasks handled by our media flow manager. Their graphs correspond to our media flows and their data exporters correspond to our media processing units. However, while the graph manager invokes data explorers, control commands for these units are actually passed through the graphs of data explorers from end-user devices. Whenever a user sends a command affecting an explorer at the end of a long path, the system can take a significant amount of time to give any feedback to the user. In contrast, we want a more direct control path to appropriate media processing units. We feel that application servers in collaboration with the resource manager can handle these control paths more efficiently. Furthermore, our approach is not limited to media streams, it can be used in asynchronous applications such as traditional messaging applications.

The KISS project[4] allows network access points (NAPs) to search for processing and network resources throughout the network. In contrast, our media flow manager selects from a pool of registered processing units. We feel that our approach is more efficient since the search time for a network or processing resource depends on a table in the MFM and not on a recursive search in a network of NAPs. Again, we do not limit our concern to media streams.

The Ninja Path [2] project sets up flows automatically. Similar to the KISS project, the processing nodes have to find each other, which we feel is less efficient. Furthermore, the source node in the Ninja project needs to know capabilities of the sink. We feel that this limitation restricts the utility and re-use of servers. This project only supports linear transformation paths. In contrast, our approach allows arbitrary graphs to be deployed dynamically on behalf of an application server.

Within international standards organizations, groups are creating standards that encourage development of servers with distributed components. ITU and IETF groups are cooperating to define the Megaco protocol [5] (ITU proposal H.248) to support development of distributed media gateway servers. The commands and events included in Megaco are geared towards IP telephony gateways. These events are not of concern to our media processing units.

The Session Description Protocol (SDP) [7], from the Network Working Group, allows people to learn about communication sessions. This protocol contains standard means of describing session participants and devices. It could be used in our description of a media flow. The Real Time Streaming Protocol (RTSP) [6], also from the Network Working Group, allows multimedia data to be streamed to endpoint devices over packet networks. Among the media server we plan to build is an RTSP media server.

Example

We have used the media flow manager (MFM) and a set of media processing units (MPUs) to develop a messaging application and a bulletin board application. Both applications allow people to post and retrieve messages through their telephones, PDAs, or PCs. People use the messaging application to send messages to specific recipients. This application also notifies people when new messages arrive. As part of its notification process, it posts the audio portion of audio/video messages to an audio message server, where they can be retrieved over telephones. Messages are retrieved with this application by downloading message files. People use the bulletin board application to post messages to a central repository. People retrieve messages from the bulletin board by downloading files or receiving media streams.

The figure shows the basic architectural components of our system—application servers (ellipse), a media flow manager, media processing units, and end-user clients. These components may execute on one or more processors. The distribution of these software elements among physical devices generally depends upon performance considerations. Distributing the tasks over multiple processors allows processing to

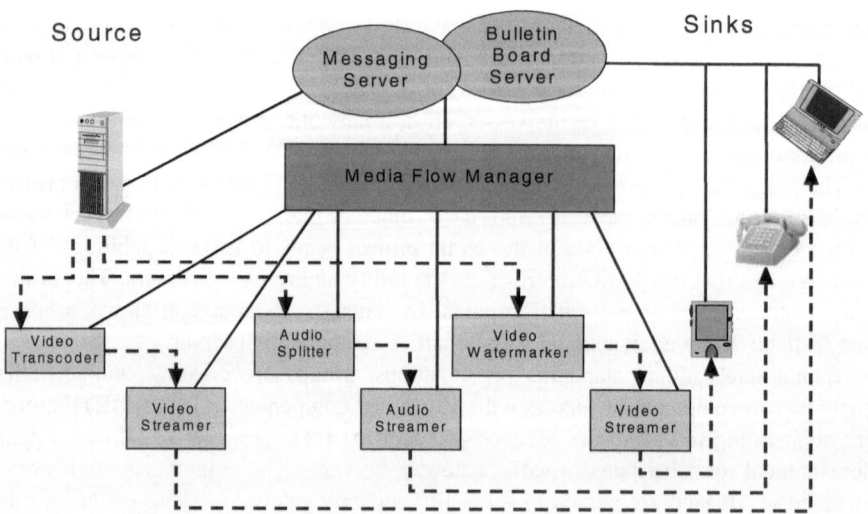

occur in parallel. However, distribution over distinct machines can also introduce communication delays among the components. Such delays can be especially important for real-time media streams.

In the figure, solid lines connecting components represents control channels. The application server and our media flow manager exchange messages using the *Media Flow Control Protocol (MFCP)*. This protocol allows applications to request media flow services, such as MPU allocation, MPU interconnection, and MPU status reports. The media flow manager also exchanges control messages with MPUs. Each MPU registers with an MFM by sending a description of its functionality and input/output requirements. The MFM maintains these descriptions and uses them to manage their resources on behalf of the applications.

Each dashed line in the figure represents a network link of a media flow. The PC is able to display MPEG-1 video streams, so a flow is established from a video store through a video transcoder to produce an MPEG-1 file. The video streamer then sends in real-time the file to the PC. The telephone is the sink of a flow that first creates a file containing the audio portion of an audio/video file and then streams the audio to the phone. The PDA receives watermarked video streams. The flow that creates this set of data takes a video file from the storage server, adds watermarking data, and then streams the resulting file to the PDA.

Both applications make use of common MPUs. The cost of developing these MPUs is, therefore, shared by the applications. Furthermore, the application servers were easily implemented. Each one needs only to interpret its specific user-level protocol.

Summary

This paper has presented a brief description of the media flow management middleware system that we are currently building. It has described how we have used this middleware to build two applications—a messaging application and a bulletin board application.

We have designed the media flow architecture and are developing middleware according to the architecture in response to an emerging trend. People are using more and more specialized, limited capability devices for digital communication. These devices not only have different capabilities to process data, but they also exchange data over multiple networks, which also have different capabilities. Hence, users are now accessing communication applications through a large number of device/network combinations.

We believe that developers need access to many specialized media servers to build communication applications for this environment,. Building a collection of media servers for each application is too expensive. Media servers need to be re-used, to be shared among multiple applications. Furthermore, different collections of these media servers can be used by single applications to accommodate user access through various device/network combinations.

We are testing our hypothesis by creating applications that use media flow managers to create and manage collections of media servers. Our current experiments will help us measure the quality of the two interfaces defined by our architecture—the media server description language and the media flow control protocol, MFCP. Future work will allow us to test the performance of media flows and the general utility of our approach.

References

1. L. Amini, J. Lepre, M. Kienzle, "Distributed Stream Control for Self-Managing Media Processing Graphs," pp. 99-102, ACM Multimedia '99, Orlando FL., October 1999.

2. S. Chandrasekaran, S. Madden, M. Ionescu, "Ninja Paths: An Architecture for Composing Services Over Wide Area Networks", http://ninja.cs.berkeley.edu/dist/ papers/path.ps.gz, Berkeley, CA.

3. A. Fox, S. Gribble, E. Brewer, E. Amir, "Adapting to Network and Client Variability via On-Demand Dynamic Distillation", pp. 160-170, ACM ASPLOS '96, MA, October 1996.

4. K. Jonas, M. Kretschmer, J. Modeker, "Get a KISS-Communication Infrastructure for Steaming Services in a Heterogeneous Environment," pp. 401-410, ACM Multimedia '98, Bristol UK.

5. 5. Megaco Protocol, work in progress, Internet Engineering Task Force Internet Draft,   January 27, 2000.

6. Real Time Streaming Protocol (RTSP), Network Working Group, Request for Comments: 2326, April 1998.

7. SDP: Session Description Protocol, Network Working Group, Request for Comments: 2327, April 1998

8. John R. Smith, Rakesh Mohan, Chung-Sheng Li, Scalable multimedia delivery for pervasive computing, pp. 131-140, ACM Multimedia '99, Orlando FL, October 1999.

Spatially Aware Local Communication in the RAUM System

Felix Hupfeld and Michael Beigl

Universität Karlsruhe, Telecooperation Office,
Vincenz-Prießnitz-Str. 1, D-76131 Karlsruhe, Germany
{hupfeld, michael}@TecO.edu

Abstract. In this paper, we propose a new paradigm for local communication between devices in Ubiquitous Computing environments, assuming a multitude of computerized everyday appliances communicating with each other to solve tasks. This paradigm is based on the concept that the location of devices is central for the communication in such a scenario. Devices define their communication scope by spatial criteria. In our paradigm no explicit addressing or identification of communication partners is used. In comparison to traditional communication methods the approach eases routing and discovery problems and can be deployed in a highly dynamic environment without centralized services. We use the term local communication as inter-device communication in a physically restricted local area. This is well distinguish from the terms telecommunication as communication over distance where location information is explicitly hidden. The communication model (RAUM) introduced is based on the observation that humans structure their environment primarily spatially. We show that spatially aware communication, is an efficient method communication in ubiquitous computing environments. We relate the communication architecture of the OSI/ISO reference architecture. An exemplary implementation that realizes a context information system is described. Based on this system several applications (Smart Doorplate, Communication with peripheral devices) have been implemented and evaluated.

Introduction

In the vision of "ubiquitous computing" devices and humans obtain access to any information everywhere [Weiser, 1991]. This vision is implemented by many research projects like Stanfords Interacive Workspace [Fox et al., 2000] or GMDs iLand [Streitz et al., 1998]. A further class of systems is based on the idea that there are services and information, that is not location bound, that should be ubiquitously accessible, e.g. database access and web-pages. For example, in the ParcTab project [Want et al., 1995] information in the Internet, e.g. a text document or e-mail at your "home server", are accessible from everywhere around the world with the help of small devices. This prototype can be considered as an ancestor of the upcoming WAP-enabled

H. Scholten and M. van Sinderen (Eds.): IDMS 2000, LNCS 1905, pp. 285-296, 2000.

devices [WAP] or Internet enabled Personal Digital Assistants (PDAs) as 3COMs PalmV, that allow Web-access or access to a virtual "home base" via Internet.

Other ubiquitous computing systems provide local information for global access. For instance in the Olivetti ActiveBadge project [Harder & Hopper, 1995] stuff was wearing electronic badges. These badges are tracked by a location system and the processed information is then presented on the Web. Using the SmartBadge [Beadle et al., 1997] even context information collected from sensors on the badge are accessible to internet users.

An example for a project addressing the research on context information is Georgia Techs Aware Home project [Kidd et al., 1999], where a house is equipped with context retrieving sensor systems. A great part of such context information is not of interest for a user or machine outside of a spatial scope, but builds the basis for "local" application. We have identified location as the most important context for communication of computer-augmented devices in ubiquitous computing [Beigl]. This kind of local communication is different from telecommunication and will be explained next. The architecture and experiments with the system will then give a closer insight into the RAUM-system.

Types of Communication

The challenges in telecommunication of the last decades were primarily to enable people to communicate over great distances independently of their location. As the prefix suggests, tele[1]communication is about remote communication. Telecommunication system abstract from the location of the communication partners.

The use of a telecommunication system arises from the need to communicate with somebody or something who or which is not physically present at the location of the caller. The communication partners (more precise the end-systems) are chosen by their **identity**, e.g. if you pick up your phone, you want to call a specific person whose identity is represented by her or his telephone number. The same applies to device telecommunication, here using the Internet. The first criterion for selecting the remote computer system is its identity represented by the IP address or the DNS name and then choosing the service. As a consequence parameters as the **physical location** of the partners are **hidden** from the user and from the end-system by the telecommunication system.

Local Communication

We regard communication between partners that are located in the same physical space as **local communication.** We consider the same physical space as what is perceived by a human as "environmental" [Montello, 1993] or local to the device. Local

---

[1] Tele- is Greek for remote

communication is carried out not only between humans but also between machines in **interactive rooms**. Such interactive rooms are equipped with (invisible) computer systems everywhere in the environment:

> Interactive rooms are places where the environment and many artifacts of the everyday live are extended with computer and communication technology.

Local communication supports this by providing a basic communication that is easily understandable to the human and flexible in its use without administrative overhead. In local communication partners (e.g. computerized artifacts) are selected according to their spatial co-location and not based on their identity. This selection process does not require any user interaction or administration. Services or information, which are requested by one device, are retrieved through communication to devices nearby. In this form of communication access to and from devices outside the local environment is not supported.

Spatially Aware Communication

The importance of the spatial relationship is obvious in human communication: If you want to talk to somebody, you stand in front of him. When you're talking, you neither want to communicate with a person behind you nor with a person in another room. Furthermore, the identity of the person you are talking with is of less importance. Your scope of communication is the person next to you.

This kind of spatial communication relationship is transferred to communication of computerized devices. As with humans, the spatial relationship defines an area of interest based on the spatial layout. As recent findings in cognitive science suggest, that for humans space is central for the understanding of interaction and communication [Kirsh, 1995]. Considering the involvement of humans for local communication, human understanding of the communication process is important. In interactive rooms application developers and users will benefit from the human understanding of communication [Brooks, 1991]. Our research suggests that such an understandable and efficient form of human-like communication is also efficient for device communication.

In spatially aware communication devices use location as their primary attribute for selecting potential communication partners. Devices describe a scope of interest in which they are willing to communicate depending on the current state of the application running on the device. By doing so the communication simulates the communication behavior of humans. Only communication from partners inside a scope is accepted. At the end of the paper we describe some example applications, that are build upon such communication paradigm.

Service Oriented Local Communication

Location awareness is important for the communication of local devices and applications, as it was shown in the examples above. Beside from building a new kind of communication system we can achieve this goal by using existing communication technology: a service-oriented telecommunication system enhanced with a lookup service or an infrastructure as in Jini or T-Spaces. As we mentioned before, the location information is hidden by these telecommunication systems. It has to be reintroduced and modeled explicitly via a service. This service must be able to track the location of each device and to select communication partners by spatial layout. See [Maaß, 1997] for a system architecture of this kind.

This solution has several drawbacks. First, the usage of a telecommunication system introduces a first indirection. The devices are mobile, thus a dynamic mapping between the device identifier and its position has to be made for routing packets. The network has to know in which subnetwork the device is located in order to forward the packets there. The central property of local communication, the device location, is hidden at first and then reintroduced explicitly through a service. This introduces (apart from the overhead caused by the lookup) a second indirection. A device position is mapped to a network address, which has to be mapped back to a position (i.e. routing information) for delivery. This makes the system inefficient and depends on dynamic mappings, which have to be set up and kept up-to-date. [Beigl, 1999] shows that the more devices take part in such a system and the more complex the system is (in terms of communication scope extension), the overhead introduced this way is rising against the spatially depended communication system we present here.

Additionally, the reintroduction of the location in a telecommunication system requires a service rendered by a central server in current systems which introduces the problems of service discovery, configuration/administration and availability.

Using telecommunication systems for local communication introduces a large overhead. A genuine location-centric local communication system foregoes the problem of routing packets to mobile devices due to the fact that the message identifies the position of the targets. Thus no techniques such as mobile IP, stemming from implicated locality of the IP address system, need to be employed to be able to route the flow of information to a device in a constantly changing network topology. It does not depend on a central service to define the spatial communication scopes, which introduces the problems of service discovery, automatic configuration and availability.

RAUM²-Architecture

The RAUM-architecture describes the communication stack and the protocol for artifacts and optional communication infrastructure. The RAUM-system presented here is based on the RAUM-architecture and allows artifacts in interactive rooms to commu-

---

[2] location-based Relation of Application objects for communicating Ubicomp event Messages. In German the word *Raum* stands for space as well as for room.

nicate either with or without infrastructure in an ad-hoc manner. The RAUM-architecture is modeled after the abstract concept of spatially dependent communication [Beigl, 1999].

The RAUM-concept describes the communication of devices according to their spatial relationship in the physical world. Devices define regions of interest in which they communicate (RAUMs). Such RAUMs describe a congruent geometric space in the physical world; the geometric space is defined as a envelop over all points defining the RAUM. Because the communication order in the RAUM-concept is bound to the physical location of the objects taking part in the communication, objects that are inside the space of interest are also inside the communication scope of the object defining the RAUM. Every device that is physically located inside the envelop of a RAUM (or more precisely: where the location is related to the set of points in the physical space) is then a member of the RAUM.

OSI/ISO No.	RAUM System Layers	Functionality
7	Application Layer	Application
4-6	RAUM Event-Layer	Reliable service, world modeling, abstract presentation of packets for application
3	RAUM-Layer	Implements the core functionality of the RAUM system
2/1	RAUM Communication-Layer	Non-reliable packet broadcast and physical transport

Fig. 1. Stack architecture of the RAUM system

The architecture of the RAUM system is layered and can be related to the one defined in the ISO/OSI model. The major functional operations for communication within the RAUM model are concentrated on layer 3, the RAUM-Layer.

The complete stack of the RAUM-System consists of 4 layers (see Fig. 1). The lowest layer is called "RAUM-Communication-Layer" and groups the physical and the DLL layer (ISO/OSI layers 1&2). The current definition of the RAUM-architecture does not specify this layer. Instead existing implementations of DLL and physical layer are used. The RAUM-Layer is responsible for the creation and dissolution of communication relationships, for device location and routing of packets. Communication relationships in the RAUM-system are created according to the definition of a spatial area, the RAUM. Note that these RAUMs are not restricted to the physical broadcast space of the medium.

The next layer, the RAUM Event-Layer, spans the layers 4 to 6 providing reliable communication and eventually offering abstract event communication services to the application, that resides at layer 7. The rest of this section will focus on the RAUM and RAUM-Event-Layer, but first the run of the RAUM-system is explained in more detail.

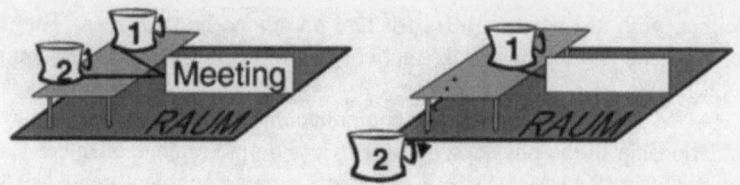

Fig. 2. Example of RAUM communication

Run of a RAUM-Communication

The communication of the RAUM-system is datagram oriented. Applications running on the devices send datagram packets to every RAUM they belong to. At the application and RAUM Event-Layer these packets are called **events**. Figure 2 shows an example run: A doorplate defines a RAUM where it wants to receive events from. After definition it gets all events from the two cups (left) inside the RAUM. As one of the cups leaves the RAUM (right), only the remaining cup keeps sending events to the doorplate.

As it is not possible to technically prevent the doorplate from hearing also events from cup 2, the RAUM protocol has to shield the application from these unwanted events. Therefore every object keeps a list of RAUMs which he is belonging to, and a list of all RAUMs it has defined. According to this list the RAUM-Layer decides if a packet has to be ignored or to be processed.

To allow a flexible handling of events adopted to the application needs we had defined three kind of RAUMs:

☐ Listener-RAUMs: With this kind of RAUM only the object that has defined the RAUM receives all events from all objects inside the RAUM. In the above example the doorplate defines such a Listener-RAUM

☐ Speaker-RAUMs: If an object defines a Speaker-RAUM all objects inside the RAUM receive all events from the speaker (the defining object). An example of a Speaker-RAUM is a Location-Beacon that constantly sends its location into a certain area.

☐ Discussion-RAUM: In the Discussion-RAUM every object inside the RAUM receives every event from all objects in the RAUM. This kind of RAUM is needed to allow negotiation with devices in an unknown environment.

RAUM-Layer

The purpose of the RAUM-Layer is to decide whether a received datagram is to be passed on for the user application or to be discarded, according to the RAUM model. As defined in the previous section, RAUM packets are accepted if the potential receiving device and the datagram packet itself are located inside the same RAUM. This is a purely geometric requirement.

To allow such a test, every packet sent must be mapable to a geometric position and every device has to know its own position in space and be aware of the defined RAUMs it is located

```
Identifier open_RAUM([Name], Rtype, Shape, Relative )
close_RAUM( Identifier )
change_RAUM( Shape )
send( Packet )
Packet receive()
```

Fig. 3. The interface to the RAUM-Layer

in. Our devices have a built-in facility to detect their geometric location via a location system (e.g. our IR-Beacons), so they know their location and can tag broadcasted datagram packets with their current position. Device position information is handled by a subsystem called the *Locator service* whose instances communicate with the RAUM Location Protocol (RLP).

The RAUMs themselves are named and defined geometrically by a sets of points. Currently, unions of spherical and cubic shapes are possible. The RAUMs are either stationary or moving with the device that is defining that RAUM. The interface to layer 3 provides upper layers with functions to open and close RAUMs. New or updated RAUM control information is broadcasted to all other devices via the RAUM Information Protocol (RIP). RAUM management is done by a subsystem called the *RAUM service*. Data is exchanged via the RAUM Data Protocol (RDP).

Routing

If devices use different media for communication or the broadcast area of the medium does not allow direct communication between devices coupling of network segments is needed. This coupling enables for example a PDA with infrared communication to interact with an ordinary network printer or an application running on a desktop computer.

Every medium with its limitations forms something like a logical Local Area Network (LAN). This is not exactly the original LAN concept, but every range-limited medium can be thought of as a network in a local area. For example, the infrared (IR) communication in a room form a LAN, as all IR transmitter-equipped devices in the room can communicate, but no information reaches other rooms. Low power radio frequency (RF) transmission and especially cable networks like Ethernet are other examples. Each LAN covers a specific geometric area, for example the room the wireless transceiver is in or the spots where devices are connected to the Ethernet. The regions covered by these LANs can overlap, and thus a certain area in a room can be supplied by several media (Figure 4).

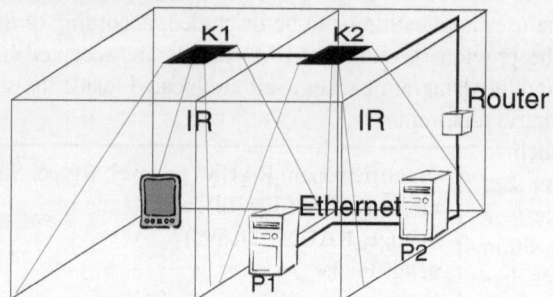

Fig. 4. Infrared and Ethernet networking

To allow the inter-medium communication mentioned above, these LANs have to be coupled with devices connected to some type of infrastructure giving them access to the other LANs. This coupling is similar to inter-LAN communication in telecommunication systems. Because there is no explicit addressing in RAUM communication location information of the communication packets is used for routing. The location information paired with the RAUM definition results in an addressing concept where the location is congruent to the address in local communication and which can be used for routing decisions. Such a router caches all communicated RAUM definitions of adjoining LANs. Furthermore the component knows the physical dimension of theses LANs, too. Using this information and the fact that RAUMs are geometrical objects it can be decided which packets from one network interface should be forwarded on the other. RAUM packets are of interest to a LAN if the described RAUM intersects with the area the LAN covers, so they must be forwarded in this case. Packets must be forwarded into a LAN when any of the devices there should receive it, for example when a RAUM covers the geometrical areas of both LANs. So the communication stack has to check all RAUMs in this LAN to see whether the packet is inside. Position information of a distinct device is needed in a LAN when a packet from it has to be routed. Therefore location information has to be associated with every packet send.

Application Examples

The RAUM system is currently used for inter-device communication in our research in Ubiquitous Computing. The RAUM stack runs on desktop computers, Palm Pilots and PIC microcontrollers [PIC]. Several routers couple the different media with an infrastructure of infrared, radio frequency, CAN and Ethernet transceivers. There are several LANs with numerous IR spots, RF areas and an Ethernet in several rooms. We built several applications, some of them are used to yield context information for applications in human computer interaction. We describe here two of them, more can be found in [Beigl &Gellersen, 1999] and in [Beigl, 1999].

The Context-Aware Doorplate

The context-aware doorplate (SmartDoorPlate) displays the name of the room, the names of the people working in that room, or events taking place there (e.g a meeting is in progress in this room). The information for recognizing a meeting in progress is inferred from the context established by MediaCups [Beigl et al., 2000]. The Media-Cup is a computer-augmented coffee cup with infrared communication, which distributes status information such as temperature and usage in specific time intervals.

The doorplate application defines a RAUM with the same shape as the room it is related to. In our set-up the plate is mounted next to the door and waits for context information emitted in events by MediaCups. These events get packed into RAUM datagrams and sent out via the infrared link. Every other infrared device within range, i.e. every other device in this LAN, receives the packets. The router device, which has a transceiver in every LAN, receives them, checks for the extension of the defined RAUMs and routes them to the appropriate LANs, here the Ethernet. The communication stack of the doorplate application receives them and puts them into the recognized "world", a storage space where the RAUM Event-Layer puts all information of interest into. Events of interest are expressed with the receive operator. The doorplate is able to track the presence of MediaCups in the room this way, and thus can infer from this context to an ongoing meeting.

Writing an application using the spatial awareness feature based on local context can be done easily in a natural way, because the communication stack holds all important information for it. When setting up this application using the RAUM system, the doorplates define a RAUM with the same spatial dimensions as the conference room they are monitoring. The cups in turn are set up to emit their data as messages sent to this RAUM. With this simple setup, using just one RAUM per conference room, it is possible to implement this application; the pseudo code of the smart doorplate is shown below:

```
MyRoomShape := Sphere (x,y,z,r)
initializeRAUM (myRaumName, myRoomShape)
world = receive ("Cup, hot, *")
while TRUE
    int rec_cups := 0
    rec_cups := count_known_objects(world)
    if rec_cups > 2 then printLCD
                            ("Meeting")
```

Communication with Peripheral Devices

A common scenario in local communication is the wireless connection between an appliance and a peripheral device. For example, a digital photo camera has to communicate with a printer or a storage medium on user's request, input devices as a mouse or a keyboard have to be bound to a computer. We built a wireless computer keyboard (AwareKeyboard) which can be used to input text on Palm Pilots. The connection between the keyboard and the Palm Pilot is location sensitive: if you put the Palm Pilot next to the keyboard, you can use it to enter text.

The Palm Pilot defines a RAUM of circular shape around his current location and filters for keyboard event messages. On keypress, the keyboard puts an event in the RAUM system, which, when the Palm Pilot is near enough and thus the keyboard is inside its RAUM, is received by the Palm Pilot:

```
initializeRAUM (PilotRAUMName, Circle (0.5 m))
events = receive ("Keyboard, KeyRelease, *")
while TRUE
    inject_into_input_queue( events )
```

Related Work

The idea of using a telecommunication system paired with a lookup service to do local communication can be found in other systems. Some of them enable devices to select their communication partner by spatial criteria, others simply pass location information to the application. The introduction of location information in a telecommunication environment is usually managed by a central location service, few systems use decentralized facilities. Leonhardt [Leonhardt & Magee, 1998] provides an architecture for acquisition of location information from divers sources. [Leonhardt, 1998] covers the aspects of location-aware telecommunication.

Jini [Jini] of Sun Microsystems is an object-oriented distributed system for inter-device communication. Jini-enabled devices register with a central lookup service where other devices can find them and thus select their partner. Physical location plays no role in this selection process, its focus lies on service compatibility.

[Maas, 1997] describes an architecture which is similar in functionality to the RAUM system. Location information is introduced in a telecommunication system via a central location component, a directory service. The directory service allows clients to retrieve communication partners by several different spatially dependent criteria such as distance, type and special conditions, devices can be located and clients can get notified when given devices meet or enter an area. Hive [Minar et al., 1999] of the Massachusetts Institute of Technology is a system for device communication modeled after the Agent paradigm. No central services are used. Devices are modeled as cells, their functionality is bundled in so-called shadows. Active Agents move between those cells and are enabled to do location dependant communication tasks by semantic models.

The routing in the RAUM system between the interconnected LANs is based on this "location addressing". The aspect of using location information in the addressing scheme in WANs (wide area networks) is studied in [Navas & Imielinski, 1997]. Here, IPv6 addressing is combined with geographical information from GPS (global positioning system) to aid in routing. [Ko & Vaidya, 1998] describes a routing scheme which uses location information as an aid in routing, too.

Conclusion and Future Work

We introduced the term local communication, contrasted it to telecommunication and showed that location-based communication is more suitable to human needs and to local communication topics. Such location-based communication is best suited for interactive rooms where a multitude of computerized artifacts are communicating with each other to carry out tasks mostly without direct user interaction.

The presented RAUM system is an example of such a location-based communication system. The architecture of the RAUM system is layered and can be related to the one defined in the ISO/OSI model. The major functional operations for communication within the RAUM model are concentrated on layer 3, the RAUM-Layer. In this layer delivery and reception of packets according to spatial areas (RAUMs) and routing of packets is handled.

The RAUM system and applications implemented on top of the system are in usage in an everyday environment since September 1999. The experiences collected with the RAUM system and with the applications indicate that the chosen communication paradigm is efficient in the given environment and allow us to identify areas of further work: Human-Computer Interaction (HCI), Appliance (Hard- and Software) design and networks. In interactive rooms HCI issues are very critical: In Appliance design the focus is on energy safe design. Energy saving protocols (e.g. [Tsaoussidis et al., 1999]) make an important contribution in that area.

Our current setup shows the advantage of using location-based communication for local applications. Further work will investigate how location-based and telecommunication can be assembled to provide the appropriate communication basis for different application tasks. Also the routing algorithm will be extended to allow for more flexible routing and load balancing. Work on the RAUM architecture will research on the lifetime of events in RAUM-communication and on supporting applications with a rule-based language to describe situation-reaction pairs.

References

[Beadle et al., 1997] H. W. P. Beadle, G. O. Maguire Jr., M. T. Smith. Location Based Personal Mobile Computing and Communication. *Proceedings of EEE/IEEE International Conference on Information, Communications and Signal Processing (ICICS) '97*, September 1997.

[Beigl, 1999] Beigl, Michael. Using spatial Co-location for Coordination in Ubiquitous Computing Envirnonments. *Handheld and Ubiquitous Computing, First International Symposium*, HUC'99, Karlsruhe.

[Beigl & Gellersen, 1999] Michael Beigl, Hans-Werner Gellersen. Ambient Telepresence. *Proceedings of the Workshop on Changing Places*. London, UK, April 1999

[Beigl et al., 2000] Michael Beigl, Hans-Werner Gellersen, Albrecht Schmidt. MediaCups: Experience with Design and Use of Computer-Augmented Everyday Objects, Computer Networks, Special Issue on Pervasive Computing, Elsevier, 2000

[Brooks, 1991] R.A. Brooks. Intelligence without representation. *Artificial Intelligence*, 47, 1991, 139-159.

[Fox et al., 2000] Armando Fox, Brad Johanson, Pat Hanrahan, and Terry Winograd, Integrating Information Appliances into an Interactive Workspace, IEEE CG&A, May/June 2000

[Harder & Hopper, 1994] Andy Harter, Andy Hopper. A Distributed Location System for the Active Office. *IEEE Network*, 8(1), Januar 1994

[Jini] http://www.sun.com/jini/

[Kidd et al., 1999] Kidd, Cory D., Robert J. Orr, Gregory D. Abowd, Christopher G. Atkeson, Irfan A. Essa, Blair MacIntyre, Elizabeth Mynatt, Thad E. Starner and Wendy Newstetter. The Aware Home: A Living Laboratory for Ubiquitous Computing Research" *Proceedings of the Second International Workshop on Cooperative Buildings*, October 1999.

[Kirsh, 1995] David Kirsh. The intelligent use of space. *Artificial Intelligence* 73(1-2), 1995, pages 31-68.

[Ko & Vaidya, 1998] Young-Bae Ko and Nitin H. Vaidya: "Location-Aided Routing (LAR) in Mobile Ad Hoc Networks", MobiCom '98, *The fourth annual ACM/IEEE international conference on Mobile computing and networking*.

[Leonhardt & Magee, 1996] Leonhardt, Ulf and Magee, Jeff. Towards a general location service for mobile environments. *Proceedings of the Third IEEE Workshop on Services in Distributed and Networked Environments*, pages 43-50, June 1996.

[Leonhardt, 1998] Leonhardt, Ulf. *Supporting Location-Awareness in Open Distributed Systems*. Doctor's thesis, Department of Computing, Imperial College of Science, Technology and Medicine, University of London.

[Maaß, 1997] Maaß, Henning. Location-aware mobile applications based on directory services. MobiCOM '97, *Proceedings of the third annual ACM/IEEE international conference on Mobile computing and networking*.

[Minar et al., 1999] Nelson Minar, Matthew Gray, Oliver Roup, Raffi Krikorian, Pattie Maes. Hive: Distributed Agents for Networking Things. *Proceedings of ASA/MA '99, the First International Symposium on Agent Systems and Applications and Third International Symposium on Mobile Agents*, 1999. http://hive.media.mit.edu/.

[Montello, 1993] D.R. Montello. Scale and Multiple Psychologies of Space. Lecture Notes in *Computer Science 716*, 1993, Seite 312-321.

[Navas & Imielinski, 1997] Julio C. Navas and Tomasz Imielinski. GeoCast - geographic addressing and routing. MobiCOM '97, *Proceedings of the third annual ACM/IEEE international conference on Mobile computing and networking*, 1997

[PIC] http://www.microchip.com/10/Lit/PICmicro/index.htm

[Streitz et al., 1998] N.A. Streitz, V. Hartkopf, H. Ishii, S. Kaplan, T. Moran. Cooperative Buildings: Integrating Information, Organization, and Architecture. *Proceedings of CoBuild '98*, Darmstadt, Germany, Lecture Notes in Computer Science 1370. Springer-Verlag, Heidelberg, ISBN 3-540-64237-4, 1998, 267 Seiten

[Tsaoussidis et al., 1999] V. Tsaoussidis, H. Badr, R. Verma. Wave and Wait Protocol (WWP): An Energy Saving Protocol for Mobile IP-Devices, *Proc. of the 7th International Conference on Network Protocols*, Toronto, Canada, 1999

[Want et al., 1995] Roy Want, Bill N. Schilit, Norman I. Adams, Rich Gold, Karin Petersen, David Goldberg, John R Ellis, Mark Weiser. An overview of the PARCTAB Ubiquitous Computing experiment. *IEEE Personal Communications*, 2(6), 1995, Seite 28-43.

[WAP] http://www.wapforum.org

[Weiser, 1991] Mark Weiser. The computer for the 21st century. *Scientific American*, September 1991, p. 94-104

The UbiCampus Project: Applying Ubiquitous Computing Technologies in a University Environment

C. Müller-Schloer[1] and P. Mähönen[2]

[1] Institut für Technische Informatik - Rechnerstrukturen und Betriebssysteme (IRB),
Universität Hannover, Appelstr. 4, D-30167 Hannover
cms@irb.uni-hannover.de
[2] VTT Electronics, Wireless Internet Laboratory, Kaitovyl 1, P.O.Box 1100,
FIN-90571 Oulu
Petri.Mahonen@vtt.fi

Abstract. The term "Ubiquitous Computing" was minted by Marc Weiser from Xerox PARC for an environment where computers are available everywhere without being recognized as such. It requires small and portable, even wearable, terminal devices, and wireless network access over multiple communication channels. The UbiCampus project is visioned by IRB and VTT to study and develop the post 3G ubiquitous networking environment for intelligent applications The UbiCampus Project aims at building and using a local ubiquitous computing environment on (a part of) the campus of the University of Hannover. The project includes transmission technology, middleware and application aspects. This paper gives a short introduction into Ubiquitous Computing, introduces the basic idea of the UbiCampus project and discusses its objectives in terms of infrastructure, applications and Jini-based service architectures.

1 Introduction

The computing landscape will change considerably within the next few years. One of the main influences will be the availability of affordable computing power anytime at any place. The traditional platforms like workstations, PCs and servers will be augmented by new, personal devices such as Personal Digital Assistants (PDAs), intelligent mobile phones (SmartPhones), palmtops, subnotebooks and even wearable computers. These platforms will be equipped with communication facilities with increasing bandwidth over stationary and wireless LANs, MANs and WANs which will eventually allow for multimedia applications. The internet protocol, especially forthcoming IPv6, will play a central role for standardization and interoperability. Another technology push is the growing momentum of common compatible software platforms and service architectures like those based on Java/Jini, Universal Plug and Play or the Home Audio and Video Interface HAVI.

Mark Weiser from Xerox PARC has minted the term "Ubiquitous Computing" for an environment where computers are available everywhere without

H. Scholten and M. van Sinderen (Eds.): IDMS 2000, LNCS 1905, pp. 297–303, 2000.

being recognized as such. Ubiquitous Computing (UC) names the third wave in computing, just now beginning. First were mainframes, each shared by lots of people. Now we are in the personal computing era, with a 1:1 relationship between user and machine. Next comes ubiquitous computing, or the age of calm technology, when technology recedes into the background of our lives.

The initial incarnation of ubiquitous computing was in the form of "tabs", "pads", and "boards" built at Xerox PARC, 1988-1994. Ubiquitous Computing helped kick off the recent boom in mobile computing research, although it is not the same thing as mobile computing, neither a superset nor a subset.

This upcoming scenario poses a number of technological challenges and questions. New ideas and prototypic implementations are needed for ubiquitous infrastructures as well as the services and applications that will use them. Ubiquitous Computing requires ubiquitous network access. Europe still stands on the forefront of wireless technologies. There is an exciting window of opportunity to build a good research and industrial base for ubiquitous computing in Europe.

This article reviews briefly the state of the art of the basic technologies needed for ubiquitous computing environments, develops a vision for an application scenario in a university environment, points out some missing technologies and discusses UbiCampus, a new project just being started at the University of Hannover (Germany) and VTT (Finland) with the goal to serve as a focal point to gain practical experience with the integration of a range of new technologies in this area.

2   The Technological Basis

It is beyond the scope of this paper to review in detail the state-of-the-art regarding all the technologies needed to build a true ubiquitous computing environment. Instead, we want to briefly discuss the main constituents needed for such systems. The future computing landscape will be characterized by the combination and integration of the three main application areas of microelectronics systems: Entertainment, Computing and Communication. Presently we are witnessing the combination and integration of once stand-alone devices like the TV-set, the telephone and the computer. This integration results in hybrid devices like the email or the internet telephone, palmtops communicating data via GSM or (sub-) notebooks with telephony facilities. From the point-of-view of ubiquitous computing, three basic technologies are key: networking and communication, devices and terminals, and the emerging glue logic in between, the middleware.

3   The UbiCampus Idea

Technological development will bring about increased chip complexity, increased performance, increased bandwidth and, building upon these basic technologies, a host of new devices and communication means. The challenge of future system design is not so much to add still another level of complexity but to manage this complexity adequately. What the user of these systems wants is a range of qualities like safety, reliability, usability, functionality, interoperability, and finally low cost and reasonable performance. He or she will not be willing to pay for technical gimmicks but rather for useful applications.

In a few years the basic technologies necessary for a ubiquitous computing environment will be available at low enough cost to assume that they will be standard equipment for teaching staff as well as for university students, just like pocket calculators or PCs today. Moreover, the methods of knowledge transfer within the universities will make extensive use of all kinds of multi media equipment. The UbiCampus project wants to provide a test environment to experimentally use existing UC technologies, to improve and augment them, and to utilize their potential for the benefit of the organization of a modern university. The project addresses the following areas:

- Establishing a UC infrastructure based on mobile terminals, a server network in the background and a ubiquitous wireless and stationary communication network.
- Establishing a service architecture based on Jini or other comparable middleware systems.
- Developing a variety of test applications for academic teaching and administration.

A ubiquitous infrastructure is only as useful as the applications it offers. Hence, UbiCampus will be application- driven. Its main emphasis is on the integration of commercially existing technology to provide high-quality services.

4   The UbiCampus Project

The UbiCampus Project aims at building and using a local ubiquitous computing environment on (a part of) the campus of the University of Hannover. It serves as a testbed for a variety of PhD projects on infrastructure, services and applications in a modern university environment. The UbiCampus project is closely linked with more content-oriented projects already ongoing at the universities of Hannover, Braunschweig and Clausthal (PROMISE - Project-oriented multimedia-based learning in EE and CS) and the Educational Technology initiative at the University of Hannover (see [3]).

4.1   Infrastructure

The UbiCampus project will be centered around the new 3000 m^2 Computer Engineering Building at the University of Hannover housing 4 Computer Engineering institutes. It is equipped with a multimedia lecture hall and a multimedia seminar room, a 100 Mbit / 1 Gbit LAN, current technology PCs, workstations and servers.

Typically, future mobile terminals will have a choice between several networks at different bandwidths and cost. This situation will be reflected within the experimental setup by the availability of (1) a wireless LAN, (2) a public network (GSM for the time being), (3) stationary Ethernet, and (4) IR links.

For local wireless communication within and in the close neighborhood of the Computer Engineering building we will install an IEEE 802.11 wireless LAN working at a data rate of 2 .. 11 Mbit/s. For close range communication, IR transmitters (IrDA) will be used.

In a real environment, there will co-exist a variety of heterogeneous mobile devices, ranging from wearables through palmtops to subnotes. It is obvious that not all of these devices are equally suited to all applications. It will be necessary to insert an adaptation layer between the terminal device and the service provider which is able to modify the I/O behavior of the application to adapt it to the abilities of the user interface. This adaptation layer might even block certain applications from being called from terminal devices with inadequate capabilities. As an example for such an adaptation we can imagine a palmtop that is perfectly suited for the IrDA- or WLAN- based download of lecture notes or transparencies but cannot be used to interact with a complex simulation program. Within the UbiCampus project we will select at least two different terminal devices: pen-operated low- end devices with a palmtop form factor, and subnotebooks with full OS support. Both are equipped with all four communication links mentioned above. A third possible choice are pen-operated workpads like the B5-size workpanel HW 90300 offered by Höft&Wessel (Fig. 1) which might be used with or without an OS being visible to the user.

Fig. 1. B5-size workpanel from Höft&Wessel with an 8.2" DSTN-Display, 32bit/200 MHz CPU and four wireless communication channels (see also [1]).

A key issue for heterogeneous devices is the portability of application software. The Java and JVM technology is a promising candidate but it remains to be seen if the limited resources available in the low-end devices are sufficient in terms of speed and memory to support this technology.

4.2 Applications

Before we will discuss the service architecture approach of UbiCampus, let's sketch a few typical applications. University life is all about knowledge transfer. UbiCampus will improve knowledge transfer by providing improved information logistics. Two patterns of information flow will be supported: information distribution (1:many) and interaction with multiple 1:1 relations. Clearly, the interactive pattern is the more challenging one but also the more desirable to break up the traditional teaching mode of '1-active:many-sleeping'. The following services will be implemented by 'assistants' or 'managers', represented by service entities in the Jini sense.

Information Distribution

The *Help Desk* is the central portal service offering assistance to find needed information and services. It includes search, overview and navigation functionality.

The *Lecture Note Distributor* offers lecture-accompanying information like textbook components, learning modules (i.e. small linked portions of knowledge, hyperlinked into a larger network of other modules) or transparencies used during the oral presentation of the lecture. The primary mode of usage is download.

The *Task Assignment Assistant* also acts as a distribution center but has the added ability to individualize the distributed information. This ability can be used for the assignment of personalized (mini-) tasks to students.

The *Presentation Setup Assistant* plays the role of a director orchestrating the cooperation of the different actors in a presentation session. It helps to select the necessary views to be presented (such as transparencies, notes, sidebars, bird's eye view, animation, simulation etc.), defines the channels to the different I/O devices (such as monitors, local or remote full screen projection or student's terminals) and limits the range of distribution (such as the class present in the auditorium).

Interaction

The interactive pattern requires the establishment of one individual channel with its own state during the whole length of the session. It can include also summarizing, evaluating or assessment functions.

The *Application Access Manager* has the very general task to mediate a connection between a user and his local GUI (graphical user interface) on the one side and an application program running on an arbitrary computer on the other. This mediation process includes the selection of a GUI suitable for the I/O capabilities available on the personal device while still fulfilling the needs of the application program. Typical applications can be e.g. a UNIX shell, an animation or a complex simulation. The *Progress Assessment Assistant* is an extension to the Task Assignment Assistant with the added ability to individually evaluate the task solutions of the students and generate some (possibly aggregated) feedback information for the lecturer. An automatic evaluation of this kind requires a formalized answering scheme like the one used in multiple choice tests.

The *Examination Manager* is a more advanced version of the Progress Assessment Assistant. It allows for computer assisted examinations and has to take into account the possibility of fraud during examinations. This is all the more complex since a ubiquitous infrastructure lends itself quite naturally to misuse for ubiquitous and computer-assisted fraud. This type of application is a very interesting learning vehicle since it requires the usage of various cryptographic techniques in conjunction with biometric sensor technology for authentication purposes. At the same time is requires more centralized services for certification and key distribution.

The *Campus Guide* is a means for navigation and orientation in the 3D world of a real campus with the help of mobile devices and infrared transmitters. It extends the idea of the SAMoA project [2] from a relatively limited exhibition space to a campus area with an extension in the range of a few kilometers. In SAMoA, situation-aware mobile computers have been used as guides through an exhibition providing information related to actual time and place. In UbiCampus, IR transmitters distributed over the university campus serve the purpose of providing a low cost equivalent of the gyro-based mechanisms used in car navigation systems.

4.3   Service Architecture

Jini provides predefined mechanisms to establish a marketplace-like environment for service providers to offer their services, and for clients to find and use these services. Lookup, discovery and join services act as mediators to establish the proper connection between clients and servers. Jini doesn't offer, however, anything beyond this low level functionality. It is therefore the task of the system architect to define a suitable architecture with a variety of generally applicable basic services and some application-specific ones related to specific functionalities.

Figure 2 shows an architecture positioning the distribution and interaction services discussed in the previous section on top of some basic services useful to some or all of the higher level ones. Possible basic services envisaged for UbiCampus are:

The *Repository Manager* knows the available data resources and manages the access to them in cooperation with the Database Service which offers the functionality of a unified view of possibly distributed and heterogeneous single databases.

The *Security Manager* knows the authorised users, their rights and public keys.

The *Print Manager* knows available print services and their status and properties.

The *Location Manager* knows and administrates the users currently active in the system. The use of that information might be restricted according to privacy and data security requirements.

On the side of the mobile terminal, the functionality will be limited to a GUI and possibly local storage. This helps to keep processor performance requirements and hence power consumption low, one of the central requirements of mobile technologies. For portability reasons, all terminal devices should support Java. A further standardization can be achieved by adopting so-called "portal"

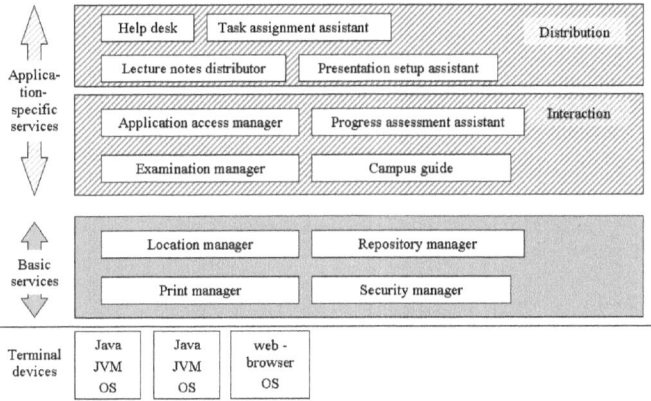

Fig. 2. Service architecture with examples of possible basic and application-specific services on top of the individual Java JVM operating system environments. Access can also be provided through an ultra-thin client based on a web browser interface portal solution.

solutions: they split the system functionality into an ultra-thin client, realized by a web browser, and the application running on a server.

5   Summary and Conclusion

The UbiCampus project is planned as an application oriented research project carried out at the University of Hannover and the Wireless Internet Laboratory of VTT, Finland. It applies Mark Weiser's idea of Ubiquitous Computing to the limited area of a university campus. The aim of UbiCampus is the integration of single technologies, available today, into a usable system architecture offering services typical for a university environment. In order to avoid being overtaken by the fast development of the necessary basic technologies, especially in the communication area, the project will heavily rely on standards.

UbiCampus combines a number of single development tasks centered around the two main project partners. The project is guided by a small group of industrial partners helping to set directions and keeping the work in line with industry development.

References

1. http://www.hoeft-wessel.de/
2. Chávez, Ide, Kirste: "SAMoA: An Experimental Platform for Situation-Aware Mobile Assistance", Proc. ARCS'99/APS'99, VDE-Verlag, 4.-7.10.99
3. http://www.etl.uni-hannover.de

Author Index

Lecture Notes in Computer Science

For information about Vols. 1–1856
please contact your bookseller or Springer-Verlag

Vol. 1893: M. Nielsen, B. Rovan (Eds.), Mathematical Foundations of Computer Science 2000. Proceedings, 2000. XIII, 710 pages. 2000.

Vol. 1894: R. Dechter (Ed.), Principles and Practice of Constraint Programming – CP 2000. Proceedings, 2000. XII, 556 pages. 2000.

Vol. 1895: F. Cuppens, Y. Deswarte, D. Gollmann, M. Waidner (Eds.), Computer Security – ESORICS 2000. Proceedings, 2000. X, 325 pages. 2000.

Vol. 1896: R. W. Hartenstein, H. Grünbacher (Eds.), Field-Programmable Logic and Applications. Proceedings, 2000. XVII, 856 pages. 2000.

Vol. 1897: J. Gutknecht, W. Weck (Eds.), Modular Programming Languages. Proceedings, 2000. XII, 299 pages. 2000.

Vol. 1898: E. Blanzieri, L. Portinale (Eds.), Advances in Case-Based Reasoning. Proceedings, 2000. XII, 530 pages. 2000. (Subseries LNAI).

Vol. 1899: H.-H. Nagel, F.J. Perales López (Eds.), Articulated Motion and Deformable Objects. Proceedings, 2000. X, 183 pages. 2000.

Vol. 1900: A. Bode, T. Ludwig, W. Karl, R. Wismüller (Eds.), Euro-Par 2000 Parallel Processing. Proceedings, 2000. XXXV, 1368 pages. 2000.

Vol. 1901: O. Etzion, P. Scheuermann (Eds.), Cooperative Information Systems. Proceedings, 2000. XI, 336 pages. 2000.

Vol. 1902: P. Sojka, I. Kopeček, K. Pala (Eds.), Text, Speech and Dialogue. Proceedings, 2000. XIII, 463 pages. 2000. (Subseries LNAI).

Vol. 1903: S. Reich, K.M. Anderson (Eds.), Open Hypermedia Systems and Structural Computing. Proceedings, 2000. VIII, 187 pages. 2000.

Vol. 1904: S.A. Cerri, D. Dochev (Eds.), Artificial Intelligence: Methodology, Systems, and Applications. Proceedings, 2000. XII, 366 pages. 2000. (Subseries LNAI).

Vol. 1905: H. Scholten, M.J. van Sinderen (Eds.), Interactive Distributed Multimedia Systems and Telecommunication Services. Proceedings, 2000. XI, 306 pages. 2000.

Vol. 1906: A. Porto, G.-C. Roman (Eds.), Coordination Languages and Models. Proceedings, 2000. IX, 353 pages. 2000.

Vol. 1907: H. Debar, L. Mé, S.F. Wu (Eds.), Recent Advances in Intrusion Detection. Proceedings, 2000. X, 227 pages. 2000.

Vol. 1908: J. Dongarra, P. Kacsuk, N. Podhorszki (Eds.), Recent Advances in Parallel Virtual Machine and Message Passing Interface. Proceedings, 2000. XV, 364 pages. 2000.

Vol. 1910: D.A. Zighed, J. Komorowski, J. Żytkow (Eds.), Principles of Data Mining and Knowledge Discovery. Proceedings, 2000. XV, 701 pages. 2000. (Subseries LNAI).

Vol. 1912: Y. Gurevich, P.W. Kutter, M. Odersky, L. Thiele (Eds.), Abstract State Machines. Proceedings, 2000. X, 381 pages. 2000.

Vol. 1913: K. Jansen, S. Khuller (Eds.), Approximation Algorithms for Combinatorial Optimization. Proceedings, 2000. IX, 275 pages. 2000.

Vol. 1914: M. Herlihy (Ed.), Distributed Computing. Proceedings, 2000. VIII, 389 pages. 2000.

Vol. 1917: M. Schoenauer, K. Deb, G. Rudolph, X. Yao, E. Lutton, J.J. Merelo, H.-P. Schwefel (Eds.), Parallel Problem Solving from Nature – PPSN VI. Proceedings, 2000. XXI, 914 pages. 2000.

Vol. 1918: D. Soudris, P. Pirsch, E. Barke (Eds.), Integrated Circuit Design. Proceedings, 2000. XII, 338 pages. 2000.

Vol. 1919: M. Ojeda-Aciego, I.P. de Guzman, G. Brewka, L. Moniz Pereira (Eds.), Logics in Artificial Intelligence. Proceedings, 2000. XI, 407 pages. 2000. (Subseries LNAI).

Vol. 1920: A.H.F. Laender, S.W. Liddle, V.C. Storey (Eds.), Conceptual Modeling – ER 2000. Proceedings, 2000. XV, 588 pages. 2000.

Vol. 1921: S.W. Liddle, H.C. Mayr, B. Thalheim (Eds.), Conceptual Modeling for E-Business and the Web. Proceedings, 2000. X, 179 pages. 2000.

Vol. 1922: J. Crowcroft, J. Roberts, M.I. Smirnov (Eds.), Quality of Future Internet Services. Proceedings, 2000. XI, 368 pages. 2000.

Vol. 1923: J. Borbinha, T. Baker (Eds.), Research and Advanced Technology for Digital Libraries. Proceedings, 2000. XVII, 513 pages. 2000.

Vol. 1924: W. Taha (Ed.), Semantics, Applications, and Implementation of Program Generation. Proceedings, 2000. VIII, 231 pages. 2000.

Vol. 1925: J. Cussens, S. Džeroski (Eds.), Learning Language in Logic. X, 301 pages 2000. (Subseries LNAI).

Vol. 1926: M. Joseph (Ed.), Formal Techniques in Real-Time and Fault-Tolerant Systems. Proceedings, 2000. X, 305 pages. 2000.

Vol. 1927: P. Thomas, H.W. Gellersen, (Eds.), Handheld and Ubiquitous Computing. Proceedings, 2000. X, 249 pages. 2000.

Vol. 1931: E. Horlait (Ed.), Mobile Agents for Telecommunication Applications. Proceedings, 2000. IX, 271 pages. 2000.

Vol. 1766: M. Jazayeri, R.G.K. Loos, D.R. Musser (Eds.), Generic Programming. Proceedings, 1998. X, 269 pages. 2000.

Vol. 1791: D. Fensel, Problem-Solving Methods. XII, 153 pages. 2000. (Subseries LNAI).

Vol. 1932: Z.W. Raś, S. Ohsuga (Eds.), Foundations of Intelligent Systems. Proceedings, 2000. XII, 646 pages. (Subseries LNAI).

Vol. 1933: R.W. Brause, E. Hanisch (Eds.), Medical Data Analysis. Proceedings, 2000. XI, 316 pages. 2000.

Vol. 1934: J.S. White (Ed.), Envisioning Machine Translation in the Information Future. Proceedings, 2000. XV, 254 pages. 2000. (Subseries LNAI).

Vol. 1937: R. Dieng, O. Corby (Eds.), Knowledge Engineering and Knowledge Management. Proceedings, 2000. XIII, 457 pages. 2000. (Subseries LNAI).

Vol. 1938: S.Rao, K.I. Sletta (Eds.), Next Generation Networks. Proceedings, 2000. XI, 392 pages. 2000.

Vol. 1939: A. Evans, S. Kent (Eds.), «UML» – The Unified Modeling Language. Proceedings, 2000. XIV, 572 pages. 2000.